AN ELUSIVE VICTORIAN

AN ELUSIVE VICTORIAN

The Evolution of Alfred Russel Wallace

MARTIN FICHMAN

THE UNIVERSITY OF CHICAGO PRESS
CHICAGO AND LONDON

Martin Fichman is professor of humanities at York University in Canada. He is the author, most recently, of *Evolutionary Theory and Victorian Culture* and *Science, Technology, and Society: A Historical Perspective*.

The University of Chicago Press, Chicago 60637
The University of Chicago Press, Ltd., London
© 2004 by The University of Chicago
All rights reserved. Published 2004
Printed in the United States of America

13 12 11 10 09 08 07 06 05 04   1 2 3 4 5

ISBN: 0-226-24613-2 (cloth)

Library of Congress Cataloging-in-Publication Data

Fichman, Martin, 1944–
    An elusive Victorian : the evolution of Alfred Russel Wallace / Martin Fichman.
        p. cm.
    Includes bibliographical references and index.
    ISBN 0-226-24613-2 (cloth : alk. paper)
        1. Wallace, Alfred Russel, 1823–1913.   2. Naturalists—England—Biography.
3. Spiritualists—England—Biography.   4. Socialists—England—Biography.   5. Natural
selection.   I. Title.
QH31 .W2 F524 2004
508'.092—dc21

                                                                        2003010949

For Ken,
And in memory of my parents

CONTENTS

ACKNOWLEDGMENTS

This book has had a long gestation period, and in the researching and writing of it I have incurred the debt of many individuals and institutions. Wallace was, and remains, an intriguing but controversial figure. In uncovering and formulating what I believe is an accurate assessment of his complex place in the history of Victorian science and culture, I have sought the knowledge, advice, and encouragement of colleagues and friends. Their views were essential, but the portrait of Wallace I present remains my interpretation: I hope it is as rich and provocative as was the subject himself.

To Jane Camerini, as passionate about Wallace as I am, I owe special thanks for her expertise, insights, and encouragement in my journey over this past decade. My lengthy discussions and communications with Charles H. Smith, a key figure in the community of Wallace scholars, have been spirited and challenging. Ruth Barton, Peter Bowler, Sherrie Lyons, and James Moore have generously shared their knowledge of the history of biology and the history of the Victorian era. Two colleagues—and friends—at York University have been unstinting of their time and patience in dealing with my questions about both Wallace and his place in Victorian history. Bernard Lightman read portions of the manuscript and has put his formidable knowledge of Victorian intellectual and cultural history at my disposal on many, many occasions. Paul Fayter has been an invaluable sounding board for my assessment of Wallace's religious and theological views. The reports of the anonymous referees for the University of Chicago Press have been exceptionally helpful to me in sharpening my argument and revising my manuscript.

York University (particularly the Faculty of Arts) and the Social Sciences and Humanities Research Council of Canada have been most generous in providing funding to support my research and travel in pursuit of Wallace.

Many archives and libraries have kindly made it possible for me to use Wallace materials in their possession. The librarians and staff of the British Library, the British Museum (Natural History), the Cambridge University Libraries, the Imperial College of Science, Technology and Medicine (London), the Linnean Society of London, the New York Public Library, the Royal Botanic Gardens (Kew), the Royal Geographical Society of London, the University of London, and the University of Toronto have been most courteous and helpful in aiding me to sift through the mass of Wallace correspondence and notebooks in their possession. Most crucially, this book would not have been possible without the cooperation of Richard Ovenden and his staff at the Special Collections Division of Edinburgh University Library; they permitted me full access to the rich repository of more than four hundred books from Wallace's personal library—most of them heavily annotated by him—now housed at Edinburgh.

I have been fortunate indeed to work with the University of Chicago Press in having this project move from conception to completion. Susan Abrams, grand friend to historians of science, encouraged me from the outset to pursue my analytical study of Wallace. Her keen intellect, critical acumen, sense of humor, and infinite patience in fielding my questions and concerns have been an inspiration to me. Christie Henry, my present editor, has been equally supportive in overseeing the final stages of publication of this book. Special thanks, too, are owed to Jennifer Howard and Yvonne Zipter of the University of Chicago Press for their administrative and copy-editing expertise.

For a project as long in the making as this one, kind friends are essential to offer support, encouragement, and the reassurance that all is worthwhile. Selma Zimmerman, cell biologist, colleague, and long-time friend, has given me the benefit of a practicing scientist's view of the history of science. My dear friends Terry and Shelly Fowler never failed to listen to anything I had to say about the book and the joys—and frustrations—it presented. Finally, Ken Setterington has been both a stern critic of any infelicities in my writing style as well as a supportive and mercifully tolerant companion through the many years it has taken me to write this book.

To all of the above, and many other friends, colleagues, and students, I owe a priceless debt of encouragement and cheer. I believe Wallace is worth the struggle it has taken me, and other historians, to assign him his proper role in the pantheon of Victorian greats.

Introduction

Alfred Russel Wallace (1823–1913) was one of the most innovative and controversial figures of the Victorian era. His thought and career were interwoven with the scientific, ethical, philosophical, social, religious, political, and economic currents that flowed in such profusion through the Victorian cultural landscape. Wallace was a prism by which a vast array of intellectual concepts and cultural forces was refracted. His receptivity to diverse ideas has tended to obscure the existence, and persistence, of certain major themes that guided Wallace through the vicissitudes of a life that spanned nine decades. Wallace challenged many of the comforting—if sometimes illusory—boundaries that were being constructed to preserve a sense of order in Victorian culture amid powerful, and confusing, changes. Some of these changes occurred within science and the scientific community. Others resulted from the momentous transformations of the mid- and late nineteenth century.

Wallace was an elusive Victorian in the sense that it was—and remains—difficult to pigeonhole him into any neat category. Many labels have been applied to him: field naturalist, biological theorist, socialist, spiritualist, theist, land nationalizationist, philosopher, and ethicist. Each of these labels is apt. But Wallace was more than the sum of his parts. Different fields of interest loomed larger at different phases of his life. His central goal of integrating these diverse concerns functioned as an anchor for Wallace's evolving intellectual and activist endeavors. His quest culminated in an overarching worldview—a teleological evolutionary cosmology. The various elements of Wallace's synthesis came together most clearly in the later decades of his life.

Wallace was hardly unique in attempting to construct a meaningful evolutionary cosmology. The Victorian era abounded in efforts to erect worldviews on evolutionary pedestals. Robert Chambers and Herbert Spencer immedi-

ately come to mind as influential evolutionary cosmologists (Secord 2000).
What set Wallace apart from the pack was his stature as a practicing scientist.
As codiscoverer with Charles Darwin of one of the most influential theories of
the nineteenth century, evolution by natural selection, Wallace was destined
for scientific superstardom. Indeed, most of the existing scholarly literature
on Wallace focuses on his science and either leaves out Wallace's "other"
interests or fails to view his political, social, theistic, and philosophical concerns in context or in depth. Consequently, Wallace's efforts to integrate his
wide-ranging interests have long eluded historical understanding. This study
demonstrates the seriously flawed nature of the conventional historiographic
portrait of Wallace as a brilliant scientist who lapsed unfortunately into theism, socialism, or spiritualism.

To many of his contemporaries, Wallace was a sage. His pronouncements
on political, social, and ethical issues struck responsive chords in individuals
and groups in Victorian society. One group, however, that remained decidedly unimpressed by Wallace's forays into these domains was the scientific
naturalists. Spearheaded by Thomas Henry Huxley, this group included John
Tyndall, William Kingdon Clifford, E. Ray Lankester, Frederick Harrison,
G. H. Lewes, and Edward Tylor (Turner 1993). This influential group was
not uninterested in political, social, ethical, or religious matters. As Adrian
Desmond's recent biography of Huxley demonstrates, Huxley and his fellow
scientific naturalists were deeply involved in such debates (Desmond 1998).
But as advocates of a specific ideal of science professionalization they were
committed to constructing a definition of value-neutral and hence "objective" science. They sought to distance professional science from the pejorative depictions of value-laden and hence "subjective" nonscience or pseudo-
science. The scientific naturalists recognized the professional gains to be had
by proclaiming the ideological neutrality of science. Huxley and his camp
could claim that they spoke as objective experts, not political or ideological
partisans. This strategy involved erecting an epistemological divide between
science and politics, ethics, religion, and other cultural forces. It also encouraged a distinction between elite and popular science. The strategy was
essentially enunciated by the end of the 1860s and served Huxley, Darwin,
and their colleagues well for several decades (Moore 1991). Such a strategy
was brilliant but disingenuous. The scientific naturalists invoked an "ideologically pure" science that concealed their own varied sociopolitical agendas
behind the banner of a rigorous professionalism.

As Bernard Lightman and others have pointed out, however, "often scientific naturalists are treated as the dominant group in the intellectual landscape
of the latter half of the nineteenth century, and their power is seen as symbolic of the triumph of a process of secularization. But we have overestimated

their influence. They were one group among many that vied for cultural authority during this period by drawing on the immense prestige provided by science" (Lightman 2001, 362–364). A number of other groups were also key players in Victorian debates about science. The North British physicists and the popularizers of science, for example, challenged the scientific naturalists' interpretation of the meaning of science. These and other individuals and groups envisioned a larger, more inclusive role for science as part of the broader culture (Turner 1993; Otter 1996; Richards 1997; Smith 1998; Lightman 2001, 355–362).

The issue of science professionalization is crucial to an assessment of Wallace's life and career. In the latter half of the nineteenth century, the definitions of professional "science" and "scientist" were far from resolved—and remained so until the early decades of the twentieth century.[1] The scientific naturalists' verdict on Wallace is both polemical and inaccurate. John Tyndall may have told Wallace that he read his spiritualist writings with "deep disappointment." And Thomas Henry Huxley was similarly disturbed by what he regarded as Wallace's forays into unworthy arenas (Wallace [1905] 1969, 2:280–281).[2] Many of Wallace's contemporaries, in contrast, regarded his effort to forge an evolutionary cosmology as entirely appropriate for a scientist. The majority of twentieth-century scholars, however, have elected to perpetuate Huxley's and his allies' portrayal of Wallace.[3]

New analytical approaches are required to establish a more realistic portrait of Wallace. Certain historians in recent years have begun to free themselves from the caricature of Wallace imposed by the scientific naturalists and many twentieth-century scholars—and my debt to them will be made clear in the course of this book.[4] Wallace's career seems eccentric or paradoxical only if we cling to the scientific naturalists' model. This study takes a different tack. It does not depict Wallace as one who departed from the norms of professionalized science precisely because what we think of as professional science today was in the process of being created in the late nineteenth century—and the process was very much contested. Instead, Wallace is approached here through the lens of the diverse, complex, and competing forces seeking to define the appropriate role of science in the broader Victorian culture.

The main theme of this study is that there is an identifiable coherence—a leitmotif—in Wallace's thought and activities. This quest for integration at the philosophical as well as practical levels, coupled with Wallace's passion to understand and change the world into which he was born, help explain his multifaceted approach. There is an underlying link—his evolutionary cosmology—that binds together Wallace's highly varied intellectual and practical pursuits. Uncovering and then exploring this link enables us to make sense of how he understood the relations among science, politics, economics, and religion.

Wallace was a formidable Victorian intellectual activist whose scientific acumen, when coupled with a maturing theistic outlook, culminated in a powerful and holistic philosophy of nature and society. Wallace was committed to science and its methodology as one of humanity's grandest achievements. But he recognized that the unbridled embrace of scientific and technological developments in the name of material "progress" was misguided and potentially destructive. Science, and its increasingly potent industrial applications, had to be tamed. What Wallace termed, with pointed irony, the "wonderful century" had striking successes and dangerous failures. Wallace's own definition of individual and societal progress was at odds with some of the most fundamental precepts of Victorian capitalism and imperialism.

By identifying and examining relations among his most fundamental convictions, Wallace's evolutionary worldview gains a compelling clarity. Many of the paradoxes and unorthodoxies of which he was habitually charged are seen to fall into a coherent pattern of belief and behavior. This pattern was not fixed but continually developing. As an evolutionary biologist, Wallace would have found naive the notion that an individual's life or the structure of a given society was static or unambiguous. During the past two decades, the increasingly sophisticated attention to the interplay between societal context and individual thought and action has dramatically enriched the historiography of science. The contextualist approach to the history of science has transformed our understanding of domains such as theory formation, ideology and science, the semantic and linguistic complexities of science terminology, religion and science, and gender issues.[5] Of all the periods whose histories have benefited from these new perspectives, the Victorian era is one of the most conspicuous.[6] A contextualist approach permits one to see Wallace as a whole human being, who was more than the sum of his (superficially) disparate parts.

THE CENTER(S) OF WALLACE'S VISION

The "strands of system" in Wallace's evolutionary worldview are diverse. They reveal tensions, both philosophical and personal, that he spent his life trying to resolve. But the notion of strands of system is crucial to an understanding of who and what Wallace was. He was aware that his efforts to forge a syncretic evolutionary teleology drew on a variety of sources, hypotheses, and observations that some of his contemporaries—and a majority of later historians—found incongruous. It took Wallace the better part of a long lifetime to transform those strands into components of a powerful evolutionary cosmology. An examination of relations among the separate strands he espoused makes Wallace's overarching vision clear.[7]

Such eclecticism inevitably created tensions in the private and public Wallace. The tension between the individual and the community is pointedly manifest in Wallace's career. It is necessary to confront directly this tension, characteristic of many of his contemporaries, in assessing Wallace's place in Victorian science and culture. This requires focusing on the powerful role of individual passion in Wallace's quest for a comprehensive worldview that merged his science with his other pursuits. Focusing on Wallace's motivations as an individual is not equivalent to psychoanalytical history, nor does it preclude sociological analysis. Rather, an emphasis on Wallace's personal coherence elucidates those decisive elements of character and those decisive events that shaped his particular evolution into a thinker of great range and accomplishment. Wallace, like many of his creative contemporaries, was a dynamic participant in the rich tapestry of Victorian cultural forces. They were real, three-dimensional people, who affected and interacted with one another in fascinating ways. Viewed thus, individuals are meeting points for influences, mobile rather than static or passive agents (Shortland and Yeo 1996, 12–14, 36–37).

Wallace made a key decision, early in his career, to incorporate science into a broader ethical framework. Once this decision was made, certain benefits and risks ensued. A major benefit for Wallace was the ability to deploy his scientific expertise on behalf of causes that he regarded as indispensable to the definition of an ethical life. A major risk was the marginalization of certain aspects of his maturing evolutionary cosmology by some influential voices in the scientific community—most notably the scientific naturalists. These risks were often of considerable cost, both professionally and financially (Soderqvist 1996, 49–53, 60–65, 70–74; Raby 2001, 218–222). Wallace concluded his autobiography with an assessment of his life's work that expresses the convictions that sustained him: "If, therefore, my books and essays have been of any use to the world—and though I cannot quite understand it, scores of people have written to me telling me so—then the losses and struggles I have had to go through have been a necessary discipline calculated to bring into action whatever faculties I possess" (Wallace [1905] 1969, 2:380). From the outset, Wallace's path was marked by a sense of mission. He sought to integrate his theoretical insights and achievements with practical efforts and strategies to ameliorate what he deemed to be the most flagrant abuses of the society in which he lived—vast social, economic, and gender inequities, unfettered industrial expansion, and increasing environmental degradation.

Wallace's particular field of scientific expertise, evolutionary biology, created both opportunities and dilemmas for him and his contemporaries. Precisely because evolutionary biology was at an interface between the natural and social sciences, it was notoriously susceptible to sociopolitical influences

and deductions of all stripes (Jones 1980; Bowler 1993). The methodological issues raised by the convoluted history of the interactions between various
models of natural science and social science also need to be considered (Cohen 1994). The challenging and contentious efforts to construct appropriate
professional boundaries for evolutionary biology speak volumes about Victorian science and culture. These efforts dramatically illustrate the complex
process by which any age attempts to define or redefine the domain of science and grasp, for itself, the ever-malleable border between scientific and
nonscientific discourse (Young 1985b).

Professionalizing science had its star performer in Huxley. He "boosted
the 'Scientist's' profile by trenching on the clergy's domain, raising the terri-
torial tension by equating authority with technical expertise." But the victory
of this new breed of scientist/professionals by the 1870s was anything but
certain. Many of "the English public schools and universities shunned science as useless and dehumanizing. . . . Oxford and Cambridge were finishing
schools for prosperous Anglicans. Against 145 Classics Fellowships at Oxford in 1870, there were four in science. The stacked odds explain Huxley's
single-minded assault on the ivy seminaries using his new-professionalized
forces" (Desmond 1998, xviii–xix). The forces arrayed against the new breed
of professional scientist were not only those of older, entrenched, and well-
to-do professions such as the clergy and law. Many of those fully sympathetic
to enhancing the prestige and power of science and scientists had reservations
about the scientific naturalists' particular strategy. Wallace had major reservations. By the early 1870s, he had emerged as one of the most outspoken
critics of the strategy of ideological neutrality. Wallace's candid insistence on
the reciprocal interaction between biology and ideology would increasingly
shape his mature evolutionary worldview (Fichman 1997). His fluid movement from sphere to sphere was troublesome to those, like Huxley, hoping
to establish a new orthodoxy. But once the Victorian period is recognized as
one of competing visions of science's proper role in culture, Wallace's putative
unorthodoxies are seen to be artifacts of historiography. Like Wallace, others
in the scientific community—as well as many elements in the wider Victorian
society—represented a much broader spectrum of response to science and
its cultural role. Science popularizers, female and working-class scientists,
Idealist philosophers, and religious thinkers were as crucial to the realities of
Victorian science as were the elite new professionals and scientific naturalists
(Livingstone 1987; Otter 1996; Gates and Shteir 1997; Brooke and Cantor
1998; Lightman 2001).

Wallace's evolutionary cosmology, with its mix of sociopolitical reformism, theism, spiritualism, and ethical philosophy, abandoned any pretext of
ideological neutrality. A central motive in Wallace's career was to unveil the

hollowness of the scientific naturalists' claims to demarcate science objectively from the broader culture. Wallace's writings from his earlier period may be seen as precursors to his later, more overt and passionate manifestos of the necessary ideological context and texture of evolutionary biology (Moore 1997, 300–303). In this sense, Wallace retained Spencer's and Francis Galton's goal of linking biology and culture while shedding their comforting armature of objective neutrality. How Wallace was able to perform that task, despite the risks and opprobrium he endured in pursuing his mission to effect cultural reform, constituted a Victorian "tract for the times." Wallace's social progressionism interacted with his biological progressionism. Together, they reinforced his contention that science could not function as a neutral blueprint for political or ethical philosophy. Reformist and theistic convictions and biological insights were equal partners in his evolutionary teleology. Science was but one element, albeit a crucial one, in his construction of a comprehensive cultural vision (Wallace [1882] 1906). Wallace's spirit and passion for using knowledge, derived from a diverse array of sources, to articulate an evolutionary cosmology that would aid him and others to "arrive at a juster conception of the . . . Life-World" remained the overarching purpose of his life's work (Wallace 1910a, 10).

Road Map for the Reader

This is not a biography of Wallace in any conventional, chronological sense. The details of Wallace's life and career have recently been the subject of two useful biographies that provide accurate chronological portraits (Raby 2001; Shermer 2002) Rather, this book is a contextualist and analytical study of Wallace's major intellectual and cultural views and activities. Given the daunting range of Wallace's interests, the focus is on the most central of Wallace's concerns: evolutionary biology (chap. 2), the philosophical and humanistic context of evolutionary theory (chap. 3), spiritualism (chap. 4), land nationalization and socialism (chap. 5), and, finally, theistic evolutionary teleology (chap. 6). Since the major goal of this work is to show how Wallace integrated these diverse concerns, the subject matter of any one chapter surfaces, to a greater or lesser degree, in other chapters. Spiritualism and socialism, for example, were regarded by Wallace—if not by all his contemporaries—as fundamentally linked. Each chapter, consequently, is not autonomous but rather merges into the kaleidoscope that was Wallace. The reader will find, for example, that Wallace's tour of North America (1886–1887) is discussed in several chapters (though in different contexts). Similarly, Wallace's relations with William James and Charles Peirce find analytical niches in several chapters.

Although evolutionary biology is the subject of chapter 2, the reader should not assume it takes pride of place among Wallace's other interests. Wallace's codiscovery of natural selection is the most well known, and among the first, of his numerous intellectual achievements. It is thus understandable why Wallace as expert field naturalist and brilliant biological theorist has been the focus of most previous historical attention. But Wallace's intellectual work was just warming up when he formulated the theory of natural selection. He went on to explore a vast array of social, political, economic, philosophical, and theological issues—as well as maintaining his intense scientific activity and productivity—in the course of his long life. The overview of Wallace's scientific achievements provided in chapter 2 is a necessary prolegomenon to the analysis of the integration of science and the broader culture that is the distinguishing quest, and enduring legacy, of Wallace's life.

If the book's plan is neither strictly chronological nor simply thematic, then what is it? This study is an investigation of how Wallace managed to make sense of apparently diverse strands of thought and action to construct a philosophy of nature and of human beings in nature that was at the least provocative and, at its best, profound. Wallace gradually succeeded in using his evolutionary insights for a wide-ranging exploration of the human condition. The book is chronological to the degree that Wallace encountered different ideas and social realities at successive periods in his long life. But his path to integrating so many elements of the intoxicating soil of Victorian culture was by no means direct or without pitfalls. We shall see the process by which the youthful naturalist evolved into a deeply philosophical and outspoken cultural critic.

This analysis provides, therefore, a framework for understanding how and why Wallace combined these diverse elements into a comprehensive evolutionary cosmology. The whole of Wallace's oeuvre is taken seriously in order to deepen and broaden our view of him. I have refrained from prejudging Wallace's forays into what are often considered unscientific and/or misguided endeavors. Instead, the book follows those paths and concerns that Wallace, himself, deemed significant. Though scarcely defending each path Wallace chose to follow, I do provide a revised, and largely sympathetic, interpretation of Wallace's personality, goals, and achievements. This book follows Wallace's evolution both as a thinker and a three-dimensional human being. The particular answers at which he arrived are significant for what they reveal about the uncertainties and ambiguities of Victorian culture. The paths Wallace pursued to ferret out those answers testify to his enduring conviction that science afforded a powerful but incomplete source of knowledge. Wallace was able to live with perplexities because he was grounded in the conviction that if this is certainly not the best of all possible worlds, it can nonetheless be made a better world. The horizon of his thinking, while powerfully informed by his scientific training and observations, encompassed a wide-ranging critical assessment of the impacts of science on the broader culture. The profound advances in science and technology of the late nineteenth century led many Victorians to embrace the "cult of progress." Others were frightened and bothered by the equally obvious flip side of progress: rampant urbanization, the dark, satanic mills that transformed the lush green landscape of Britain, and the creation of an industrial underclass. Wallace sought to reconcile his humanistic beliefs with his scientific investigations. He embarked on a personal and public quest to articulate an evolutionary cosmology that would ensure the dignity of all individuals in an ethical sociopolitical order appropriate to a new age.

———•◆•———

NOTES

1. The professionalization of science in the second half of the nineteenth century is a subject currently undergoing intense historiographic and sociological analysis. For an introduction to this rapidly growing body of analysis, see Collini 1991; Turner 1993; Yeo 1993; A. Secord 1994a; Lightman 1997; Barton 1998a; Desmond 2001; Waller 2001.

2. Interestingly, some recent studies of both Tyndall and Huxley emphasize the ambiguities about spiritualism and materialism in their own professional and personal lives. See Barton 1987; Lightman 1987.

3. George 1964; Schwartz 1984; Shermer 1994, 2001. Although Shermer provides an accurate account of Wallace's commitment to spiritualism—and acknowledges the influence of spiritualist beliefs on his evolutionary theory—Shermer concludes that Wallace's "supernaturalism" drove him, ultimately, to "pseudoscientific fool-hardiness"; more to the point, Shermer argues that such historical reconstructions as his on Wallace will help to "illuminate how and why perfectly reasonable and rational scientists come to believe in the reality of the paranormal and supernatural" (Shermer 1984, 83). For Shermer (and the others cited here), the (assumed) schism between Wallace as scientist and Wallace as nonscientist looms large.

4. Scarpelli 1992; Smith 1992; Moore 1997; Vetter 1999; Berry 2002; Camerini 2002; Jones 2002; Peck 2003, 6–12, 19–22. Wallace's antivaccination activities are not dealt with at length in my book. He was part of a considerable portion of the population in both Europe and North America who were opposed to mandatory vaccination programs. Although Wallace's name and formidable power of argumentation became an important tool for the antivaccination movement, his active involvement in the campaign was rather limited. His writings, however, generated considerable interest among both the proponents and detractors of antivaccination. Wallace's sophisticated statistics-based critique of the medical efficacy of, and dubious public health safeguards relating to, vaccination are discussed in Smith (1991, 202–216) and Scarpelli (1992).

5. The recent contextualist literature on the history of science is vast. I cite the following seminal works to indicate the range of historiographic activity, though they represent only the tip of the iceberg: Shapin 1994; Pickering 1995; Kuklick and Kohler

1996; Shortland and Yeo 1996; Brooke and Cantor 1998; Golinski 1998; Harding 1998; Lenoir 1998; Gieryn 1999; Schiebinger 1999.

6. A representative sample of the growing contexualist literature on Victorian intellectual and cultural history includes Cooter 1984; Alter 1987; Russett 1989; Stafford 1989; Desmond 1989, 1998; Collini 1991; Pratt 1992; Bowler 1993; Turner 1993; Yeo 1993; Allen 1994; Jardine, Secord, and Spary 1996; West 1996; Lightman 1997; Smith 1998; Winter 1998; Yanni 1999.

7. I owe the term "strands of system" to Douglas R. Anderson; see his *Strands of System* (1995, esp. 26–30, 56–67). Anderson's approach to Peirce is methodologically similar to my analysis of Wallace, and his study of Peirce has provided me with several suggestive insights. The parallels between certain aspects of Peirce's life, character, and efforts to construct a syncretic philosophy to those of Wallace are striking. The Wallace-Peirce connection is discussed in chapter 3.

CHAPTER 2

The Making of a Victorian Naturalist

The process by which Wallace became one of the greatest Victorian naturalists is central to understanding his evolutionary worldview. The concepts and attitudes that would guide Wallace's quest for an integrative framework linking social, political, religious, philosophical, and scientific issues had their genesis in his earliest encounters with the natural world.

LEARNING LAND SURVEYING

In the summer of 1837, Wallace went to join his brother William, a land surveyor, in Bedfordshire to acquire the rudiments of surveying and mapping. The year before, Parliament had terminated the traditional right of farmers to pay the charges assessed (tithes) on the land they worked in kind. Instead, rent charges, based both on the assumed productive potential of the land and on the farmer's average actual yield, were put into effect. Surveying in the British Isles became, almost overnight, a trade in high demand. Tithe owners delighted in, and tenant farmers fulminated against, the precise surveys now required to assess accurately the new rent charges of field properties. These new survey maps were crucial legal documents in a new economic world (Kain and Prince 1985; Moore 1997, 301). Wallace's surveying expertise would prove extremely valuable when he later tackled the more philosophical questions of animal and plant distribution (Camerini 1993).

Wallace's surveying career, in addition to its regular income, afforded him the joy of working daily in the open countryside. It was also his introduction to the science of geology. William, like most land surveyors, had acquired a tolerable knowledge of geological principles. Wallace soon became familiar with the fossils abundant in the chalk and gravel in the regions in which he worked. He also became adept at calculating the areas of the fields he and William surveyed (Wallace [1905] 1969, 1:108–109; Bennett 1987). This technical knowledge of mapping was to serve Wallace well in his later travels in the Amazon Basin and the Malay Archipelago. Wallace's work made him look with renewed interest at the paltry mathematics he had learned in grammar school. He purchased some cheap elementary books on mechanics and on optics published by the Society for the Diffusion of Useful Knowledge. It was these basic texts that "thus laid the foundation for that interest in physical science and acquaintance with its general principles which have remained with me throughout my life." Wallace had ample opportunity to take solitary rambles in the British countryside. It was in this setting that he "first began to feel the influence of nature and to wish to know more of the various flowers, shrubs, and trees I daily met with, but of which for the most part I did not even know the English (much less Latin) names." The embryonic evolutionist hardly realized that there was such a science as systematic botany. He didn't know that every flower and weed he encountered had already been accurately described and systematically classified. Since William took scarcely any interest in the native plants and animals constantly encountered in their travels, Wallace found little encouragement for pursuing the matter further. His introduction to the scientific literature of British flora and fauna would not come until several years later (Wallace [1905] 1969, 1:110–111).

In 1838, the Wallace brothers moved from Barton to Turvey, some twenty miles away. William had obtained the commission for a survey for tithe commutation of another parish. That work lasted until the summer of 1838. The two then relocated to Silsoe and, shortly thereafter, to Leighton Buzzard to undertake a survey of the parish of Soulbury (Wallace [1905] 1969, 1:117, 129–131). These experiences provided Wallace with one of his first direct encounters with the impact of nineteenth-century industrialization on rural Britain. Soulbury parish was crossed by the river Ouse and its tributaries. Parallel with the Ouse was the Grand Junction Canal, which then carried most of the heavy goods from the manufacturing districts of the Midlands to London. Following the same general direction but on higher ground a half mile west was the London and Birmingham Railway, then under construction. Most of the earthwork had been completed and many of the rail bridges either built or nearly so. Wallace regarded the entire region "as enlivened by the work going on." The Grand Junction Canal itself was in the midst of substantial modernization. It needed upgrading to carry the ever-increasing trade caused by the rapid growth of London and the boom in agricultural prosperity during the first third of the nineteenth century. To supply sufficient water for the locks of the canal, and the barges carrying the precious cargo, it was necessary to construct steam engines to pump up the water at each lock from the lower to the higher level. Wallace had never before seen a steam engine. He took the greatest delight in examining these monuments of modernity, all erected by the celebrated firm of Boulton and Watt. William's technical expertise permitted him to explain to his brother how these engines worked. Alfred gained his "first insight into some of the more important applications of the science of mechanics and physics." Wallace, like many of his contemporaries, failed to comprehend the implications of the revolutionary expansion of railways all over Britain (and elsewhere). He did not perceive that the canals would soon yield their economic status to the railways.

During the next two years (1839–1840), the two brothers' surveying pursuits took them to the Welsh border country. Wallace realized that part of the function of survey maps was to facilitate the enclosure of traditional commons land. The broader ethical, political, legal, and historical aspects of land ownership and distribution at that time, however, did not seem particularly problematic to Wallace. He "certainly thought it a pity to enclose a wild, picturesque, boggy, and barren moor, but I took it for granted that there was *some* right and reason in it." It would take several decades for him to conclude that enclosures were "unjust, unwise, and cruel" (Wallace [1905] 1969, 1:132–134, 140–150, 158). But the firsthand knowledge of rural conditions he was gaining would prove invaluable when Wallace later became an ardent advocate of land nationalization.

In the summer and autumn of 1841, Alfred and William surveyed parishes in Brecknockshire and Shropshire. The Brecon Beacons, the highest mountain range in South Wales, are the source of the tributaries that feed into the river Usk. Wallace was exhilarated to work amid the picturesque environs of the Usk valley. To be once again living near his childhood home was an added pleasure. Wallace delighted in hearing, though never learned, the Welsh language. The Beacons presented him with his first contact with the phenomena of subaerial denudation. He did not fully appreciate their scientific significance until some years later when he had studied Charles Lyell's *Principles of Geology*. Almost the whole of this region is of Old Red Sandstone formation. Subjected to centuries of rain, frost, and snow, the sandstone gives these mountains their characteristic rounded summits and surrounding extensive gently undulating plains. As Wallace put it in his autobiography, "We obtain an excellent illustration of how nature works in moulding he earth's surface by a process so slow as to be to us almost imperceptible." Wallace also saw for the first time one of those strange relics of antiquity, a huge erect slab of old red sandstone. This produced an impression that, more than six decades later, was "still clear and vivid" (Wallace [1905] 1969, 1:160–167).

Late in 1841, the two brothers settled in Neath (in Glamorganshire,

Wales) to survey and prepare corrected maps of the district. After complet-
ing that task, their remaining work, including some minor architectural and
engineering projects, was not onerous. Wallace found ample time to savor
the delights of the Welsh moors and mountains. William was often away,
seeking additional employment or engaged in some minor business ventures
in various parts of the country. Alfred thus was left frequently to his own
devices. Having learned the use of the sextant in surveying, he began reading
books on astronomy. He constructed a simple telescope with which he was
able to observe details of the moon's surface as well as Jupiter's satellites.
These modest ventures led to another of Wallace's life-long pursuits. He
would keep abreast of the "grand onward march of astronomical discovery"
(Wallace [1905] 1969, 1:191–192). In later years, Wallace's knowledge of
astronomy became quite sophisticated. He took an active part in the late-
nineteenth-century debates on the possibility of life on Mars and on the
astronomical factors that affected global climatic change. The years spent
surveying served the critical function of broadening Wallace's interest and
skills relating to science. He acquired expertise that he otherwise would not
have had. But Wallace's early entry into the workforce deprived him of any
higher formal education. His lack of exposure to university life set him apart
from many of those who later became his colleagues as an adult.

NATURE'S MYSTERIES

The most profound aspect of Wallace's surveying was his "first introduction
to the variety, the beauty, and the mystery of nature as manifested in the
vegetable kingdom." His botanical pursuits at this period were amateurish.
William often characterized his brother's ruminations in this area as worthless, being of no obvious practical value. They were, however, precious to
Alfred. Given his ignorance of natural history, any exposure to field study
was instructive. Out of these tentative steps emerged Wallace's passion for
the subject that would become his life's work: evolutionary biology. The purchase in 1841 of a shilling paper-covered book published by the Society for
the Diffusion of Useful Knowledge was Wallace's initiation into the world
of systematic botany. The pamphlet contained an outline of the structure of
plants and a short description of their various parts and organs. It included
a good description of a dozen of the most common of the natural orders of
British plants. The impact of this inexpensive tract on Wallace's scientific
development was momentous. Wallace recalled that "this little book was a
revelation to me, and for a year my constant companion. On Sundays I would
stroll in the fields and woods, learning the various parts and organs of any
flowers I could gather, and then trying how many of them belonged to any

of the orders described in my book. Great was my delight when I found that I could identify a Crucifer, an Umbellifer, and a Labiate; and as one after another the different orders were recognized, I began to realize for the first time the order that underlay all the variety of nature." Wallace's self-directed botanizing quickly revealed that many of the plants he encountered, including some with the most beautiful and curious flowers, were not described in his cherished book. Wallace was in a quandary. The absence of formal education had left him ignorant of the existence of any suitable text of British floras.

Wallace happened by chance to come upon an old issue of the *Gardener's Chronicle*. He read it with growing excitement, particularly the advertisements and reviews of books. Wallace asked a friend for more issues. He found in one of them a notice of the fourth edition of John Lindley's *Elements of Botany*. Described as a comprehensive work, it contained descriptions of all the natural orders of plants and was illustrated by numerous excellent woodcuts. Lindley was professor of botany in the newly founded University of London and Fellow of the Royal Society. Wallace thought this just what he required to advance his botanical studies. The book's ten-shilling price tag was a challenge, however. Wallace was "rather frightened" at the prospect of such a major purchase. He was "always very short of cash; but happening to have [just then] so much in my possession, and feeling that I *must* have some book to go on with," Wallace ordered it at the bookstore of a Mr. Hayward. Lindley's book was excellent. It did have one grave shortcoming. There were scarcely any references to uniquely British species. Wallace was still at a loss as to the names of the plants he was observing daily.

Wallace asked Hayward if he knew of any book that could rectify this defect. The bookseller at once declared that he had a copy of John C. Loudon's comprehensive *Encyclopedia of Plants* (1829). Hayward said he would be glad to lend it, so that Wallace could copy the characters of the British species into the Lindley volume. For the next several weeks, Wallace spent all his leisure time copying the descriptions of the British species and genera into the (fortunately) broad margins of Lindley's work. Wallace was able to identify and copy the majority of plants he had encountered. There were several plants for which Wallace could not assign an unambiguous species classification. Such ambiguities were merely a minor annoyance at the time. But they plagued his thoughts. This practical difficulty in classifying plants was a catalyst that later provoked Wallace to assert that the taxonomic distinction between species and varieties was often an arbitrary, rather than a "natural," property. His collecting and identifying British specimens was both a pleasure and an education. The delight produced by the descriptions of exotic plants was even more decisive in Wallace's path to glory. Even in words and drawings, the wondrous plants exerted a "weird and mysterious charm . . .

which, I believe, had its share in producing that longing for the tropics which a few years later was satisfied in the equatorial forests of the Amazon (Wallace [1905] 1969, 1:192–195). When Wallace left Britain for South America in 1848, he had acquired—on his own initiative—an impressive knowledge of the general principles of plant systematics.

ON HIS OWN IN LEICESTER

Nineteenth-century British surveying was bound to suffer some diminution in its original capacity to absorb experienced practitioners. William had, for some time, attempted to shield his brother from the difficulties in finding remunerative work in that trade. In January 1844, William was forced to tell Alfred, then just turning twenty-one, that he could no longer provide him with any more surveying jobs. He encouraged Wallace to go to London to try to obtain a new type of employment. Wallace lived briefly in the capital, sharing his other brother John's lodgings. With surveying and engineering no longer options, Wallace decided to try for a post at some school. He could teach English, surveying, elementary drawing, and other subjects with which he had some familiarity. He applied, successfully, for a post at the Collegiate School at Leicester. There he taught, for some thirty or forty pounds annual salary (an amount with which he "was quite satisfied"), English, drawing, and some surveying and arithmetic. At Leicester, Wallace availed himself of the mathematical knowledge of his headmaster, the Reverend Abraham Hill. Hill had been a "rather high Cambridge wrangler." He provided Wallace with books on algebra, trigonometry, and calculus. Alfred taught himself some of the rudiments of those subjects. But the major benefit of Hill's encouragement was Wallace's fascination with "the ever-growing complexity of the higher mathematics as exhibiting powers of the human mind so very far above my own" (Wallace [1905] 1969, 1:229–232).

Leicester also possessed a fine town library. Wallace spent several hours each day reading through its collections. He encountered two works that were to exert decisive influences on his career: Alexander von Humboldt's *Personal Narrative of Travels in South America* and Thomas Malthus's *Essay on the Principle of Population*. Von Humboldt's vivid description of the tropics provoked an intense desire in Wallace to travel to those regions, a desire earlier whetted by his reading (probably in 1842) Darwin's *Voyage of the "Beagle"* (McKinney 1972, 5). Humboldt's and Darwin's books affected Wallace deeply and immediately, given his developing passion for natural history. The full impact of Malthus's *Essay* was delayed. When it hit, in 1858, Malthus's work provided Wallace with a major clue to the problem of the origin of species. At Leicester, Wallace met Henry Walter Bates, the entomologist and his future companion in the Amazon. Bates was born in Leicester in 1825, making him two years Wallace's junior. Although an excellent student, Bates, like Wallace, was compelled for family reasons to terminate his formal schooling in 1838. Bates was apprenticed to a local hosiery manufacturer at the tender age of thirteen. He had a prodigious capacity for work, both physical and intellectual. After laboring thirteen hours each day, Bates attended the Leicester Mechanics' Institute at night. There he excelled in Greek, Latin, French, drawing, and composition. Like Wallace, Bates was, and remained throughout his life, an avid reader in a wide variety of fields. Bates was particularly fond of reading Homer in the original and Gibbon's *Decline and Fall of the Roman Empire.*

More significant for the Wallace-Bates's relationship was Bates's passion for entomology. He combed the nearby woods of Charnwood Forest for specimens whenever the opportunity arose. Bates's scientific expertise was already in evidence. At the age of eighteen, he had a short paper on beetles published in the first issue of the *Zoologist* in 1843 (McKinney 1970). Wallace first met Bates at the Leicester library in late 1844 or early 1845. Local libraries were part of the growing, nonmainstream, scientific networks of early nineteenth-century Britain. Their books and opportunities for personal contacts afforded intellectual stimulation for those financially or socially excluded from the existing elitist British universities.[1] Bates introduced Wallace to beetle collecting. Bates's extensive personal collection was all the more amazing to Wallace when he learned that "the great number and variety of beetles, [with] their many strange forms and often beautiful markings or colouring, [had] almost all been collected around Leicester." Bates's friendship provided Wallace with a more direct stimulus than either Darwin or von Humboldt. Equipped with collecting bottle, pins, and a storage box, as well as James F. Stephens's *Manual of British Coleoptera, or Beetles,* Wallace accelerated his pursuit of biological knowledge (Wallace [1905] 1969, 1:232–238).

At Leicester, Wallace heard his first lectures on mesmerism, given by Spencer Hall. Along with spiritualism, this was to become a major, and controversial, field of enquiry for him.[2] Wallace had read George Combe's essay *The Constitution of Man* (1828). Combes's combination of phrenological and progressivist ideas seemed to be corroborated by the mesmeric experiments Wallace witnessed at Leicester and later repeated on his pupils (Wallace [1905] 1969, 1:234–236). Wallace's growing interest in psychical phenomena must be viewed alongside his youthful disavowal of traditional religious institutions. The secular rationalism to which he had been exposed in London was reinforced by public lectures on David Friedrich Strauss's *Life of Jesus* (1835). Further evidence of Wallace's youthful antipathy toward certain prevalent religious doctrines occurs in his annotations to William Swainson's *Treatise on the Geography and Classification of Animals* (1835). He purchased a copy of Swainson in 1842. Wallace dismissed as ridiculous Swainson's attempt to reconcile a literal interpretation of Scripture with modern geology and zoology (McKinney 1972, 6). Wallace rejected special creation. At this period, he regarded empirical investigations and evidence, not natural theology, as the initial path toward the elucidation of the problems of natural history. But Wallace's broadly theistic sympathies lurked in the background of his thought. Among the aspects of life in Leicester he regarded most highly, was "the opportunity of hearing almost every Sunday one of the most impressive and eloquent preachers I have ever met with—Dr. John Brown." Wallace stated that "Brown was one of the few Church of England clergymen who preached extempore, and he did it admirably so that it was a continual pleasure to listen to him." Wallace had become too convinced of the incredibility of large portions of the Bible and of the "absence of sense or reason in many of the doctrines of orthodox religion" to be persuaded of the literal truth of Brown's eloquent sermons. He remained, nonetheless, open to the return "to some form of religious belief" (Wallace [1905] 1969, 1:240).

EARLY CAREER CHOICES

The totally unexpected death of William in February 1845 forced Alfred to return to Neath. With his surviving brother John, he wound up William's business affairs. Wallace left Leicester with considerable regret. Although he regarded the year he spent there as perhaps the most important of his early life, he realized he had neither vocation nor any deep commitment to teaching grade school. He drew from his varied, and often financially precarious, experiences the optimistic moral that adversity has its rewards. Wallace viewed his hardships as

> really useful to me . . . and an important factor in moulding my character and determining my work in life. Had my father been a moderately rich man and had supplied me with a good wardrobe and ample pocket-money; had my brother obtained a partnership in some firm in a populous town or city, or had established himself in his profession, I might never have turned to nature as the solace and enjoyment of my solitary hours, my whole life would have been differently shaped, and though I should, no doubt, have given some attention to science, it seems very unlikely that I should have ever undertaken what at that time seemed rather a wild scheme, a journey to the almost unknown forests of the Amazon in order to observe nature and make a living by collecting.

Apart from the design and construction of a local Mechanics' Institute, Wallace found the work at Neath unpleasant. He determined to give it up as soon as alternative employment became available. He consoled himself, meanwhile, with entomological and botanical collecting. Wallace began to focus his attention on central questions of philosophical biology. He read extensively in those works that dealt, either explicitly or implicitly, with evolution, the origin of species, the geographical distribution of animals and plants, and the difference between species and varieties. Wallace devoured Charles Lyell's *Principles of Geology, Vestiges of the Natural History of Creation* (by the then anonymous author Robert Chambers), James Cowles Pritchard's *Researches into the Physical History of Man* (1813), William Lawrence's *Lectures on Comparative Anatomy, Physiology, Zoology, and the Natural History of Man,* and, for the second time, Darwin's *Voyage of the "Beagle"* (Wallace [1905] 1969, 1:197, 239–240, 254–257).

The year 1845–1846 in Neath, as in Britain generally, was one of heady speculation in railways and their construction. Wallace asked John to give up his job as a journeyman carpenter in London and move to Neath. The two brothers, Wallace suggested, could set up some type of profitable business. This was not an odd request, given their combined practical experience and general knowledge of architecture, building, and engineering, as well as surveying. John agreed. He joined Alfred in January 1846, where they both lodged with Mr. Sims (Alfred's sister Fanny's future husband). When their mother indicated that she would like to live with them, Alfred and John rented a small cottage. Less than a mile from the center of Neath, the cottage had "a pretty view across the valley . . . and the fine Drumau mountain." The brothers' chief work in 1846 was the survey of their own parish of Llantwit-juxta-Neath, for the purpose of a new valuation of the tithe commutation. That work was completed smoothly. Wallace was then told that it was he who would be responsible for collecting the payments from farmers (who could afterward deduct it from their annual rents). This proved to be a highly unpalatable task for Wallace, since many of the farmers were very poor. His discomfort was compounded by the fact that some of the local farmers could not speak English. They could not even understand why they had to make the payments. Wallace often had to deal with farmers who, simply and adamantly, refused to pay him the amounts (sometimes quite small) owed.

The net result was another experience that "disgusted me with business, and made me more than ever disposed to give it all up if I could but get anything else to do." Wallace consoled himself by giving lectures on elementary physics at the Mechanics' Institute to interested workers. In 1895, a half-century later, Wallace received a letter from one of those workers, Matthew

Jones. Jones asked if the author of *Island Life* and the *Malay Archipelago* was the same Mr. Wallace who taught the evening classes at Neath in 1845–1846. He added, "I have often had a desire to know, as I benefited more while in your class . . . than ever I was taught at school. I have often wished I knew how to thank you for the good I and others received from your teaching." This letter is a tribute both to the powerful impact of the Mechanics' Institutes in Victorian Britain and to Wallace's teaching abilities. At Neath, Wallace indulged his growing fascination with all aspects of natural history. John proved far more supportive of his brother's nascent scientific proclivities than William had been. He often accompanied Wallace on lengthy tours of the geologically striking countryside around Neath. It was on one of these nature walks (June 1846), that Wallace obtained his first (and only) specimen of one of the most beautiful British beetles, *Trichius fasciatus* (Wallace [1905] 1969, 1:241–247, 251).

Wallace and Bates had begun corresponding on various subjects but chiefly on their mutual passion of collecting insects. They exchanged specimens by mail. During the summer of 1847, Bates paid Wallace a week's visit at Neath. In his autobiography, Wallace states that "it must have been at this time that we talked over a proposed collecting journey to the tropics, but had not then decided where to go." Fortunately, Bates's widow retained some of these early letters of Wallace to her husband and returned them for the preparation of *My Life*. Wallace quoted a few passages from the letters. Since they do not appear in Marchant's edition of Wallace's letters, these passages are extremely important. They provide excellent clues to what Wallace was thinking about in this momentous period prior to his and Bates's decision to embark on their Amazon travels (Wallace [1905] 1969, 1:254–257). That decision was to alter Wallace's life and career profoundly.

THE LURE OF THE TROPICS

One letter to Bates (11 April 1846) evokes Wallace's increasingly urgent desire to visit, and collect in, the tropics. Commenting on Bates's favorable opinion of Lyell's *Geology*, an opinion he fully shared, Wallace noted that he had recently reread Darwin's *Beagle Journal*. The second reading reinforced his opinion that "as the Journal of a scientific traveller, it is second only to Humboldt's 'Personal Narrative. . . .' [Darwin] is an ardent admirer and most able supporter of Mr. Lyell's views." Wallace declared that he envied Bates for having friends living near him who shared his natural history pursuits. Wallace said that he knew of no one else in Neath who studied even one branch of natural history. "I am quite alone in this respect," he complained. With the exception of his correspondence with Bates, Wallace was a solitary seeker in nature's domain. His reading of Humboldt, in particular, assumed greater importance as one of the major determinants in Wallace's career choice at this turning point in his life (Wallace [1905] 1969, 1:256).

Wallace's sister Fanny also played a role in his decision to embark for South America. In 1834, when the Wallace family's finances had not yet deteriorated completely, Fanny had attended a school in France to perfect her knowledge of the language. She worked in England for several years as a governess and schoolmistress. In 1844, Fanny obtained a better-paying post at an episcopal college in Georgia. She then was appointed headmistress at a private school near Montgomery, Alabama (Wallace [1905] 1969, 1:72). Fanny came back to England in 1847. To commemorate her return from America, she invited her two surviving brothers, Alfred and John, to join her for a week's vacation in Paris (Wallace [1905] 1969, 1:72, 223, 256). Alfred enjoyed the visit, especially since he had an excellent interpreter. The pleasures of the French capital were not the only positive aspects of the trip for Wallace. He made several visits to the museums of the Jardin des Plantes.

In his last letter to Bates before their imminent voyage, Wallace compared the holdings of beetles and butterflies of the Jardin des Plantes with the collections in the insect room at the British Museum. He boasted to Bates that he was way beyond the mere phase of studying local collections. "I should like," Wallace asserted, "to take some one family to study thoroughly, principally with a view to the theory of the origin of species. By that means, I am strongly of [the] opinion that some definite results might be arrived at." Wallace concluded this letter, which exudes excitement, by referring to Lorenz Oken's *Elements of Physiophilosophy* (Wallace read the English translation of 1847). He wanted "very much to see [it]. . . . There is a review of it in the *Athenaeum*. It contains some remarkable views on my favourite subject—the variations, arrangements, distribution, etc., of species" (Wallace [1905] 1969, 1:256–257). What finally triggered the two naturalists to undertake their scientific odyssey was their reading of William H. Edwards's just-published *Voyage up the River Amazon* (1847). Edwards's provocative description of the beauty and grandeur of tropical vegetation proved irresistible. Wallace and Bates decided that the Amazon Basin, with Pará as their headquarters, was the most appropriate region for their expedition. Edwards's statement that traveling and living expenses there were moderate clinched matters. Wallace and Bates contacted Edward Doubleday, entomologist and a curator of the butterfly collections of the British Museum. Doubleday assured them the whole of northern Brazil was comparatively unknown and a collection of novel species of insects, land shells, birds, and mammals would easily pay their expenses. The latter factor was essential. Wallace and Bates, unlike either von Humboldt (a man of independent means) or Darwin and Huxley (both of whom were officially attached to naval surveys), were without financial support. They had to rely on the sale of their specimens. But Wallace and Bates were not, primarily, professional collectors. Their travels were motivated, rather, by a "true love for the objects of" their affection (Camerini 1996, 47).

After a further study of South American animal and plant holdings at the British Museum, Wallace and Bates made arrangements with an agent, Samuel Stevens, to receive and sell their collections. Stevens was an enthusiastic collector of British Coleoptera and Lepidoptera. He was also the brother of John Crace Stevens, the influential natural history auctioneer of King Street, Covent Garden. Samuel Stevens and Wallace were to enjoy a mutually advantageous and wholly amicable business and personal relationship for the next fifteen years. Stevens served the crucial function of being Wallace's highly talented and trusted guardian of the immense collections he sent to England from South America and the Malay Archipelago. Absent from England, except for the brief period (1852–1854) between his two major expeditions, Wallace came to rely completely on Stevens. He preserved those specimens that were intended for Wallace's private collections and, equally important, arranged for the sale of Wallace's exotica at the most advantageous prices the London market would yield. Stevens thereby kept Wallace supplied with cash during his long years abroad. He provided Wallace with the provisions and other stores required for the laborious and often dangerous expeditions in South America and the Malay regions. Finally, Stevens kept Wallace informed on matters of general scientific interest (Wallace [1905] 1969, 1:264–266). The rich and complex relationship between Stevens and Wallace was a major component of Wallace's development as a scientist. Wallace, and Bates, had "tapped into an existing set of resources—the practical skills needed for collecting, the knowledge base for identifying what to collect, and the critical link through an agent to collectors of natural objects, whose money made their trip possible. They, [thus], participated in the culture of collecting—the networks of correspondence, publishing, specimen trading, and equipment manufacture—that made their venture financially" and materially possible (Camerini 1996, 48, 62–64).

CONTEXTS OF VICTORIAN EXPLORATION

A fortuitous meeting with Edwards, who happened to be in London, put the finishing touches on Bates's and Wallace's plans. He wrote letters of introduction for them to some of his American friends in Pará. Bates's parents invited the two prospective scientific adventurers to spend a final week with them at Leicester before sailing off. At Leicester, Wallace visited some of his old friends from his teaching days there. More important, he practiced shooting and skinning birds. These techniques perfected, Wallace would use them to great advantage in South America as well as the Malay Archipelago. He also visited the wild district of Charnwood Forest, which had been so significant a force in Bates's development as a naturalist. Everything in order, the two companions left Leicester by stagecoach for a "cold and rather miserable journey" to Liverpool. They made one stop at Chatsworth, to see its palm and orchid houses, then considered the finest specimens of their kind in England. In Liverpool, Wallace and Bates called on J. G. Smith, who had collected butterflies at Pará and Pernambuco. Smith invited the two to dine with him. He showed them his collections and gave them extremely useful information about the country and peoples of the Amazon basin. Smith also spoke of the great natural beauty of the region. His verbal landscape strengthened Wallace's and Bates's vision of what they expected to encounter on their arrival in South America. The two naturalists left England on 26 April 1848 aboard the *Mischief,* a relatively small but fast sailing ship destined for Pará (Wallace [1905] 1969, 1:265–267).

When Wallace embarked for Pará (now Belém, Brazil) on 26 April 1848 he was a gifted but untested naturalist. He had a clear objective. Wallace wanted to investigate all aspects of the "species question" (Wallace [1905] 1969, 1:257). When he returned to England from the Malay Archipelago fourteen years later, he was a biologist of established reputation. Wallace was one of many Victorian naturalists whose careers benefited significantly from voyages of exploration at early phases of those careers. Not only Englishmen but also numerous Europeans were guided by the writings and vision of von Humboldt. The Prussian polymath's own voyages of exploration to many regions of the globe in the early nineteenth century provided a model for aspiring naturalists. The Humboldtian ethos eloquently argued for the importance of expeditions that observed, measured, and collected everything they encountered. It became a touchstone for nineteenth-century scientific advances in fields as diverse as climatology, zoology, botany, and physical and human geography. Humboldt provided a standard and a center of gravity for the proliferation of scientific exploring expeditions beginning in the middle third of the century.[3] Wallace considered von Humboldt's *Personal Narrative* "the first book that gave me a desire to visit the tropics" (Wallace [1905] 1969, 1:232).

For Wallace, Darwin, Huxley, Joseph Dalton Hooker, Bates, and numerous others, training in the field served as a rite of passage en route to becoming scientists of eminence. They came from different social backgrounds and possessed varying agendas. But all were united by the social worlds and institutions that shaped the development of nineteenth-century natural history research. The twin mantras of Victorian Britain, colonialism and capitalist industrialization, established the context for a shared network of relationships, opportunities, and obligations within which the naturalist-scientists lived and functioned. The connection between natural history and the Royal Navy was not as significant in Wallace's career as it was for Hooker, Huxley, and many others. Wallace's travels and field explorations were, nonetheless, situated in the common milieu that defined mid-Victorian scientific exploration. Wallace's activities, particularly in the second of his two major voyages of discovery, were inseparable from the worldwide British colonial network as well as the lucrative trade in naturalists' specimens shipped back to London.[4] His voyages to the Amazon and the Malay Archipelago were scientific and cultural rites of passage for him. They were also, in the deepest sense, journeys of personal exploration and development.

SOUTH AMERICA: TRAVELS PERSONAL AND SCIENTIFIC

Wallace reached Pará on 28 May 1848 and remained in South America for four years. He recounted his experiences in *A Narrative of Travels on the Amazon and Rio Negro* (1853). In this, his first major book, Wallace described his often-risky journeys and vividly recorded his observations on the flora and fauna of the Amazon basin (Wallace 1853a). Wallace's interest in the region's human inhabitants was no less keen. Much of the *Narrative* is devoted to a detailed account of the life and customs of the residents of the cities, as well as of the native tribes he encountered in traveling through the interior of the continent.

Despite the novelty of Pará, Wallace was at first disappointed. "The weather was not so hot," he explained, "the people were not so peculiar, the vegetation was not so striking, as the glowing picture I had conjured up in my imagination, and had been brooding over during the tedium of a sea-voyage. . . . [During] the first week of our residence in Pará, though constantly in the forest in the neighbourhood of the city, I did not see a single humming-bird, parrot, or monkey." The naturalist's trade had to be learned by patience and experience. One of the major achievements of the Amazon travels was Wallace's transformation from a novice into an expert naturalist. He acquired a formidable ability to discern the various peculiarities of different regions—"the costume of the people, the strange forms of vegetation and the novelty of the animal world." After only two months of collecting at Pará, Wallace and Bates were able to send their first specimens back to England. It was a staggering total of more than 1,300 species of insects (Wallace 1853a, 3–4, 34).

The coexplorers traveled together for two years, with brief independent forays, in the environs of Pará, along the Tocantins River and up the Amazon as far as Barra (today Manaus). At Barra, where the Rio Negro joins the Amazon, the two decided to separate permanently in order to maximize their collections. Wallace went on to explore the Rio Negro, the relatively unknown Uaupes, and other northern tributaries; Bates continued along the Upper Amazon. The journey up the Uaupes was one of the high points of Wallace's South American sojourn. He wrote, nearly sixty years later, that so far as he had heard, "no English traveller had to this day ascended the Uaupes River so far as I did, and no collector has stayed at any time at Javita, or has even passed through it" (Marchant [1916] 1975, 23–24). The map that Wallace constructed remained the most accurate one until the first years of the twentieth century (McKinney 1972, 14; Knapp 1999). It detailed the course and width of the river for its first four hundred miles and the location of the various Indian tribes inhabiting its banks. Wallace also specified the location of the most important vegetable products of the surrounding forest.[5]

It was on the Uaupes that Wallace had his first encounter with "man in a state of nature—with absolute uncontaminated savages!" Unlike the half-civilized tribes among whom he had lived previously, the Uaupes Indians were in "every detail . . . original and self-sustaining as are the wild animals of the forests, absolutely independent of civilization, and who could and did live their own lives in their own way, as they had done for countless generations before America was discovered." The appearance and behavior of these Indians left an indelible impression on Wallace. "I could not have believed," he declared, "that there would be so much difference in the aspect of the same people in their native state and when living under European supervision. The true denizen of the Amazonian forests, like the forest itself, is unique and not to be forgotten" (Wallace 1853b, 190–94; [1905] 1969, 1:288).

Wallace's fascination with the Amazonian aborigines did not preclude a critical response to the culture of the half-civilized and urban inhabitants. In district after district, he noted that the "indolent disposition of the people . . . will prevent the capabilities of this fine country from being developed till European or North American colonies are formed." Despite Wallace's profound love of unspoiled nature, he shared (at this period in his life) the prevalent Victorian conviction that nineteenth-century European civilization defined the standard by which all cultures should be measured. His condemnation of the widespread practice of slavery in Brazil is chauvinistic as well as moralistic. "Can it be right," Wallace asked, "to keep a number of our fellow creatures in a state of adult infancy,—of unthinking childhood? It is the responsibility and self-dependence of manhood that calls forth the highest powers and energies of our race. It is the struggle for existence, the 'battle of life,' which exercises the moral faculties and calls forth the latent sparks of genius. The hope of gain, the love of power, the desire of fame and approbation, excite to noble deeds, and call into action all those faculties which are the distinctive attributes of man" (Wallace 1853a, 55, 83).

A Narrative of Travels is the first sustained example of Wallace's skillful interweaving of biological theory with personal reflections, a mode he would employ to even greater advantage in *The Malay Archipelago* (1869). As a contribution to the scientific literature, *Narrative* is less impressive for one obvious reason. The ship Wallace had taken for his return voyage to England in 1852 sank. This resulted in the loss of his extensive private collection of insects and birds. The majority of his sketches, drawings, notes, and journals also ended up in the ocean depths. Fortunately, Wallace's agent Stevens had insured the greater part of the Amazonian collection, so Wallace's finances were not completely devastated (Camerini 1997, 369). That he was able to write the *Narrative,* the *Palm Trees of the Amazon and Their Uses* (1853), and several technical papers, using only the meager materials he salvaged from the burning vessel, is testimony to Wallace's growing mastery as a naturalist. Some careful sketches of the Amazonian species of palms and fishes, his diary while on the Rio Negro, some notes for maps of that river and the Uaupes, plus the letters he had sent home were the only treasures that made their way with Wallace back to England (Wallace [1905] 1969, 1:305–306, 313–314).[6] Despite the shortcomings of the *Narrative* from the standpoint of precise data and documentation, Wallace did set forth observations and ideas on geographical distribution and speciation that were to be fully developed in his later work.

THE PERPLEXITIES OF GEOGRAPHICAL DISTRIBUTION

The study of the geographical distribution of animals and plants was a familiar one at the time of Wallace's voyage. Explanations of distributional data were generally embedded within the framework of the argument from design. This argument received an influential rendition in the 1830s with the publication of the eight *Bridgewater Treatises* (Gillispie [1951] 1959, 209–216). Most naturalists believed that the multiplicity of species, their detailed—at times apparently perfect—adaptations to their particular environments, and the succession of organic forms in time accorded with the wisdom and foresight of a Creator God. Eighteenth- and early-nineteenth-century advances in geology had seriously eroded the authority of literal biblical interpretation. But the tradition of natural theology remained strong, especially in Great Britain. William Paley, William Whewell, Hugh Miller, William Buckland, and Adam Sedgwick were among the powerful voices for divine providence in the course of nature (Gillispie [1951] 1959, 219–222; Brooke and Cantor 1998, 141–167).

By the 1840s and early 1850s, however, the orthodox theory of special creation had lost much of its luster. The idea that each species had been uniquely and expressly created to occupy its particular niche in the environment was becoming less tenable. A consensus remained, nonetheless, that the development of life was part of a harmonious and divinely inspired plan worked out through the agency of secondary (natural) causes. Baden Powell, a Liberal Anglican clergyman-scientist, declared in 1855 that "the term 'creation' indeed, especially as respects new species, seems now, by common consent, to be adopted among geologists as a mere *term of convenience,* to signify simply the fact of origination of a particular form of animal or vegetable life, without implying anything as to the *precise mode* of such origination—as simply involving the assertion that a period can be assigned at which that species appears, and before which we have no evidence of its appearance" (Powell 1855, 399, 476–481). Powell later was one of seven well-known contributors to the controversial *Essays and Reviews,* published in 1860, which argued for the compatibility of evolution and religion—an extremely touchy issue since the appearance the year before of *Origin of Species* with its detailed public exposition of natural selection.

In the *Narrative,* Wallace was stating received opinion when he noted that "countries possessing a climate and soil very similar, may differ almost entirely in their productions. Thus Europe and North America have scarcely an animal in common in the temperate zone; and South America contrasts equally with the opposite coast of Africa; while Australia differs almost entirely in its productions from districts under the same parallel of latitude in South Africa and South America." However, on the assumption that the characteristic fauna of a region was the direct "product" of environmental conditions, this dissimilarity of the faunas of ecologically identical areas was problematical. For those regions separated by great oceans or mountain barriers, it could plausibly be argued that the present faunas were (in some unspecified manner) the product of history as well as ecology. Faunal differences would be expected, maintained by water and land barriers to the dispersion and intermingling of species. Lyell had clearly laid out these general anomalies of geographical distribution in *Principles of Geology,* a book that profoundly influenced Wallace's thinking and that he probably took with him to South America in 1848 (McKinney 1972, 20–21). But as Wallace had discovered in his Amazon travels, places "not more than fifty or a hundred miles apart often have species of insects and birds at the one, which are not found at the other" (Wallace 1853a, 326–327). The existence of several closely related but not identical species in adjacent areas of practically identical climate and topography was unexpected. Wallace had also been much struck by the fact that rivers, though generally easily passable by birds and insects, frequently acted as sharp demarcations between closely related species. The two beautiful butterflies *Callithea sapphira* and *C. leprieuri,* collected by him along the Amazon, are each restricted to one bank though separated only by the expanse of the river. A similar localization of the monkeys of the Amazon region emphasized the anomaly to Wallace. There should not have been a number of slightly different species in a given ecological niche (Wallace 1853a, 328–329).

The recognition that the distribution of closely allied species was often marked by precise boundaries is the most important consequence of Wallace's Amazon field studies. Thereafter, he insisted on the need to specify the exact locale at which species and varieties were collected. This rigor was not the practice of naturalists then. Vague designations such as "Amazon" or even "South America" abounded in the standard catalogs of species. Wallace himself was not aware of the need for precise notation of locale when he began collecting (Wallace [1905] 1969, 1:377). He did not, in the *Narrative,* invoke his new distributional data to support explicitly a theory of evolution. But Wallace was fully aware of their significance in suggesting that closely allied species in adjacent areas resulted from an earlier isolation of populations from an original stock. Chance migration across river barriers, for example, would have set up the conditions for subsequent variation of the now isolated populations. Continued variation would, over time, result in the formation of distinct species (McKinney 1972, 24–26). In one passage, Wallace directly attacked the hypothesis that adaptation to conditions was the determining factor in the distributional patterns of species:

> In all works on Natural History, we constantly find details of the marvelous adaptation of animals to their food, their habits, and the localities in which they are found. But naturalists are now beginning to look beyond this, and to see that there must be some other principle regulating the infinitely varied forms of animal life. It must strike every one, that the numbers of birds and insects of different groups, having scarcely any resemblance to each other, which yet feed on the same food and inhabit the same localities, cannot have been so differently constructed and adorned for that purpose alone. Thus the goat-suckers, the swallows, the tyrant fly-catchers, and the jacamars, all use the same kind of food, and procure it in the same manner (they all capture insects on the wing), yet how entirely different is the structure and the whole appearance of these birds!
>
> (Wallace 1853a, 58)

Though no evolutionary explanation was put forward, the *Narrative of Travels* signaled the direction Wallace's ideas were to take.

THE MALAY ARCHIPELAGO: WALLACE AT THE PERIPHERY

Although the solution to the problem of the origin of species still eluded him, Wallace returned from South America with a more sophisticated understanding of major conceptual issues in natural history. In London, he prepared the manuscripts for *A Narrative of Travels on the Amazon and Rio Negro* and *Palm Trees of the Amazon and Their Uses,* both published in 1853. His South American collections made Wallace's name known to the leading members of the zoological and entomological societies. Wallace attended meetings of both groups assiduously. He presented several detailed papers on Amazonian butterflies, monkeys, and fishes, as well as one titled "On the Insects Used for Food by the Indians of the Amazon." During this period he also studied the extensive insect and bird collections of the British Museum and the Linnean Society and the botanical collections at the Kew Herbarium.

Wallace was committed to another voyage as the best means of securing his reputation as a naturalist and of providing the data required for resolving the species problem. He decided on an expedition to the Malay Archipelago. The collections in London indicated that the archipelago was promisingly rich in the number and variety of its species. The fact that the natural history of the region, with the exception of the island of Java, was relatively unexplored was an intellectual invitation to Wallace. It was also a financial inducement. Wallace knew that specimens of the archipelago's lesser-known fauna would find a ready market in Europe. While Wallace labored, collected, and shipped back his magnificent specimens from the Malay Archipelago, his agent and friend Stevens worked wonders in London. Stevens used his contacts in the burgeoning specimen trade to execute sales that were rewarding to himself and to Wallace. Stevens's business savvy was sufficient to cover expenses for Wallace during his eight-year Malay sojourn. He also made shrewd investments with part of the surplus from the London sales. These provided Wallace with a small income for the first few years on his return to London in 1862 (George 1979).

Start-up funds for a voyage to the East were another matter. In 1854, Wallace's private resources were meager. Through the intercession of Sir Roderick Murchison, president of the Royal Geographical Society, Wallace gained free passage on the steamer *Euxine,* which left England in March 1854. Disembarking at Alexandria, he proceeded overland to Suez, where he boarded the steamer *Bengal* and arrived in Singapore on 20 April 1854. Thus began "the eight years of wandering throughout the Malay Archipelago, which constituted the central and controlling incident" of Wallace's life. He traveled nearly 14,000 miles, collected more than 125,000 specimens, and formulated those ideas that became the basis of evolutionary biology (Wallace [1905] 1969, 1:320–327, 331–332, 336). These travels also resulted in Wallace's *The Malay Archipelago: The Land of the Orang-utan, and the Bird of Paradise; a Narrative of Travel, with Studies of Man and Nature* (Wallace [1869] 1962), regarded as one of the masterpieces of Victorian travel writing (Raby 2001, 199).

Wallace's "eight years of wandering" were not haphazard or solitary. They took place within the framework of an established colonial society in the East Indies. It was the culture, both ideological and physical, created by generations of European explorers, traders, engineers, missionaries, and doctors that facilitated Wallace's arduous travels, collecting, and achievements in the Malay Archipelago. Through the efforts of the British, Dutch, Danish, and German contacts he had, or made, Wallace secured housing, directions to collecting sites, translators, servants, and introductions to local rulers (Camerini 1997, 370–371). The most notable instance of his "British connection" was Wallace's introduction to Sir James Brooke, the English rajah of Sarawak (in Borneo). Wallace first called on him in late September 1854. Brooke received him "most cordially, and offered me every assistance at Sarawak . . . [he] has given me a letter to his nephew, Capt. Brooke, to make me at home [during the rajah's frequent official absences]. . . . I look forward with much interest to see what he has done and how he governs" (Marchant [1916] 1975, 42–43). Wallace developed a very cordial relationship with Sir James. As he wrote his mother in late 1855:

> You will see that I am spending a second Christmas Day with the
> Rajah . . . I have lived a month with the Dyaks and have been a jour-
> ney about sixty miles into the interior. They are a very kind, simple
> and hospitable people, and I do not wonder at the great interest Sir
> J. Brooke takes in them. In moral character, they are far superior to
> either Malays or Chinese, for though head-taking has been a cus-
> tom among them it is only as a trophy in war. In their own villages
> crimes are very rare. Ever since Sir J. has been here, more than twelve
> years, in a large population there has been but one case of murder
> in a Dyak tribe, and that one was committed by a stranger who had
> been adopted into the tribe. . . . I have now seen a good deal of Sir
> James, and the more I see of him the more I admire him. With the
> highest talents for government he combines the greatest goodness of
> heart and gentleness of manner. At the same time he has such con-
> fidence and determination, that he has put down with the greatest

ease some conspiracies of one or two Malay chiefs against him. It is a unique case in the history of the world, for a European gentleman to rule over two conflicting races of semi-savages with their own consent, without any means of coercion, and depending solely upon them for protection and support, and at the same time to introduce the benefits of civilisation and check all crime and semi-barbarous practices. . . . Under his government, "running amuck," so frequent in all other Malay countries, has never taken place. The people are never taxed but with their own consent, and Sir J.'s private fortune has been spent in the government and improvement of the country; yet this is the man who has been accused of injuring other parties for his own private interests, and of wholesale murder and butchery to secure his government!

(Marchant [1916] 1975, 48–49)

Wallace's letter is more revealing than he realized. It displays his usual observational acuity and his keen interest in sociopolitical affairs. It also shows Wallace's gentle character and his predilection for failing to detect in those whom he respected shortcomings obvious to others. This particular trait was to lead Wallace later into a series of contentious defenses of fraudulent spiritualist mediums in England. Wallace's praise of Sir James betrays traces of "Rule Britannia." His generally positive response to British colonial rule during the 1850s did not, however, preclude a critical ethnological outlook that was not typical of many Eurocentric Victorian travelers. "The more I see of uncivilized people," Wallace affirmed, "the better I think of human nature on the whole, and the essential differences between civilized and savage man seem to disappear. . . . I can safely say that in any part of Europe where the same opportunities for crime and disturbance existed, things would not go so smoothly as they do here. We sleep with open doors, and go about constantly unarmed" (Wallace [1905] 1969, 1:342–343). Wallace was still convinced of the superiority of European technical and scientific prowess. He was less convinced of the superior status of European cultural mores. However, Wallace's crusades to expose the darker side of British imperialism, both at home and abroad, lay in the future. Now, he delighted in beholding and interpreting the spectacular natural history of the Malay Archipelago.

WALLACE'S BOND WITH INDIGENOUS PEOPLES

Important as the Malay colonial network was, Wallace's brilliant scientific fieldwork would not have been possible without the aid of the indigenous peoples. Crucial to his collecting endeavors was the availability of help in carrying

out such specific, and often tedious, functions as locating and procuring animals and processing specimens for on-site description and transport back
to England. Wallace was also wholly dependent on servants and laborers for
carrying his precious specimens and personal belongings, building temporary housing, communicating with local inhabitants, and procuring food and
other necessary supplies. He would likely have perished during one of his frequent bouts with illness without the aid of faithful attendants. They secured
whatever remedies were available and nursed him back to some semblance
of health and vigor (Camerini 1996, 53). Wallace had first employed as his
assistant a sixteen-year old English boy, Charles Allen, who accompanied
him to the East in 1854. Bright and hard working, Charles did not live up
to Wallace's expectations as a scientific traveling companion. Charles proved
careless in his personal habits and his work in preparing and mounting Wallace's ever-increasing collections of butterflies, beetles, and birds (Marchant
[1916] 1975, 39, 41).

Exasperated, Wallace wrote his sister Fanny from Borneo (25 June 1855),
asking her to confer with Stevens on a replacement for Charles. Stevens had a
possible new assistant in mind. Wallace entreated Fanny to "let me know what
you think of him. Do not tell me merely that he is a 'very nice young man.' Of
course he is. So is Charles a very nice boy, but I could not be troubled with
another like him for any consideration whatever. . . . Ask [Stevens's suggested
assistant] whether he can live on rice and salt fish for a week on occasion . . .
whether he likes the hottest weather in England—whether he is too delicate to
skin a stinking animal—whether he can walk twenty miles a day—whether he
can work, for there is sometimes as hard work in collecting as in anything. Can
he draw (not copy)? Can he speak French? Does he write a good hand? Can
he make anything? Can he saw a piece of board straight? (Charles cannot, and
every bit of carpenter work I have to do myself.)" Stevens retracted his suggestion. When Wallace left Sarawak in 1856, he was at a loss. Charles stayed
behind to work with the bishop of Sarawak, training to become a teacher. He
and Wallace remained friends, and Charles periodically rejoined Wallace's
expeditions (Wallace [1869] 1962, 417–419; Marchant [1916] 1975, 46–
50). Wallace recorded in his field journal that each of his more than sixty
separate Malay journeys necessarily involved preparation and significant loss
of time on a number of occasions. It can be inferred that he was referring to
trouble with finding suitable help.[7]

One native who did fulfill all Wallace's requirements was a Malay boy
named Ali. Hired as Wallace's personal servant in 1855, Ali turned out to be
Wallace's most trusted and valued companion. Initially, Ali was to serve as
Wallace's cook and handyman and to help him learn the Malay language. As
Ali's personal and intellectual qualities revealed themselves, Wallace came to

rely on him in many other respects. He taught Ali how to shoot birds, to skin them properly, and to mount the skins expertly. Ali was an adept boatman and hiker. His talents were essential in surmounting the numerous difficulties and dangers Wallace encountered in his travels into unchartered regions. Ali became familiar with Wallace's needs and habits. When additional native help was required, Ali taught them those duties that he knew were essential to the success of the evolutionist's research and observations.

During Wallace's stay in Ternate, Ali got married. Since his wife continued to live with her family, Ali's role as Wallace's "faithful companion of almost all my journeyings among the islands of the far East" remained undiminished until Wallace's final departure from Singapore in 1862 (Wallace [1905] 1969, 1:382–383). The bond forged between Wallace and Ali was powerful. Theirs was a complex relationship that straddled the elusive boundary category of servant and master in the colonial East Indies (Taylor 1983, 17–19, 69–71). When the young Harvard zoologist Thomas Barbour traveled to the Dutch East Indies in 1907, he encountered a "wizened old Malay man" on the island of Ternate. The Malay introduced himself by saying "I am Ali Wallace." Barbour "knew at once that there stood before me Wallace's faithful companion of many years." Barbour took Ali's photograph and sent it to Wallace on his return home. Wallace subsequently wrote Barbour "a delightful letter acknowledging it and reminiscing over the time when Ali had saved his life, nursing him through a terrific attack of malaria" (Barbour 1950, 36). Ali was more than Wallace's headman. He was Wallace's teacher in the native language and manners and, in the truest sense, a friend and companion during years of thrilling adventure. When Wallace left the Malay Archipelago, he gave Ali presents of money, guns and ammunition, and sundry provisions and tools. By local standards, Ali had become a wealthy man through Wallace's gifts. These gifts were not only tokens of the sincere respect, trust, and gratitude on Wallace's part for all that Ali had done to make his sojourn in the Dutch East Indies as comfortable and productive as possible. They also represent Wallace's acknowledgment of the vital role played by Ali in the realm of knowledge-making itself (Camerini 1996, 60).

THE MALAY ODYSSEY: CONTINUED AND CONCLUDED

Wallace's lengthy stay in the Malay Archipelago exposed him to prolonged and intimate contact with the varied peoples making up that melting pot of races. Chinese, Malays, Dyaks, Portuguese, Dutch, English, and the occasional American constituted a human tapestry as rich and intriguing as the complex faunal assemblages of birds, monkeys, butterflies, and fish. "The streets of Singapore," Wallace wrote his brother-in-law Thomas Sims in

March 1856, "on a fine day are as crowded and busy as Tottenham Road, and from the variety of nations and occupations far more interesting. I am more convinced than ever that no one can appreciate a new country in a short visit. After two years . . . I only now begin to understand Singapore and to marvel at the life and bustle, the varied occupations, and strange population, on a spot that so short a time ago was an uninhabited jungle" (Marchant [1916] 1975, 50–51).

In another letter to Fanny's husband (Timor, 15 March 1861), Wallace elaborated in a postscript on his religious views. He knew these differed from the traditional faith of his sister and brother-in-law. "In my early youth I heard," Wallace explained,

> as ninety-nine-hundreths of the world do, only the evidence on one side, and became impressed with a veneration for religion that has left some traces even to this day. I have since heard and read much on both sides, and pondered much upon the matter in all its bearings. . . . I have [also] wandered among men of many races and many religions. I have studied man and nature in all its aspects, and I have sought after truth. In my solitude I have pondered much on the incomprehensible subjects of space, eternity, life and death. I think I have fairly heard and fairly weighed the evidence on both sides, and I remain an *utter disbeliever* in almost all that you consider the most sacred truths. . . . I am thankful I can see much to admire in all religions. To the mass of mankind religion of some kind is a necessity. But whether there be a God and whatever be His nature, whether we have an immortal soul or not, or whatever may be our state after death, I can have no fear of having to suffer for the study of nature and the search for truth, or believe that those will be better off in a future state who have lived in the belief of doctrines inculcated from childhood, and which are to them rather a matter of blind faith than intelligent conviction.

Wallace quickly added to Thomas: "This for yourself; show the *letter only* to my mother" (Marchant [1916] 1975, 65–67). His Malay travels reinforced Wallace's disavowal of adherence to any traditional orthodox religion. But Wallace's emerging evolutionary worldview was compatible with a broader spiritual and teleological framework that would become more overt on his return to England. Wallace's eight-year sojourn in the Malay Archipelago is of crucial importance in the history of biology. It cemented his reputation as one of the premier scientists of the Victorian era. But Wallace was never a closeted scientist. His Malay travels were as much a personal journey of discovery as it was a scientific odyssey. From 1854 to 1862 Wallace was exposed to an exotic realm, the creatures of which—humans, animals, and plants—sparked his imagination in numerous directions. Although the "main object of all [his] journeys was to obtain specimens of natural history, both for my private collection and to supply duplicates to museums and amateurs," Wallace had other motives (Wallace [1869] 1962, xi).

Wallace's evocative account of those eight years was published in 1869 (fig. 1). He dedicated *The Malay Archipelago* to Darwin. Its subtitle is, *The Land of the Orang-Utan and the Bird of Paradise; a Narrative of Travel with Studies of Man and Nature.* The subtitle and the text itself indicate that Wallace was already pondering what he would later term the question of *Man's Place in the Universe* ([1903] 1907) and *The World of Life* (1910a). Wallace explained the six-year delay in publishing the full account of his travels as necessary, given the immediate and immense task of studying, cataloging, and interpreting theoretically his abundant collections of fauna and flora. He wrote nearly thirty technical articles for the transactions and proceedings of the major scientific societies of London, including the Linnean, the Zoological, and the Entomological, in the years following his return to England. *The Malay Archipelago,* touching as it did on the social, political, and cultural

Figure 1. Photograph of Wallace at age forty-six, taken when *The Malay Archipelago* was published.

characteristics of the countries and peoples he encountered and lived among, is a masterly piece of Victorian travel writing. It went through ten editions, was immediately translated into German, French, and Danish, and exerted a major influence on a whole generation of tropical naturalists (Marchant [1916] 1975, 201). Although Wallace provided extensive descriptions of the natural history of those island countries, it is significant that the concluding chapter of the book is entitled "The Races of Man in the Malay Archipelago." Unraveling the mysteries of the human species was an enduring thread of Wallace's life and career. He concluded the preface to *The Malay Archipelago* with the hope that those "who have been in any way interested in my travels and collections, may derive from the perusal of my book some faint reflexion of the pleasures I myself enjoyed amid the scenes and objects it describes" (Wallace [1869] 1962, xii).

WALLACE AND LITERATURE

One such reader was the novelist Joseph Conrad. Conrad regarded *The Malay Archipelago* as his favorite book (Smith 1991, 455; Raby 2001, 199; Camerini 2002, 14–15). And Lyell, in a letter to Wallace (13 March 1869), declared that "nothing equal to it has come out since Darwin's 'Voyage of the *Beagle.*' . . . The history of the Mias [Orang-utans] is very well done. I am not yet through the first volume, but my wife is deep in the second and much taken with it. It is so rare to be able to depend on the scientific knowledge and accuracy of those who have so much of the wonderful to relate" (Marchant [1916] 1975, 287). Darwin's reaction was also highly enthusiastic (Marchant [1916] 1975, 194).

A letter written by Wallace to his schoolfellow and oldest friend George Silk, near the end of his Malay travels, encapsulated the richness of Wallace's personality. In his quiet and dignified way, Wallace dined heartily at the feast of human experience:

> The Amazon now seem[s] to me quite unreal—a sort of former ex-
> istence or long-ago dream. Malays and Papuans, beetles and birds,
> are what now occupy my thoughts, mixed with financial calcula-
> tions and hopes for a happy future in old England, where I may live
> in solitude and seclusion, except from a few choice friends. . . . Tell
> me about yourself, your own private doings, your health, your visits,
> your new and old acquaintances. . . . But, above all, tell me what you
> read. . . . Follow the advice in [the] *Family Herald* Article on 'Hap-
> piness', *Ride a Hobby,* and you will assuredly find happiness in it,
> as I do. Let ethnology be your hobby, as you seem already to have

put your foot in the stirrup, but *ride it hard.* If I live to return I shall
come out strong on Malay and Papuan races. . . . You must read
'Pritchard' through, and Lawrence's 'Lectures on Man' carefully;
but I am convinced no man can be a good ethnologist who does not
travel, and not *travel* merely, but reside, as I do, months and years
with each race. . . . When I went to New Guinea, I took an old copy
of 'Tristram Shandy', which I read through about three times. It is an
annoying and, you will perhaps say, a very gross book; but there are
passages in it that have never been surpassed. . . . I have lately read a
good number of Dumas's wonderful novels, and they *are* wonderful,
but often very careless and some quite unfinished. 'The memoirs of
a Physician' is a wonderful wild mixture of history, science, and romance; the second part, the Queen's Necklace, being the most wonderful and, perhaps, the most true.

(Wallace [1905] 1969, 1:365–367)

Wallace was an insatiable reader. He devoured innumerable scientific works,
as well as an eclectic array of political tracts, philosophical treatises, religious
texts, and, above all, novels and other fictional genres. Starting from his
childhood, when his father had been appointed librarian to the Hertford
Town Library, Wallace consumed all the books he could get. His daughter
Violet recalled that her father was never without a book in hand. She and her
brother William were especially impressed by their father's ability to quote
at delightful length passages from Lewis Carroll's *Alice through the Looking-
Glass,* which became a family classic (Marchant [1916] 1975, 11, 350).

Wallace's fascination with literature is fundamental to his life and work.
He was an eloquent stylist. His scientific works are models of clarity and force.
His works on sociopolitical and cultural matters are equally passionate and
compelling, if not always convincing to all readers. Wallace's zest for works on
all subjects prevented his scientific theorization from myopia and profoundly
enhanced his complex appreciation of this earthly "paradise." Two sections,
in particular, of *The Malay Archipelago* testify to Wallace's literary as well
as scientific prowess. The first dealt with his quest to find the mysterious
orangutan. The second described his pursuit of the several varieties of birds
of paradise. The opening paragraph of the first chapter entices the reader:

If we look at a globe or a map of the Eastern hemisphere, we shall
perceive between Asia and Australia a number of large and small islands, forming a connected group distinct from those great masses
of land, and having little connexion with either of them. Situated
upon the Equator, and bathed by the tepid water of the great tropical

oceans, this region enjoys a climate more uniformly hot and moist
than almost any other part of the globe, and teems with natural productions which are elsewhere unknown. The richest of fruits and the
most precious of spices are here indigenous. It produces the giant
flowers of the Rafflesia, the great green-winged Ornithoptera (princes
among the butterfly tribes), the man-like Orang-Utan, and the gorgeous Birds of Paradise. It is inhabited by a peculiar and interesting
race of mankind—the Malay, found nowhere beyond the limits of this
insular tract, which has hence been named the Malay Archipelago.
To the ordinary Englishman this is perhaps the least known part of
the globe.

(Wallace [1869] 1962, 1)

In March 1855, Wallace left Sarawak to go to the coal works that were
being opened near the Simunjon River, east of Sarawak. He lodged comfort-
ably at the house of an English engineer named Coulson. Finding the locale
particularly rich in insect species, Wallace had a small house built with two
rooms and a verandah. He lived there for the next nine months. Simunjon
proved to be the most successful site for Wallace's insect collecting. Owing
to the clearing of virgin forests for the construction of coal pits and a railway
line, Wallace found Simunjon an insect collector's dream, with twenty square
miles of country filled with decaying trees, irresistible sites for a myriad of
beetles, wasps, butterflies, and other insect species. One of Wallace's chief
reasons for coming to Simunjon was to "see the Orang-utan (or great man-
like ape of Borneo) in his native habitat." Wallace hoped "to study his habits,
and obtain good specimens of the different varieties and species of both sexes,
and of the adult and young animals. In all these objects I succeeded beyond
my expectations" (Wallace [1869] 1962, 30).

ORANGUTANS AND BIRDS OF PARADISE

Wallace's account of his first sightings and subsequent contacts with the
orangutans (or *mias,* as they were called by the natives) is intriguing. It depicts the complex relationship of the specimen collector to the object of his
quest. Capturing orangutans was a tricky undertaking. They were large and
powerful. One big adult male orang, when speared by one of Wallace's aides,
seized the man. "In an instant," Wallace recorded, "the orang got hold of
the man's arm, which he seized in his mouth, making his teeth meet in the
flesh above the elbow, which he tore and lacerated in a dreadful manner.
Had not the others been close behind the man would have been . . . killed,
as he was quite powerless; but they soon destroyed the creature with their

spears and choppers. The man remained ill for a long time, and never fully recovered the use of his arm." Wallace, with his rifle, never faced that type of bodily danger. He killed, skinned, beheaded and prepared for shipment back to England many excellent specimens of orangutans. These were quickly bought by ardent purchasers, including the British Museum and the Museum at Derby.

Wallace was acutely interested in observing the habits and social structures of these evolutionary cousins of humanity. When he shot and killed one adult female, he found her young infant nearby. It was crying and terrified. Wallace carried the infant to his home. He nursed it back to health and enjoyed its companionship for several months. The naturalist who had slaughtered numerous adult orangs now found himself feeding the infant with "rice-water from a bottle with a quill in the cork, which after a few trials it learned to suck very well." As Wallace's adopted infant grew healthier, he "fitted up a little box for a cradle, with a soft mat for it to lie upon, which was changed and washed every day." Wallace soon found it necessary to wash the little *mias* as well. He noted that it "enjoyed the wiping and rubbing dry amazingly, and when I brushed its hair seemed to be perfectly happy, lying quite still with its arms and legs stretched out while I thoroughly brushed the long [orange] hair of its back and arms." A few weeks later, Wallace obtained a young harelip monkey (*Macacus cynomolgus*). The two youngsters became good friends. They afforded Wallace the delight of observing closely two similar species with very different habits and mannerisms. Despite his best efforts at nursing and medical attention, Wallace's pets succumbed to a fatal fever. Saddened but stalwart, Wallace spent his remaining weeks in Simunjon and its environs carefully observing and recording in detail the daily habits of the orangutan. He killed more specimens and prepared their skins and skeletons for London's eager market (Wallace [1869] 1962, 26–49). Wallace's firsthand and intensive study and knowledge of the orangutan made him almost unique among naturalists of the period. Despite their great appeal for collectors and museums, the orangutan in its native habitats remained relatively unstudied until the mid-twentieth century (van Oosterzee 1997, 184). The role played by his close contact with orangutans in Wallace's developing evolutionary cosmology is complex. Despite his playing with and nursing of these creatures, Wallace still seems to have regarded the gulf between these Great Apes—the "highest" representatives of the nonhuman world—and humans as conceptually unbridgeable. Did these observations of the orangutan during the 1850s plant one of the seeds that would prompt Wallace to declare in the 1860s that human evolution was guided by factors other than natural selection and contribute to his refusal to leave the divine out of the history of *Homo sapiens'* development (Berry 2002, 191)? At the very least, there are hints in the 1856

"On the Habits of the Orang-utan of Borneo" that Wallace's observations of these apes were leading him to invoke more explicitly the concept of design in nature (see chap. 4).[8]

The other major quest, indeed obsession, for Wallace was to see and obtain specimens of the magnificent birds of paradise (*Paradiseidae*). Though scarcely larger than crows, these gloriously plumed birds, Wallace exclaimed, "really deserve [their] name." He thought them "one of the most beautiful and most wonderful of living things." Certain species of birds of paradise were endemic to the Aru Islands. Other species were collected by Wallace in trips to the Molucca Islands and mainland New Guinea (Camerini 2002, 107). Wallace gained detailed knowledge of the habits of the various species of birds of paradise from firsthand observation and from the accounts of his hired hunters and conversations with natives of the region. Lest one underestimate the rigors confronting the tropical naturalist, Wallace provided in *The Malay Archipelago* a vivid if discomfiting account for the armchair traveler:

> At our first stopping-place sand-flies were very abundant at night,
> penetrating to every part of the body, and producing a more lasting
> irritation than mosquitoes. My feet and ankles especially suffered,
> and were completely covered with little red swollen specks, which
> tormented me horribly. . . . After a month's incessant punishment,
> those useful members rebelled against such treatment and broke into
> open insurrection, throwing out numerous inflamed ulcers, which
> were very painful, and stopped me from walking. So I found my-
> self confined to the house, with no immediate prospect of leaving
> it. Wounds or sores in the feet are especially difficult to heal in hot
> climates, and I therefore dreaded them more than any other illness.
> The confinement was very annoying, as the fine hot weather was ex-
> cellent for insects, of which I had every promise of obtaining a fine
> collection; and it is only by daily and unremitting search that the
> smaller kinds, and the rare and more interesting specimens, can be
> obtained. When I crawled down to the river-side to bathe, I often saw
> the blue-winged Papilio ulysses, or some other equally rare and beau-
> tiful insect; but there was nothing for it but patience, and to return
> quietly to my bird-skinning, or whatever other work I had indoors.
> The stings and bites and ceaseless irritation caused by these pests of
> the tropical forests would be borne uncomplainingly; but to be kept
> prisoner by them in so rich and unexplored a country, where rare
> and beautiful creatures are to be met with in every forest ramble—a
> country reached by such a long and tedious voyage, and which might

not in the present century be again visited for the same purpose—is a
punishment too severe for a naturalist to pass over in silence.

Wallace's greatest consolation during these ordeals was the specimens his as-
sistants brought back to him daily during his long confinement. The sight of
these birds of paradise "was quite a relief to my mind . . . , for I could hardly
have torn myself away from Aru had I not obtained specimens." The psychological, as well as scientific, significance of Wallace's success in obtaining
knowledge and specimens of these glorious birds was profound. As the first
English naturalist to view the birds of paradise in their native forests, Wallace's
experience stimulated him "to continue my researches in the Moluccas and
New Guinea for nearly five years longer. It is still the portion of my travels
to which I look back with the most complete satisfaction" (Wallace [1869]
1962, 353–354, 369).

The first European voyagers to the region, in search of precious spices
such as cloves and nutmeg, had been given dried skins of "birds so strange
and beautiful as to excite the admiration even of those wealth-seeking rovers."
The early natural history of the family was shrouded in myth and ignorance
of their actual characteristics. The Malays called them "God's birds." The
Portuguese traders, seeing no feet or wings on the dried specimens and not
venturing to obtain any authentic, observational knowledge of the living creatures, named them "birds of the sun"—speculating that they somehow lived
only in the air, always turning toward the sun. It was the Dutch travelers
who first denominated them *Avis paradiseus,* or paradise bird. As late as the
1760s, when Linnaeus named the largest species of the family *Paradisea apoda*
(the footless paradise bird), no perfectly complete specimen had been seen
in Europe. When Wallace arrived in the Aru Islands in 1855, even the Malay
traders called them *Burong mati,* or dead birds. This indicated that actual
sightings of the living birds were still rare (Wallace [1869] 1962, 419–420).

During his five years of searching, Wallace was able to observe only five
species. He offered two reasons for the elusiveness of the birds of paradise,
the first of which was almost mystical: "It seems as if Nature had taken precautions," Wallace suggested, "that these her choicest treasures should not be
made too common, and thus be undervalued." Wallace considered it somewhat of a inherent challenge by "Nature" that the birds of paradise, whose
"exquisite beauty of form and colour, and strange developments of plumage
are calculated to excite the wonder and admiration of the most civilized and
the most intellectual of mankind, and to furnish inexhaustible materials for
study to the naturalist, and for speculation to the philosopher," should be
witnessed in their actual habitats by only the most dedicated and intrepid

seekers. The second reason was far more pragmatic. Birds of paradise, like other rare and desirable commodities, lead "the inhabitants of uncivilized countries to conceal minerals or other natural products with which they may become acquainted, from the fear of being obliged to pay increased tribute, or of bringing upon themselves a new and oppressive labour" (Wallace [1869] 1962, 439–440). Wallace's observation was more astute than he first realized. When he did, finally, succeed in obtaining excellent specimens of birds of paradise, they were probably worth as much as any bird or insect on the planet to the well-heeled English collectors to whom Stevens sold Wallace's yield (Quammen 1997, 85). Charles Allen, who rejoined Wallace for some of these searches, met with innumerable obstacles by the indigenous peoples in his quest for specimens of these birds.

But what a sight they were when actually observed in their native habitat. Wallace's description of the king bird of paradise (the *Paradisea regia* of Linneaus) is rhapsodic:

> This lovely little bird is only about six and a half inches long . . . [but its] head, throat, and entire upper surface are of the richest glossy crimson red, shading to orange-crimson on the forehead, where the feathers extend beyond the nostrils more than half-way down the beak. The plumage is excessively brilliant, shining in certain lights with a metallic or glassy lustre. The breast and belly are pure silky white, between which colour and the red of the throat there is a broad band of rich metallic green. . . . From each side of the body beneath the wing springs a tuft of broad delicate feathers about an inch and a half long, of an ashy colour, but tipped with a broad band of emerald green, bordered within by a narrow line of buff. These plumes are concealed beneath the wing, but when the bird pleases, can be raised and spread out so as to form an elegant semicircular fan on each shoulder. But another ornament sill more extraordinary, and if possible more beautiful, adorns this little bird. The two middle tail feathers are modified into very slender wire-like shafts, nearly six inches long, each of which bears at the extremity, on the inner side only, a web of an emerald green colour, which is coiled up into a perfect spiral disc, and produces a most singular and charming effect. The bill is orange yellow, and the feet and legs of a fine cobalt blue.

In stark contrast to the glorious male, Wallace noted that the "female of this little gem is such a plainly coloured bird, that it can at first sight hardly be believed to belong to the same species" (Wallace [1869] 1962, 426–427). The precise evolutionary explanation for the widespread difference in appearance between the male and female of many species of animals was to be a matter of major import to Wallace for the remainder of his career. His views on the causes and significance of sexual differences were integral to his conceptualization of evolutionary theory and natural selection. They would also be dominant, and controversial, components of his speculations regarding the evolution and future prospects of the human species.

MORE QUESTIONS ABOUT THE HUMAN SPECIES

The Malay Archipelago is at once a scientific treatise, a popularized tract on travel and discovery, and a detailed account of observations of the natural world, including the human species. Its complex and compelling narrative structure constitutes a candid but entirely public diary of eight critical years in Wallace's life. Written some six years after the completion of those travels (1862), it necessarily involved much reflection and introspection on his part. Wallace pondered matters cultural, political, and ethnological, as well as those relating more strictly to the floral, faunal, and geological peculiarities of that region of the globe. The fact that the book went through ten editions (the last revision appeared in 1891 and was reprinted six times up to 1922) testifies to its enduring appeal to a wide readership. It chronicled a crucial phase in Wallace's evolution as a theoretical scientist, an astute observer of natural phenomena, and an increasingly committed political and social activist.

Wallace ended his wide-ranging survey with a final chapter devoted to "The Races of Man in the Malay Archipelago." Many of the previous chapters had dealt, directly or indirectly, with the various individuals and tribes that Wallace encountered during his eight-year sojourn. The final chapter was intended to "conclude this account of my Eastern travels with a short statement of my views as to the races of man which inhabit the various parts of the archipelago, their chief physical and mental characteristics, their affinities with each other and with surrounding tribes, their migrations, and their probable origin." In keeping with his overarching goal of establishing biogeography as a science treating all species, including *Homo sapiens*, Wallace declared that "two strongly contrasted races inhabit the Archipelago," the Malays and the Papuans. These two major human groups were distinguishable by their geographical distribution as well as their intellectual and social characteristics. The Malays occupied almost exclusively the larger western half of the archipelago. The Papuans, "whose headquarters are New Guinea and several of the adjacent islands," occupied the eastern regions. In a manner similar to that in which he described the precise locale of animal and plant species in the Amazon basin, Wallace pinpointed notable human distributional patterns. He observed that in the area demarcated by the major geographical division of the Malays and Papuans "are found tribes who are also intermediate in their chief characteristics." He immediately added that "it is sometimes a nice point to determine whether they belong to one or the other race, or have been formed by a mixture of the two."

Wallace deemed the Malays "undoubtedly the most important of these two [major] races." It was a broad racial category, in which he included the Javanese, the Bugis (inhabitants of Celebes), the Moluccan-Malays, as well as the "more savage" Malay tribes (the Dyaks of Borneo, the Battaks of Sumatra, and the Jakuns of the Malay Peninsula). Wallace's reasoning is explicitly Eurocentric. The Malays, he asserted, are the most civilized of the inhabitants of the archipelago because they have "come most in contact with Europeans, and [therefore] alone [have] any place in history." Despite their diversity of language and dialects, Wallace considered the Malay race to "present a considerable uniformity of physical and mental characteristics." Malays shared a common light reddish-brown skin color, with hair being invariably black and straight and of a rather coarse texture. The constancy of their hair attributes was so striking that Wallace asserted that "any lighter tint, or any wave or curl in it, is an almost certain proof of the admixture of some foreign blood." Wallace's extensive observations of the Malays, over so long and intimate a period among them, led him to conclude that their mental and moral traits were nearly as uniform as their physical appearance. The Malay, he declared, is impassive, reserved, diffident, undemonstrative, slow and deliberate in speech, and "particularly sensitive to breaches of etiquette, or any interference with the personal liberty of himself or another." As an example of the latter trait, Wallace mentioned that he often found it difficult to get one Malay servant to waken another: "He will call as loud as he can, but will hardly touch, much less shake, his comrade. I have frequently had to waken a hard sleeper myself when on a land or sea journey." Wallace described the "higher classes" of Malays as "exceedingly polite, [having] all the quiet ease and dignity of the best-bred Europeans." But the Malays had a darker side to their character, "a reckless cruelty and contempt of human life." This dual nature, Wallace inferred, had given rise to the disparate accounts of Malaysian temperament recorded by previous travelers to the archipelago. Wallace regarded the intellect of the Malays as "rather deficient. They are incapable of anything beyond the simplest combinations of ideas, and have little taste or energy for the acquirement of knowledge." What "civilization" the Malays possessed, Wallace argued, could not have been indigenous but was likely due (in addition to European contact) to many of them having "been converted to the Mahometan or Brahminical religions."

The other major race of the archipelago, the Papuan, Wallace regarded as "in many respects the very opposite of the Malay." The skin color of the

Papuans was a deep sooty-brown or black, sometimes approaching "the jet-black of some negro races." Their hair was harsh, dry, and very curly, growing outward in adults to the "frizzled mop which is the Papuans' pride and glory." In stature, the Papuans decidedly surpassed the Malays. They were as tall, or sometimes taller, than the average European. Despite their darker hue, Wallace regarded the facial characteristics of the Papuan as being more similar to European appearance than were the Malays. He asserted that the facial traits of the Papuan and their characteristic frizzy hair (which covered their head as well as many other parts of their bodies) were sufficiently distinctive to enable observers to distinguish "at a glance" between individuals belonging to either of the two main races of the region. The moral characteristics of the Papuan appeared, to Wallace, "to separate him as distinctly from the Malay as do his form and features." The Papuans were impulsive and demonstrative in speech and action, bold, noisy, joyous, and laughter loving. Their superior skills in domestic decoration and carvings compared to the Malays led Wallace to assert that the Papuans had a greater artistic sensibility. Like the Malays, however, the Papuan character had its dark side. He remarked that they "seem very deficient . . . in the affections and moral sentiments, [and] are often violent and cruel." Wallace attributed these latter traits, interestingly enough, to the "intellect of this race," which he was inclined to "rate . . . somewhat higher than that of the Malays." The harsher moral traits and more severe discipline of the Papuans as compared to the "listless and apathetic" character of the Malays, Wallace suggested, "may be chiefly due to that greater vigour and energy of mind which always, sooner or later, leads to the rebellion of the weaker against the stronger—the people against their rulers, the slave against their master, or the child against its parent" (Wallace [1869] 1962, 446–450).

In a passage such as this, was Wallace describing the Papuans? Or the inhabitants of his native Britain? By the time the *Malay Archipelago* was published (1869), Wallace had already commenced his path toward social and political activism. This would lead him increasingly to advocate the reform of British society, raising the poor at the expense of the rich. It was the publication of *Malay* that brought him into personal contact with John Stuart Mill. In 1870, Mill encouraged Wallace to become a member of the General Committee of the Land Tenure Reform Association. Wallace was to remain in that association until the formation of the more radical Land Nationalisation Society in 1880 provided him with a broader forum for his goal of sociopolitical transformation (Marchant [1916] 1975, 382). John (later Lord) Morley was another reader of Wallace's book who pondered its cultural speculations. As the new editor of the *Fortnightly Review*, Morley was eager to make evolution the *Fortnightly*'s creed (Desmond 1998, 367). He would clearly have been responsive to the evolutionary framework of Wallace's book. But Morley, like Mill, also was deeply interested in political issues.

Some thirty years after first reading the *Malay Archipelago*, Morley wrote Wallace (in a letter dated 31 October 1900) that "in older days I often mused upon a passage of yours in the 'Malay Archipelago,' contrasting the condition of certain types of savage life with that of life in a modern industrial city." Morley was responding to a letter written a week earlier by Wallace (20 October 1900). By then, Wallace was an avowed socialist. He appealed to Morley as one of the few politicians "left to us, who . . . is able to become the leader of the English people in their struggle for freedom against the monopolists of land, capital, and political power." Wallace alluded to a recent speech in which Morley intimated that, "if the choice for this country were between Imperialism and Socialism, [he would be] inclined to think the latter the less evil of the two." Lord Morley told Wallace that he did not suppose he would ever become a convert to Wallace's socialist cause. But he added that he always remembered "J. S. Mill's observation, after recapitulating the evils to be apprehended from Socialism, that he would face them in spite of all, if the only alternative to Socialism were our present state" (Marchant [1916] 1975, 395–396). Land nationalization and socialism belong to later phases of Wallace's career (see chap. 5), but that two such scientifically literate figures as Mill and Morley were also impressed with Wallace's political interpolations in the *Malay Archipelago* testifies to the powerful and multilevel impact that Wallace's account exerted on its readers in 1869.

Rich in political allusion as the final chapter is, Wallace's main purpose was to demonstrate that a line could be drawn that divided the two main races of the archipelago into separate geographical, as well as ethnological and cultural, entities. Wallace announced, with obvious satisfaction, that the human boundary he proposed "is on the whole almost as well defined and strongly contrasted, as is the corresponding zoological division of the Archipelago, into an Indo-Malayan and Austro-Malayan region." That the human and animal boundary demarcations did not agree exactly was, for Wallace, neither problematic nor unexpected. The human line is not coincident with what has come to be known as "Wallace's Line" but lies further east. Wallace's Line, which refers to the faunal division of the Malay Archipelago, is one of the most disputed topics in biogeography. The term itself was first coined by Huxley in 1868 (Camerini 1993, 700). If Wallace's Line is highly contentious, the human boundary is even more so. As Wallace (and others) pointed out, the human species has migratory abilities and interbreeding propensities that differ from those of most other animal species. What Wallace found conclusive for his main argument was the "remarkable fact, and something more than a mere coincidence," that the dividing lines "approach each other so closely as they do." Both these boundaries, he asserted, were "true and natural" and reflected the operation of evolutionary forces amid the constraints of a relatively stable, but not static, geological earth history (Wallace [1869] 1962, 452–455).

Using the extensive observational data from his field journals, and benefiting from the additional years of analytic scrutiny of that data after he returned to England, Wallace forged the final chapter of the *Malay Archipelago* into a concise summary of his views on the nature and distinctions of the diverse representatives of the human species (Brooks 1984, 59). But while the book came out in 1869, Wallace's Malay odyssey actually ended in 1862. On leaving Singapore in late January of that year, Wallace returned via Bombay, Suez and Alexandria in Egypt, Malta, Marseilles, and thence across France. He arrived in London in the spring (Wallace [1905] 1969, 1:382–385). The thirty-nine-year-old traveler came home a celebrated if controversial figure.

TOWARD AN EVOLUTIONARY BIOGEOGRAPHY

In the two decades following his return to England, Wallace's scientific prestige grew rapidly. He was elected president of the Entomological Society of London (1870–1872) and president of the Biology Section of the British Association for the Advancement of Science (BAAS [1876]). Wallace's presidential address (sec. D, biology) to the BAAS meeting in Glasgow on 6 September 1876 was divided into two parts. Both the first, "On Some Relations of Living Things to Their Environment," and the second, "Rise and Progress of Modern Views as to the Antiquity and Origin of Man," were reprinted in *Nature* the following day (Wallace 1876a). Most significant, this period marked the publication of two of Wallace's greatest works. The *Geographical Distribution of Animals* ([1876] 1962) and *Island Life* ([1880] 1892) are seminal contributions to the modern science of biogeography. Wallace contributed a constant stream of major theoretical and descriptive articles to the leading natural science journals of the period. His subjects ranged from ornithological classification to the comparative antiquity of continents to the origin of species and genera to the significance of glacial epochs. He also was a prolific reviewer of books on a vast array of topics, including his famous reviews of Darwin's *The Expression of the Emotions in Man and Animals* (1872) and Lyell's *Principles of Geology* (10th ed., 1867–1868) and *The Geological Evidences of the Antiquity of Man* (4th ed., 1873).[9]

In the 1860s, Wallace published a series of important articles on natural history and evolutionary biology, as well as *The Malay Archipelago*. But he had yet to produce a major book that synthesized the factual scope and theoretical power of his varied contributions to evolutionary science. At the urging of his two close friends, the eminent ornithologists Alfred Newton and Philip Lutley Sclater, Wallace was persuaded to undertake a comprehensive analysis of the global distributional patterns of animals. The result would be the classic *Geographical Distribution of Animals*. It was Wallace's technical magnum opus and stands as one of the masterpieces of Victorian biology.

Since antiquity, naturalists had been aware that different regions of the globe housed distinct and characteristic fauna and flora. The differences in organic beings were assumed to be due to varied climates and physical conditions. It was Georges-Louis LeClerc, comte de Buffon who, in the mid-eighteenth century, definitively challenged the adequacy of these traditional explanations. He pointed out that the tropical regions of the Old and New World, regions of practically identical ecology, differed strikingly in their indigenous mammals. By 1820, Buffon's observation had been broadened, notably by von Humboldt and the Swiss botanist Augustin Pyrame de Candolle, to include most other animals and plants. Wallace was, of course, familiar with the work of von Humboldt. He would have also been familiar with the ideas of Candolle, which were summarized by both Lyell and William Swainson. Wallace had purchased a copy of Swainson's *A Treatise on the Geography and Classification of Animals* (1835) in 1842. He made copious annotations in the Swainson book, many of which were objections to Swainson's overriding goal of harmonizing geology and zoology with biblical literalism. But Wallace was alert to Swainson's useful zoogeographical observations and summaries (Wallace [1905] 1969, 1:354; McKinney 1972, 6). "The great work of Lyell," which had furnished Wallace "with the main features of the succession of species in time," had become a bible of different sorts to Wallace (Wallace [1905] 1969, 1:355). Thus Wallace, like many naturalists of the early 1850s, now recognized that any regions, even of identical ecology, separated by barriers (such as mountains or oceans) would have distinct and characteristic organisms. Most important, it was understood that the present distributional patterns of animals and plants were determined by historical factors (past changes, both organic and geological) as well as by existing ecological conditions (Browne 1983). It was Wallace's genius to combine the data of geographical distribution with the concept of successive appearance of species in the famous 1855 essay "On the Law Which Has Regulated the Introduction of New Species." Although he was informed by Stevens soon after that piece was published that "several naturalists express[ed] regret that I was 'theorizing,' when what we had to do was to collect more facts," Wallace was determined to forge ahead (Wallace [1905] 1969, 1:355). In a series of articles in the late 1850s and early 1860s, in *The Geographical Distribution of Animals* ([1876] 1962), and in *Island Life* ([1880] 1892), Wallace succeeded in incorporating the manifold data of animal distribution into a unified theory.

He explained both existing and past zoological features of the various continents and islands on the basis of geological history and the dispersal and evolution of animals. Wallace's synthesis of zoology, geology, and evolution by natural selection established a causal framework for zoogeography, which was central to the formulation of evolutionary biogeography.

One of the major goals of nineteenth-century biogeography was the determination of a set, or sets, of regions that accurately described distributional patterns. An early and influential schema, limited to the world's flora, was that proposed by Candolle. His *Essai elementaire de geographie botanique* (1820) divided the globe into twenty regions, each of which possessed a characteristic (or endemic) flora. During the following four decades, the concept of biogeographical regions gained increased acceptance, although the number and boundaries of such regions varied with different authors. Zoological regions were crucial elements in the development of Wallace's evolutionary thought. He used the theoretical concept and visual map images of mammalian regions in formulating his views on evolutionary descent. Wallace became increasingly devoted to establishing precise global regions that would provide the conceptual and visual means to bring rigor and order to the multitude of details presented by the study of zoological (and, by implication, botanical) geography (Camerini 1993).

In 1858, P. L. Sclater, who became secretary of the Zoological Society of London, theorized that the earth was divided into six great ornithological regions: (1) Palearctic (Europe, northern Asia to Japan, and Africa north of the Atlas mountains); (2) Ethiopian or Western Paleotropical (Africa south of the Atlas mountains, southern Arabia, and Madagascar); (3) Indian or Central Paleotropical (India, southern Asia, and the western half of the Malay Archipelago); (4) Australian or Eastern Paleotropical (the eastern half of the Malay Archipelago, Australia, New Zealand, and most of the Pacific Islands); (5)Nearctic (Greenland and North America to northern Mexico); and (6) Neotropical (southern Mexico, South America, and the West Indies). Each of these six regions was, Sclater claimed, inhabited by a distinct set of bird populations (Sclater 1858, 130). Using the extensive data from his travels in the Malay Archipelago, Wallace argued that Sclater's assignment of the western half of the archipelago to the Indian ornithological region and the eastern half to the Australian region was valid in every branch of zoology (Wallace 1860). Surprisingly, the striking differences in the fauna between the eastern and western halves of the archipelago—marsupials, for instance, are confined to the eastern half—seemed to be precisely demarcated by the Strait of Lombock. This strait between the islands of Bali and Lombock, merely fifteen miles wide, marked the limits and abruptly separated two of the great zoological regions of the globe. The lack of any significant ecological differences between the two halves of the archipelago was seized on by Wallace. He asserted that the faunal dissimilarities were the result of past geological configurations different from those of the present. Separate evolutionary histories underlay the distributional patterns of the now proximate regions.

WALLACE'S LINE

The theory that the earth's surface had undergone significant changes in time was a central tenet of the new geology, enshrined in Lyell's *Principles of Geology* (1830–1833). Wallace's 1855 essay "On the Law" drew freely on geological speculation in explaining curious distributional phenomena. In his 1857 essay on the distribution of animals in the Aru Islands, Wallace argued that the shallow seas separating the various islands of the eastern half of the Malay Archipelago implied past land connections between them. Now, in the 1860 essay on the "Zoological Geography of the Malay Archipelago," Wallace tied these observations together and offered explicit explanations. He suggested that the faunal similarity of the eastern islands of the Malay Archipelago to New Guinea and Australia implied a former "great Pacific continent," of which the present islands and Australia are the surviving fragments. Analogously, the faunal similarity of the western islands—including Borneo, Java, and Sumatra—to southern Asia argued for a past "extension of Asia as far to the south and east as the Straits of Macassar and Lombock." In support of this view, Wallace noted that a "vast submarine plain unites together the apparently disjointed parts of the Indian zoological region . . . so completely that an elevation of only 300 feet would nearly double the extent of tropical Asia." Most significant, that plain terminates abruptly in the deep sea of the Moluccas and the Strait of Lombock—that is, at the limit of the Indian region.

The two halves of the archipelago, Wallace concluded, despite their present proximity, belonged to "regions more distinct and contrasted than any other of the great zoological divisions of the globe." South America and Africa, separated by the vast expanse of the Atlantic, seemed to Wallace not as dramatically different as the Indian and Australian regions. Further, the sharp contrasts between the faunas of the latter two are "almost unimpaired at the very limits of their respective districts; so that in a few hours we may experience an amount of zoological difference which only weeks or even months of travel will give us in any other part of the world!" Wallace cited the presence of elephants, monkeys, orangutans, pheasants, and trogons in the Indian region against the marsupials, parrots, and birds of paradise of the Australian region (Wallace 1860, 172, 174, 178–179). The boundary between the two regions, later known as Wallace's Line, was tentatively fixed by Wallace as coinciding with the deep sea separating Borneo and Celebes in the north to the strait between Bali and Lombock in the south. Wallace himself later proposed slightly different depictions of the precise location of the boundary. The actual location of Wallace's Line has been a source of debate among scientists to the present day (George 1981).

Wallace's Line found quick acceptance in the biogeographical literature and was widely adopted by zoogeographers. (Floral distributions present different issues and anomalies, and plant geographers found Wallace's boundary less useful.) The allure of a slender line separating marsupials from tigers, and honeyeaters and cockatoos from barbets and trogons, was potent not only to scientists but also to laypersons. Wallace's Line was enthusiastically noted in the popular literature of the period. Ernst Haeckel outdid all his contemporaries when he asserted in 1893 that when crossing "the narrow but deep Lombok Strait we go with a single step from the Present Era to the Mesozoicum." It was inevitable that such striking assertions would call forth objections and attempted refutations. After 1890, doubts were frequently expressed as to the validity of the specific line posited by Wallace. As the distributional data became better known, Wallace himself was less positive about the exact demarcation described by the faunal line that had come to bear his name. Defenders and opponents continue to battle it out, and an impartial study of the status of Wallace's Line is still lacking. The issues involved in locating and assessing the significance of any precise borderline between different biogeographical regions are highly complex. Wallace was fully aware of these complexities, both conceptual and observational. His attempts to posit present faunal boundaries that reflected past distributional patterns testify to the boldness of his evolutionary vision. There was a relative paucity of data, both faunal and geological, in the late nineteenth century. This compelled Wallace and others to select arbitrarily a number of "indicator" species. They then based the outlines of biogeographic regions and subregions on the distribution of these species. Although many of the details of his original boundaries have necessarily been altered as more abundant data became available, Wallace's insights were fundamental to the subsequent development of the science of biogeography (Mayr 1976, 626–628, 642–643).

CONTINENTS, OCEANS, AND LAND BRIDGES

"On the Zoological Geography of the Malay Archipelago" (1860) is a culmination of Wallace's early views on geographical distribution. It placed him, around 1860, clearly within the continental extensionist tradition. The majority of naturalists invoked postulated past land extensions of greater or lesser extent to account for present similarities of plants and animals in regions now separated by tracts of water. In 1846, Edward Forbes had postulated the existence of five past land bridges to account for similarities between British and various continental floras and faunas. He extended this idea to include other continental extensions of great magnitude. He championed an ancient Atlantic continent to explain the observed relationship between the species of the Azores, Madeira, and the Canary Islands and those of North America and Europe (Forbes 1846). Not all naturalists initially accepted each of Forbes's postulated extensions. But an important precedent had been set for subsequent investigations and theoretical explanations of geographical (particularly disjunctive) distribution. Wallace endorsed the extensionist hypothesis in his 1860 essay. He stated that it was specifically those cases in which islands possessed a rich and varied fauna closely allied with that of adjacent islands or continents that forced the conclusion that a "geologically recent disruption [had] taken place." Conversely, the distinctness of the faunas of regions now separated by seas, no matter how narrow, implied the lack of any land connection in, at least, the recent geological past (Wallace 1860, 182–83).

The sophisticated association of geographical distribution with geological changes stamps Wallace's 1860 essay as a seminal work in the evolutionist tradition. He regarded it as laying the theoretical foundation for his later work in zoogeography, and in many respects it did. One aspect of that essay, however, is conspicuous by its absence from Wallace's later theories. He ceased to utilize major continental extensions to explain present distributional anomalies. The conversion of Wallace to a position that made him a forceful opponent of the extensionist tradition and the preeminent defender of the doctrine of the permanence of the continents and oceans was a crucial development in nineteenth-century evolutionary science (Fichman 1977). Wallace's analysis of global distribution data gradually convinced him that Sclater's system had theoretical as well as descriptive significance. The assumption of the general permanence of the earth's topography, with distinct and geologically enduring continents and oceans, provided Wallace with a *vera causa* for the existence of Sclater's six well-defined zoogeographic regions. He was to devote a great portion of his subsequent scientific work and polemicization to the establishment of the thesis that the present distribution of the earth's biota reflected migration and dispersal over a relatively fixed surface rather than any major alteration or movement of that surface itself in time (Wallace 1892b).

Wallace's essay "On the Physical Geography of the Malay Archipelago" (1863) signaled the first major departure from his previous position. He now stressed that land connections could be inferred only in special instances where the geological evidence, as well as distributional data, was overwhelming (Wallace 1863, 227). Former land connections between widely separated regions, as well as between adjacent lands separated by deep seas, he deemed unlikely. Since Wallace held the six zoogeographical regions to represent fundamental geophysical *and* evolutionary divisions, any connections between them in recent geological epochs became inadmissible. Henceforth, Wallace would restrict explanations of distributional patterns to past migrations across land and sea masses similar in their general outlines to the present oceans and continents (Wallace 1863, 226–227, 233).

ANOMALOUS DISTRIBUTIONS

The 1863 essay elicited the approval of both Darwin and Lyell. Darwin thought it "an epitome of the whole theory of geographical distribution" (Lyell 1867–1868, 2:346–353; Marchant [1916] 1975, 132). Their reactions are not surprising. Wallace was for the first time fully embracing the hypothesis of the general permanence of the oceans and continents that underlay Lyell's geological uniformitarianism. Wallace's stand was made explicit the following year in "On Some Anomalies in Zoological and Botanical Geography" (Wallace 1864a). This essay treated several cases of apparently anomalous distributional patterns that had been advanced as objections to Wallace's extension of Sclater's ornithological regions. Wallace refined his argument that the six regions represented "a true Zoological and Botanical division of the earth." Unlike the various schemes proposed by naturalists that were generally intended to apply only to a particular group of organisms, Wallace claimed that Sclater's divisions were "well adapted to become the foundation for a general System of Ontological regions" (Wallace 1864a, 111–113). There had been a long tradition of descriptive biogeography in the eighteenth and early nineteenth centuries, but most such systems were ahistoric (Larson 1986). It was precisely the historical framework of Sclater's regions that enabled Wallace to seize on their import for his new conviction that evolutionary theory depended directly on the premise of continental permanence. In 1899, Sclater asserted that his original six regions had become elevated to the rank of biogeographical orthodoxy, particularly after Wallace had endorsed them unequivocally in his 1876 treatise (Sclater and Sclater 1899, chap. 1). Despite the extensive debates on the most accurate set of biogeographical regions, Sclater's claim and his assessment of Wallace is cogent. The Royal Geographical Society's *Atlas of Zoogeography* of 1911 was largely predicated on Sclater's regions as developed by Wallace (Bartholomew, Clarke, and Grimshaw 1911, 4–12). That Huxley was a friend of Sclater's, and keenly interested in the emerging science of biogeography

(Bowler 1996, 384–386, 391–394; Desmond 1998, 457, 465), provides yet another link in the complex professional and personal relationship between Wallace and Huxley.

In the 1863 essay, Wallace sought to indicate "how Zoological and Botanical regions are formed, or why organic existences come to be grouped geographically at all." He advanced five premises that he claimed must underlie any discussion of geographical distribution. First, all species have a tendency "to diffuse themselves over a wide area, some one or more in each group being actually found to have so spread" and become "dominant species." Second, there exist "barriers, checking or absolutely forbidding that diffusion." Third, there has been a continual and "progressive change or replacement of species, by allied forms," throughout the earth's history. Fourth, there has been a gradual change in certain features of the earth's surface leading "to the destruction of old and the formation of new barriers." And last, natural selection entails that these changes of climate and physical conditions will often "favour the diffusion and increase of one group, and lead to the extinction or decrease of another." Given this explanation of the formation of the earth's zoogeographical regions, Wallace then indicated how cases of anomalous distribution could be resolved within the framework of his theoretical model. Minor land connections still provided the key to certain anomalies. But Wallace was pressing hard for the antiextensionist position.

Bates provided Wallace with a serious test for the hypothesis that biogeographical regions coincided with the present distribution of land and ocean masses. Bates had shown that portions of the insect fauna of Chile and much of temperate South America showed little similarity to that of tropical North America. On Wallace's schema, there should have been one Neotropical fauna including all of South America, Mexico, and the West Indies. More disturbing were the marked insect affinities between South America and the Australian Region, especially Tasmania and New Zealand (a resemblance that Hooker had shown to characterize also the distribution of plants). Wallace once again generalized from his studies in the Malay Archipelago. There, although there are two distinct zoological regions, certain areas show a mixture of species from the Indo-Malay and Austro-Malay subregions. Wallace suggested that in some cases—such as the predominance of certain genera of Oriental (Indian) rather than Australian insects in New Guinea and the Moluccas (Spice Islands)—the original population had been overwhelmed or, in the extreme, exterminated by immigrants from the adjacent region. "The result," he declared, "is a mixture of races in which the foreign element is in excess; but naturalists need not be bound by the same rule as politicians, and may be permitted to recognise the just claims of the more ancient inhabitants, and to raise up fallen nationalities. The aborigines and not the invaders must be looked upon as the rightful owners of the soil, and should determine the position of their country in our system of Zoological geography" (Wallace 1864a, 114–115, 118–119). Wallace, as did so many other Victorian scientists, drew freely on the metaphors of imperialism, conquest, and stronger versus weaker populations of humans and other species (Browne 1992; Pratt 1992; Ritvo 1997). As Wallace's commitment to socialism and other egalitarian reformist schemes deepened in the later decades of the nineteenth century, however, these imperialist metaphors would assume a less prominent place in his writings.

Since the greater part of southern (temperate) South America was known to be of a more recent date geologically than the tropical mass, Wallace argued it would first have been subject to immigration from the tropics. This would account for the fact that the birds, mammals, and reptiles of temperate South America are modifications of indigenous Neotropical species. But Wallace then emphasized that insects and plants had greater powers of dispersal by "what may be called the adventitious aid of the glacial period and of floating ice" as well as by transoceanic migration. They could easily have traveled the greater distances from the temperate regions of North America or from Australia and Antarctic lands. Being already suited to a temperate climate, these latter would have been capable of establishing themselves successfully in competition with immigrants from the tropical region. The Neotropical region, Wallace concluded, thus retains its fundamental biogeographical status despite the instances of marked plant and insect affinities with more distant temperate regions (Wallace 1864a).

THE PERMANENCE OF CONTINENTS AND OCEANS

Implicit in these arguments is the doctrine of the general permanence of the great features of the globe. Lyell recognized this when he cited Wallace's solution of the problem of affinities between certain Australian and South American organisms. These regions, Lyell maintained, could not have had a free land connection since the Pliocene or even Miocene epochs (Lyell 1867–1868, 2:335–338). For those anomalous cases that had seemed to require past land connections, Wallace now offered a new framework for investigation: "Though the details of the distribution of the different groups may differ, there will always be more or less general agreement in this respect, because the great physical features of the earth—those which have longest maintained themselves unchanged—wide oceans, lofty mountains, extensive deserts— will have forbidden the intermingling or migration of all groups alike, during long periods of time. The great primary divisions of the Earth for purposes of Natural History, should, therefore, correspond with the great permanent features of the earth's surface—those that have undergone least change in recent geological periods."

Wallace's conception of zoogeography was predicated on—indeed rendered intelligible by—the fundamentally distinct character of the six main regions. This necessitated, almost axiomatically for him, that the oceans and continents had occupied their present positions at least within the period of development of present species. Wallace was fully aware of the difficulties, conceptual as well as practical, attendant on any attempt to establish a system of biogeographical regions that would be valid for all animals and plants. He was to become, however, more insistent that his expanded version of Sclater's system afforded the prime, and most natural, model for the study of organic distribution (Wallace 1864a, 122–123).

THE EMERGENT SYNTHESIS:
GEOGRAPHICAL DISTRIBUTION OF ANIMALS

By 1868–1869, Wallace had all the elements for a comprehensive treatment of zoogeography: the permanence of the oceans and continents (with auxiliary minor changes in physical geography), methods of dispersal and migration of organisms, glaciation, and, of course, evolution by natural selection. The resulting treatise, *The Geographical Distribution of Animals,* appeared in 1876 (Wallace [1876] 1962). It was recognized at once as a landmark in the science of zoogeography as well as a strategic contribution to evolutionary theory (George 1964, 123). Hooker and Darwin declared that it was Wallace who provided the strongest arguments for the theory of the general permanence of the oceans and continents and most effectively demolished the views of extensionists and continental mobilists (Darwin and Seward 1903, 2:28; Huxley 1918, 2:224–225).

Given Wallace's core belief that evolutionary theory held crucial implications for an understanding of human nature and culture, it is significant that he chose not to include mankind in the text of *Geographical Distribution.* There were valid reasons for this decision. Individual species were excluded from the book because Wallace considered them too numerous to provide the basis for any manageable distributional analysis. Moreover, because species represent the most recent evolutionary modifications, he deemed them less indicative than genera—"the natural groups of species"—of those fundamental distributional patterns connected with the more permanent features of the earth's history. Wallace noted that to treat the genus *Homo* zoogeographically would yield the uninformative statement "universally distributed." To deal, in contrast, with the distribution of the "varieties" or "races" of man would have violated the major methodological premise of the work. Moreover, for

Wallace, anthropology had now become "a science by itself [which] it seems better to omit . . . altogether from a zoological work, than to treat it in a necessarily superficial manner" (Wallace [1876] 1962, 1:vii–ix). Wallace's decision to omit man from this particular treatise made strategic sense in that he permitted readers to focus on his central objective, geographical distribution of organisms generally. In any event, Wallace authored numerous other works that were central to the late-nineteenth- and early-twentieth-century debates about the bearing of evolutionary theory on human matters (Hawkins 1997).

Wallace intended his 1876 treatise, as he did nearly all his writings, for the nonscientific reader as well as the professional scientist. His target audience included anyone "capable of understanding Lyell's 'Principles,' or Darwin's 'Origin.' " *Geographical Distribution* is an excellent specimen of Victorian science writing that is at once lucid and rigorous. The interweaving of fact, theory, and descriptive prose, which Wallace had practiced in *A Narrative of Travels on the Amazon and Rio Negro* and *The Malay Archipelago,* is perfected here. Particularly notable is the series of plates Wallace had executed to illustrate the physical aspect and characteristic fauna of the more important zoogeographical subregions. Their purpose was to "make the book more intelligible to those readers who have no special knowledge of systematic zoology, and to whom most of the names with which its pages are often crowded must necessarily be unmeaning" (Wallace [1876] 1962, 1:x–xii).

Part 1, "The Principles and General Phenomena of Distribution," explained why different regions possessed distinct and characteristic fauna. Wallace showed why, for example, parts of South Africa have lions, antelopes, zebras, and giraffes, while climatically similar parts of Australia house only kangaroos, wombats, phalangers, and mice. He invoked evolutionary change in conjunction with geographic "isolation by the most effectual and most permanent barriers" (Wallace [1876] 1962, 1:7–8, 11–14). Wallace also placed great stress on the effects of glaciation in bringing about present global distribution patterns (Marchant [1916] 1975, 203). Paleontological data were crucial in the development of evolutionary theory. Wallace devoted the second part of *Geographical Distribution* to a detailed analysis of the distribution of extinct mammals, with brief comments on extinct birds, reptiles, insects, and land and freshwater mollusks. Since the distribution of animal fossils is not identical with the distribution of living forms allied to them, Wallace argued convincingly that it was possible to reconstruct past migration routes in order to locate the probable origin of existing genera and families (Wallace [1876] 1962, 1:42–43, 57, 107–108). The third and fourth parts of *Geographical Distribution* dealt with the faunal characteristics of the zoogeographical regions and with the present range of each of the families and genera of vertebrates, insects, and mollusks. This analysis constituted the core of Wallace's argument. He drew on a vast array of data from the past and present distributions of animals to substantiate his thesis that all the chief types of animal life appear to have originated in the great northern continents and then migrated southward into the unoccupied continents of the Southern Hemisphere (Wallace [1876] 1962, 1:173–174). Wallace's use of the concept of northern origins of the earth's fauna, with their subsequent migration and "conquest" of the southern land masses, is one more example of the potency of the metaphor of imperialism in the late Victorian era (Pratt 1992; Ritvo 1997, 334–335, 349–350). Wallace's thesis of northern origins also drew on a strong, if controversial, nineteenth-century explanatory tradition in paleonotology and geology (Bowler 1996, 394–418). The latter source is more significant for the overall strategy of *Geographical Distribution* than is the metaphorical function of the concept.

Wallace had laid out the principles that were to guide zoogeographical research and theory formulation for nearly a century. Many details of his broad synthesis, however, have been successfully challenged. The boundaries between the major zoogeographical regions and subregions have necessarily been revised as more accurate and extensive distributional data have been forthcoming and the methods of analyzing those data made more sophisticated. The border between the Oriental and Australian regions, the most famous example, has been shifted repeatedly by zoogeographers. It is no longer considered to be defined by Wallace's original line. Wallace himself had offered his boundary as provisional. He stated that the precise limits between regions, when not formed by oceans, were somewhat arbitrary and "will be, not a defined line but a neutral territory of greater or less width, within which the forms of both regions will intermingle" (Wallace [1876] 1962, 1:184). Similarly, increased fossil evidence has required modifications in Wallace's reconstruction of past continental configurations (and connections) and the former distribution and migrations of animals. The thesis of the northern origin of the major orders and families of mammals has been effectively criticized. The higher primates—the Old World monkeys, apes, and the ancestors of man, for example—are now thought to have emerged most probably in Africa and thence spread across the globe (Foley 1995). Most significant, the recent compelling evidence for continental drift has provided a radical alternative to Wallace's explanation for the similarities between the fauna and flora of the southern continents. Given these qualifications, it remains true that the general principles advanced in *Geographical Distribution* were fundamental to the development of the science of zoogeography. And Wallace's position as the leading student of animal distribution was confirmed, four years later, with the appearance of *Island Life* ([1880] 1892).

THE ARGUMENT COMPLETED: *ISLAND LIFE*

The fauna and flora of islands had intrigued and puzzled naturalists at least since the appearance of J. R. Forster's *Observations Made during a Voyage Round the World* (1778). Insular data, especially from the Galapagos Islands and the Malay Archipelago, were crucial for Wallace's initial formulations of evolutionary theory and biogeography. *Island Life* is an analysis of the distributional phenomena presented by islands in their complex relation to each other and to continents (Wallace [1880] 1892). Although much of the first part of the book follows directly from the tenets of *Geographical Distribution of Animals, Island Life* is a powerful extension of Wallace's biogeographical system. Darwin considered it the best book Wallace published. Hooker thought it "an immense advance [which] . . . brushed away more cobwebs that have obscured the subject than any other" treatise (Marchant [1916] 1975, 252, 289–90). Aside from the brilliance of Wallace's treatise, on a subject to which Hooker had devoted a great part of his own career, Hooker would have been delighted by Wallace's dedication. *Island Life* bore the following inscription: "To Sir Joseph Dalton Hooker, who, more than any other writer, has advanced our knowledge of the geographical distribution of plants, and especially of insular floras, I dedicate this volume; on a kindred subject, as a token of admiration and regard." The plan of *Geographical Distribution* had required that Wallace discuss mainly genera and the higher orders of animals. *Island Life,* in contrast, focused on species and included important discussions of phytogeography (plant distribution). It provided a more comprehensive scope for Wallace's theorization.

Island Life marked the completion of Wallace's most innovative contributions to biogeography. Although he continued to refine details, the fundamental principles of geographical distribution had been established. Wallace's synthesis of geological and climatic data, modes of migration and dispersal of organisms, and evolutionary adaptation and divergence provided a framework that continues to guide biogeographical studies. Until the mid-twentieth century, this framework was allied to the doctrine of continental and oceanic permanence, which Wallace insisted was "the only solid basis for any general study of the geographical distribution of animals [and plants]" (Wallace [1905] 1969, 2:386). Many recent accounts of the history of biogeography, however, are colored by the perception that until the general acceptance of the theory of continental drift in the early 1970s, the science of geographical distribution was bereft of any accurate geophysical foundation.[10] The hypothesis of the large-scale displacement of continents in the earth's history was given an early formulation in Alfred Wegener's *The Origin of Continents and Oceans* (1915). Drift theory was rejected in Wegener's lifetime and subjected to bitter debates among geologists, paleontologists, and climatologists until the 1960s (Oreskes 1999). The articulation of the empirically confirmed doctrine of plate tectonics and seafloor spreading in the late 1960s gave drift theory (in a form radically changed from Wegener's conception) decisive corroboration (Le Grand 1988). Recent histories of biogeography that are inspired by such hindsight are highly problematic. They assume that we are supposed to "know" that the debates over dispersal mechanisms of animals and plants and sunken land bridges between continents—debates to which Wallace's theories were crucial—would eventually be incorporated into the drift paradigm. This is Whig history at its worst. The more appropriate historiographical approach is to look forward from Wallace and his contemporaries, rather than backward from current scientific consensus. Wallace's contributions to the late-nineteenth- and early-twentieth-century debates regarding causal mechanisms for observed distributional patterns retain their profound historical significance (Fichman 1977; Bowler 1996, 371–372). Many influential scientists, such as William Diller Matthew in *Climate and Evolution* (1915), championed Wallace's theories during the first half of the twentieth century (Rainger 1991, chap. 8). Wallace made biogeography one of the most impressive applications of the theory of evolution.

NATURAL HISTORY AND BEYOND

Wallace's London residency served him well in establishing his scientific reputation and his network of friends and colleagues. But it had significant drawbacks. Wallace was unable to secure any permanent employment that would provide him with a steady income. Ironically, Wallace at this point was unconvinced of the possibility of earning anything substantial either by lecturing or writing. He felt that the "experience of my first work on 'The Amazon' did not encourage me to think that I could write anything that would much more than pay expenses." Wallace also disliked the "confinement" of London life. He regarded it as uncongenial to his nature and a certain recipe for a shortened life span (Wallace [1905] 1969, 1:414–416).

When the government decided to establish a branch of the South Kensington museum in Bethnal Green (in East London) in 1869, to combine art and natural history for public benefit and instruction, Wallace campaigned for the directorship. He had the support of Lyell, Huxley, and other influential colleagues. However, when the museum was built and opened in 1872, the South Kensington authorities decided Bethnal Green could be managed from their headquarters. No director was chosen. This experience had a decisive effect on the course of Wallace's life. In anticipation of gaining the post, and to commence the "country life" for which he so yearned, he purchased land

and built a home (1871–1872) in the picturesque village of Grays, some twenty miles from London. The site on which the home was constructed, by Wallace himself with the assistance of an architect and some friends and contract laborers, afforded beautiful views of the Thames and the Kent hills. The house was ready for Wallace, his wife Annie, and their children to move into in March 1872. Wallace "began to take that pleasure in gardening, and especially in growing uncommon and interesting as well as beautiful plants, which in various places, under many difficulties and with mingled failures and successes, has been a delight and solace to me ever since" (Wallace [1905] 1969, 1:415–416, 2:90–93).

In 1876, Wallace sold the house and land at Grays. Though still pleasant, it was no longer suitable for his growing family. The Wallaces moved first to Dorking and then, in 1878, to a larger house in Croydon. In the same year, Wallace applied to be superintendent of Epping Forest. The forest had recently been acquired by the Corporation of London to assure its protection and improvement while preserving its "natural aspect," in accordance with the Act of Parliament that restored that immense tract to the public. Wallace's failure to get the post was disappointing. It was also a relief. He was now "free to do literary work which I should certainly not have done if I had had permanent employment so engrossing and interesting as that at Epping. In that case I should not have gone to lecture in America, and should not have written 'Darwinism,' perhaps none of my later books, and very few of the articles contained in my 'Studies.' This body of literary and popular scientific work is, perhaps, what I was best fitted to perform." Wallace's candidacy for the Epping position was supported by the presidents of most of the natural history societies in London, by a number of members of Parliament, and by numerous residents near the forest and in London. He was rejected by the city merchants and tradesmen with whom the actual decision lay on the grounds that a "practical man" was needed. They wanted someone who would implement their plans to build a large hotel and an amusement park. Wallace's article "Epping Forest" in the 1 November 1878 issue of the *Fortnightly Review*, which argued for utilizing parts of the unwooded tracts of land as "an experiment in illustration of the geographic distribution of plants [which] would have been both unique and educational," did little to recommend him to the London commercial czars (Wallace 1878b, [1905] 1969, 1:416–417, 2:101–102; Raby 2001, 218–221). Wallace's environmental concerns, made public at this juncture, would become an increasing component of his activist politics in the years to come.

Wallace's failure to get either the Bethnal Green or Epping positions did enable him to devote himself more fully to his scientific and other interests. The question arises, of course, as to just how Wallace and his family managed to live in such comparative comfort despite these setbacks. How did Wallace escape the straitened circumstances of so many members of the industrial society whose gaps between rich and poor he dedicated much of his later life to exposing? Wallace continued to enjoy some profits from the sale of his Amazonian and, especially, Malay collections. The *Malay Archipelago* was a financial as well as critical success, as were many of his books. Wallace's employment as an examiner for a variety of educational institutions provided the steadiest, if rather modest, source of income. In 1870, Bates informed Wallace of a vacant post as examiner in physical geography at the science and art department housed in one of the magnificent new complex of buildings at South Kensington. These temples of science owed much to Huxley's successful politicking (Desmond 1998, 394–397). Wallace applied and was appointed in 1871. He also became an examiner in physical geography and geology for the Indian Civil Engineering College and for the Royal Geographical Society in the same year. Thus began yet another career for Wallace. This lasted, except for his one-year lecture tour of North America (1886–1887), until 1897. A number of other scientists of repute but meager means, including Bates, worked as examiners. They each had more than a thousand papers to grade during a three-week period per year. The work, though arduous and repetitive to the point of drudgery, yielded an annual income of £50 to £60. Wallace and his coexaminers were able to exchange highly amusing examples of answers that "exhibited every possible degree of ignorance of the subject[s] . . . and thus contributed a little hilarity to our otherwise strictly business meetings." For the question "Mention the natural habitat of the horse [and] the elephant," Wallace recorded the following answer: "The habit of the horse is plowing, the elephant goes to shows." Wallace's three decades as an examiner left him with a decidedly critical assessment of the English educational establishment. For each student who did well, numerous others displayed complete ignorance of their subject matter (Wallace [1905] 1969, 2:406–418).

In Wallace's opinion, this sorry state of affairs was the result of cultural and systemic factors. Many of the teachers themselves were ignorant in the subjects they were charged with instructing. In order to justify the existence of various government departments, it was necessary to demonstrate a certain success rate. Wallace complained with great disdain that "hence the 'passes' are brought up to good general average, however bad the bulk of the papers may be; and people are deluded by the idea that because a person has passed in Physiography [the former Physical Geography] he has a good general knowledge of the whole subject, whereas many pass who are quite unfit to teach any portion of it to the smallest child. My own conclusion is that all these examinations are an enormous waste of public money, with no useful result whatever" (Wallace [1905] 1969, 2:412, 416–417). Wallace wrote a number of articles calling for the reform of British education. But this was one field in which he simply could not devote the prodigious expenditure of intellectual and activist energy that characterized his other chosen fields of battle. Wallace's educational philosophy drew heavily on his own experiences. Largely self-taught in his early years, economic and social circumstances forced him to learn his science and all other disciplines outside the groves of academe. Wallace found the English "public school" system and its Oxbridge culmination a path reserved only for the most privileged members of society.

In the first issue (1870) of *Nature,* which he assisted in founding, Wallace specified that the "broad principle I go upon is this,—that the State has no moral right to apply funds raised by the taxation of all its members to any purpose which is not directly available for the benefit of all. . . . I uphold national education, but I object absolutely to all sectional or class education" (Wallace 1870b). Wallace's views were expressed in a letter to the editor. In order to counteract its arguments, which could be taken as critiques of educational (particularly science) reforms then being endorsed by Huxley and his allies, Wallace's letter was prefaced by a lead article with the same title. This piece "Government Aid to Science" attacked his views and, Wallace felt, misrepresented his actual position. Most annoyingly, the article omitted dealing "with the main ethical question which I raised." Wallace felt strongly enough about this preemptive strike to reprint his letter in full, with explanatory comments, in his autobiography (Wallace [1905] 1969, 2:54–60). Wallace was scarcely a foe of more widespread public (in the modern sense of the term) education in sciences and arts. But his growing antiestablishment sentiments put him on a collision course with Huxley in this instance. Wallace was on more comfortable ground when he focused on the role museums could play in educating the public about natural history. He wrote several articles that set forth principles of museum design and collections display. Wallace's proposals were radical at the time but have since come to be widely adopted. One of his major objectives was that the displays of animals and plants should mimic natural distributional patterns in realistic settings. This idea has been influential since Wallace first put it forward (Wallace 1869a). His writings on museums reinforce Wallace's belief in an individual's active self-involvement, rather than passive recipience, in his or her own education.

Wallace contributed a constant stream of articles to *Nature.* The journal, which had a magazine format similar to the *Reader* or *Saturday Review,* was intended as a broad cultural forum but with a central emphasis on science. Under the guidance of men like Huxley and its liberal editor Norman Lockyer, the moderately priced journal was an immediate success. Despite some competition, including an Oxford-based monthly called the *Academy,* it became the leading popular voice for the professionalizing London scientific elite (Desmond 1998, 372). The coterie publishing *Nature* welcomed Wallace, at this point, as a powerful champion for the dissemination of the virtues of science among the educated general readers who made up its target audience. *Nature* was an influential platform for Wallace's expertise as scientist, scientific book reviewer, and science popularizer. He had become a Victorian naturalist of the first rank.

———•◆•———

NOTES

1. Another example of nonmainstream meeting places (the pub) for the discussion of science is described by Anne Secord (1994b); for more forums for popular science discussion, see Greg Myers 1994.

2. On mesmerism generally in this period, see Alison Winter (1998).

3. For the general impact of von Humboldt, see Lewis Pyenson and Susan Sheets-Pyenson (1999, 258–260); however, detailed studies of the specific, rather than general, influence of the Humboldtian ethos on individual naturalists would be welcome additions to the scholarly literature.

4. Camerini (1997, 354–377) gives an excellent general account of early Victorians in the field. See also her more detailed "Wallace in the Field," in Kuklick and Kohler (1996, 44–65). Both essays are indispensable for understanding the sociopolitical and cultural context of Wallace's early career. For additional perspectives, see Philip Rehbock (1983); Janet Browne (1992); David Allen (1994); Nicolaas Rupke (1994); Nicholas Jardine, James Secord, and Emma Spary (1996).

5. Wallace [1905] 1969, 1:316–320. The map is reprinted facing p. 320 and had appeared, of course, a half century before in his 1853 article "On the Rio Negro" (Wallace 1853b). Wallace's original hand-drawn and colored map, from which these lithographs were printed, shows great care and skill in its execution. It is mounted on linen and now housed at the Royal Geographical Society of London (Camerini 2002, 65–67).

6. Brazilian ichthyologist Mônica de Toledo-Piza Ragazzo has just published her edition of Wallace's Rio Negro fish drawings, entitled *Peixes do Rio Negro/Fishes of the Rio Negro* (2002), which features plates of all two hundred plus drawings (accompanied by Wallace's original notes), with all auxiliary text, introductory comments, tables, and appendices—offered in both Portuguese and English. Apart from the value of the work as a historical document and contribution to biodiversity studies, it also highlights an aspect of Wallace's talents that is usually overlooked: his skill as a sketch artist.

7. Alfred Russel Wallace, MS. Journal, pp. 29, 34; cited in Camerini 1996, 54n. 26. Wallace's original Malay journals and his field notebooks are housed at the Linnean Society Archives in London. Transcripts of these original documents, which are in Wallace's ostensibly clear but often difficult to decipher handwriting, are also available at the Linnean Society. The transcripts are useful for an initial overview of each document but must be checked against the originals for any obvious misreadings or word omissions.

8. For detailed discussions of Wallace's orangutan experiences, see Penny van Oosterzee (1997, chap. 10), and Gavan Daws and Marty Fujita (1999).

9. Charles Smith (1991, 490–502) lists more than 150 such entries in his authoritative bibliography of Wallace's writings.

10. Noteworthy exceptions include Frankel (1981, 1984); Greene (1982); Browne (1983); Secord (1986); and Laudan (1987).

CHAPTER 3

Wallace's Evolutionary Philosophy

Wallace's evolutionary philosophy was central to his life and career. His philosophical interests have, however, not been accorded the full recognition they warrant. Wallace's scientific and sociopolitical activities have generally been considered separate if not incongruous pursuits. An examination of his evolutionary philosophy ties together these diverse aspects of his life. Wallace made no exceptional contributions to the discipline of philosophy in the Victorian period. But he followed philosophical controversies closely. Wallace's interpretation of the different philosophical schools of the nineteenth century was crucial to the elaboration of his evolutionary worldview. Empiricism, positivism, idealism, realism, theism, and pragmatism all fascinated him. An analysis of his first writings on evolution, particularly the famous 1855 and 1858 essays, reveals that epistemological and metaphysical concerns were crucial to Wallace from the beginning of his career. Those essays laid the groundwork for the full exposition of evolutionary theory discussed in chapter 2. The focus on these early essays in this chapter highlights their epistemological significance. By contextualizing Wallace's evolutionism within the broader intellectual climate, the foundations of his critical analysis of Victorian culture become clear.

READING CHAMBERS'S *VESTIGES*

In a letter written on 9 November 1845, Wallace asked whether Bates had "read 'Vestiges of the Natural History of Creation,' or is it out of your line?" Bates's response indicated that he did not think much of the *Vestiges* as a work of any real scientific merit. Wallace replied on 28 December 1845 that he had "rather a more favourable opinion of the 'Vestiges'" (Wallace [1905] 1969,

1:254; McKinney 1969). *Vestiges of the Natural History of Creation,* by the Edinburgh publisher Robert Chambers, had a remarkable, and contentious, impact on Victorian scientific and popular readers from the instant of its publication in October 1844 (Yeo 1984; Secord 2000). It has been credited with effecting, more than any other previous work on evolution, a "sea change" in the fortunes of that doctrine. *Vestiges* brought "transmutation . . . off the streets, out of the shabby [medical] dissecting theatres, and into the drawing rooms [of Britain]. No longer was it to be the province of socialist revolutionaries and republican physicians." As *Vestiges* forced its way into the realm of the scientific elite, opinions were strongly divided. Many questioned its central evolutionary thesis, its theoretical and empirical merits, and its potential impact on the wavering religious beliefs of an unsettled Chartist age. Hooker thought it "good value for a hack-work, despite its egregious blunders." Darwin was inclined to be charitable toward its main thesis (for obvious reasons) but was extremely uncomfortable about its technical inadequacies. Adam Sedgwick's review bristled with "unmitigated contempt, scorn, and ridicule" (Desmond and Moore 1991, 320–322). Huxley granted that *Vestiges* was cleverly crafted, even a piece of brilliant journalism. But he thought Chambers's science secondhand and "loathed the book's blundering pretensions." Part of Huxley's negative response was due, doubtless, to his jealousy of the book's considerable financial rewards for its author. Huxley at this period was embarking on his grand crusade to increase the comparatively meager social and financial recognition accorded to professional scientists. That Chambers, the very antithesis of Huxley's vision of the new, culturally respected man of science, was reaping such significant monetary sums must have been particularly galling (Desmond 1998, 193). *Vestiges* was without question a major bestseller. By 1860, it had sold more than 20,000 copies in eleven British editions plus editions in the United States, Germany, and the Netherlands. For each revised edition, Chambers incorporated many of the criticisms leveled against the first edition and gradually increased the general scientific repute of his treatise (Williams 1971). The first and following editions of the *Vestiges,* with their clear statement of the developmental hypothesis, exerted a decisive influence on Wallace's emerging evolutionary speculations (McKinney 1972, 5–6, 9–12, 21, 40–42, 50–53, 84, 95, 147–148).

Wallace's reply to Bates depicted the impact of the *Vestiges* in unambiguous language. In contrast to the majority of his contemporaries, the young Wallace did not regard Chambers's specific enunciation of the developmental thesis as "a hasty generalization." Rather, he saw it as

> an ingenious hypothesis strongly supported by some striking facts and analogies, but which remains to be proved by more facts & the

additional light which future researches may throw upon the subject. It at all events furnishes a subject for every observer of nature
to turn his attention to; every fact he observes must make either for
or against it, and it thus serves both as an incitement to the collection of facts, & an object to which to apply them when collected. I
would observe that many eminent writers give great support to the
theory of the progressive development of animals & plants. There
is a very interesting & philosophical work bearing directly on the
subject—Lawrence's "Lectures on Man"—delivered before the Royal
Coll[ege] of Surgeons, & now published in a cheap form. The great
object of these lectures is to illustrate the different races of mankind
& the manner in which they probably originated—and he arrives at
the conclusion[,] as does also Mr. Pritchard [sic] in his work on the
Physical history of man, that the varieties of the Human race have
not proceeded from any external causes, but have been produced by
the development of certain distinctive peculiarities in some Individuals which have become propagated through an entire race.

This letter reveals crucial insights into Wallace's first ruminations on that
"question of questions," ruminations that would inform his specific path
toward discovery of the theory of natural selection. Wallace was unencum-
bered by any prior theoretical prejudices against Chambers's version of the
developmental hypothesis. He recognized it for what it was: a provocative
statement of a hypothesis that would require substantial empirical data to
either prove or discredit it. Wallace incorporated the *Vestiges* along with his
other readings and fieldwork in natural history into the basis for a viable
and rigorous research program. The youthful surveyor and amateur natural
historian was no naive Baconian. He already had a theoretical framework that
his future empirical observations and fieldwork would either strengthen or
render defective (Kleiner 1985).

The December 1845 letter clarifies two enduring concerns of Wallace.
From the outset, he was intrigued by the implications evolutionism held
for the species *Homo sapiens*. Whatever Wallace's fascination with beetles
and butterflies, the human implications of evolution were always in the fore-
ground of his thought. Wallace was also obsessed with the relationship between species and varieties. He declared "that a permanent peculiarity not
produced by external causes is a characteristic of 'species' and not of mere 'variety,' and thus, if the theory of the 'Vestiges' is accepted, the Negro, the Red
Indian, and the European are distinct species of the genus Homo." Wallace
would alter this opinion in a famous 1864 paper, in which he argued that all
the "races" of mankind derive from a single ancestral species (1864b). But

Vestiges reinforced his view that humans were the product of similar forces that accounted for the development of all animal and plant species. The letter to Bates concluded with a provocative statement about the relationship of species to varieties:

> An animal which differs from another by some decided & perma-
> nent character, however slight which difference is undiminished by
> propagation & unchanged by climate & external circumstances, (like
> the negro) is invariably considered as a distinct species—while one
> which is not propagated so as to form a distinct race, but is produced
> more frequently from the parent stock (like the Albino) is generally[,]
> if the difference is not very striking, considered a variety,—now I
> consider both these to be equally distinct *species*, & I would only con-
> sider those to be varieties whose differences are produced by External
> causes & which therefore are not propagated as a distinct race.
>
> In how many cases in the animal world & particularly among In-
> sects are the differences between species far less than those between
> varieties, so consid[ere]d neither however being produced by Exter-
> nal circumstances. . . . How well too does this theory account for
> those excessively rare species whose Existence seems almost a mys-
> tery. They may be produced by more common species at intervals in
> the same manner as the Albino is from European parents.
>
> (McKinney 1969)[1]

The December 1845 letter is indispensable for assessing Wallace's earliest views on the "species question." Although his concept of variation would later be modified, this letter clearly indicated that Wallace was aware that viable definitions held a major key to uncovering the origin of species (Bowler 1976). The letter also evokes Wallace's state of mind as he pondered with Bates whether, and where, they should embark for a prolonged journey to the tropics. Wallace emphasized to Bates, if only to raise the latter's estimate of Chambers's book, that he had "heard that 'Cosmos[,]' celebrated work of the venerable Humboldt, supports in almost every particular its theories. . . . This work I have a great desire to read, but fear I shall not have an opportunity at present. Read Lawrence's work—it is well worth it" (McKinney 1969, 372–373). Wallace later stated in his autobiography that one of the main reasons he chose to include these and several other letters to and from Bates in *My Life* was to show that "at this early period, only about four years after I had begun to take any interest in natural history, I was already speculating upon the origin of species, and taking note of everything bearing upon it that came in my way. It also serves to show . . . my appreciation of the 'Vestiges,' a book which, in my opinion, has always been undervalued" (Wallace [1905] 1969, 1:255). Wallace is surely correct to stress his independent path toward the discovery of natural selection (Kottler 1985).

ROBERT OWEN'S IMPACT

The impact of Robert Owen's social and political views on the philosophical development of the young Wallace at this period was even more profound. It was Owen's fundamental principle that a person's character was formed by a combination of heredity and environment (including education and family life). Wallace rejected the standard arguments against Owenite teaching, that it was immoral and denied the cherished precept of free will. Wallace never denied that heredity was fundamental to the evolutionary process; as a budding transformationist he could scarcely have done so. Rather, he agreed with Owen (fig. 2) that "character" can, and must, be improved by beneficent social and political environmental conditioning. Wallace pointed to Owen's twenty-six-year experiment at New Lanark as indisputable confirmation of the validity of Owenite theories (Wallace [1905] 1969, 1:89–91). As Greta Jones has recently demonstrated, however, Owen's influence on Wallace was even more extensive than is usually thought (Jones 2002). Owenite socialism contributed in subtle ways to Wallace's particular formulation of natural selection itself, particularly in the 1858 essay. Though Wallace would continue to refine his conception of the precise role played by natural selection, most notably in human evolution, three ideas of Owen proved highly relevant to Wallace's original formulation. These included, in addition to the belief in the educability of mankind, the notion of "home colonization" (the foundation of self-supporting communities run on socialist principles) and a form of anti-Malthusianism ("man can produce more than he can consume") (Jones 2002, 80).

Given Wallace's specific assertion in his autobiography that it was his "recollection [of] Malthus's 'Principles of Population' " on the island of Ternate in 1858 that ignited a spark leading to his enunciation of natural selection (Wallace [1905] 1969, 1:361–363), the claim that Owenite anti-Malthusianism played a critical role would at first seem contradictory to accepted historiography of that crucial event in Wallace's life. And Wallace's "Malthusian moment" has been powerfully described by James Moore (1997). But Moore has, significantly, interpreted Wallace's moment of inspiration in the broader context of the socialist legacy in Wallace's scientific work. This alerts us to the fact that there were several possible readings of Malthus available in the early nineteenth century (Benton 1995). Spencer, Darwin, David Ricardo, and many other thinkers of various philosophical and political stripes could,

Figure 2. Drawing of Robert Owen at age sixty-three by Ebenezer Morley (1834).

and did, read Malthus through quite different lenses. Malthus is certainly present in Wallace's 1858 paper, but it is Malthus read by an Owenite. Population pressure was recognized by Wallace as one major factor intensifying selection. But he also cited the enormous importance of "some alteration of physical conditions" in habitat, environment, climate, and behavior in effecting species evolution. For Wallace, as for Owen, the pressure of population is not just a unilinear force crashing against limited subsistence. Particularly in human societies, the expansion of population may be mediated by inventions that increase food supply and by various forms of social organization that can reduce an imbalance between population pressures and available resources (Jones 2002, 86–95).

Wallace's reading of Malthus, therefore, was complex. He regarded *Principles of Population* as dealing with the "problems of philosophical biology"— in short, a work he "greatly admired for its masterly summary of facts and

logical induction to conclusions" (Wallace [1905] 1969, 1:232). Malthus
was part of Wallace's philosophical education, just as Owen was. Indeed,
both men—albeit in quite different ways—were philosophical mentors in his
self-education. In one sense, Malthus as read through Owenite lenses alerted
Wallace to the epistemological distinction between physical and cultural evolution. His exposure to Owenite principles never led him to espouse any
version of Lamarckism. But Owen's emphasis on individual ethical reform
of "character" as a corollary to collectivist political reform rendered Wallace
intellectually open to considering factors other than strictly physical causation in the evolutionary process. One of the more intriguing by-products of
Owenite thinking for Wallace's conception of natural selection is that it suggested that natural selection alone might not be wholly adequate to account
for all aspects of evolution, especially that of humans. It is of no little interest
that Owen was publicly converted to spiritualism in 1853. The link between
Owenism and spiritualism suggests that Wallace may have imbibed something
more than sociopolitical messages from Owen and Owenite teachings (Jones
2002, 77). While living with his brother John in London for a few months in
London in 1837, Wallace heard Owen lecture at the Hall of Science near Tottenham Court Road. Wallace declared he was struck by "his tall spare figure,
very lofty head, and highly benevolent countenance and mode of speaking"
(Wallace [1905] 1969, 1:79, 87, 104; Raby 2001, 13–14). His fascination
with Owen did not preclude Wallace's quest for additional insights. He soon
came under the rather different spell of Herbert Spencer's individualist political economy. But Wallace never ceased to regard Owen "as my first teacher
in the philosophy of human nature and my first guide through the labyrinth
of social science. He influenced my character more than I then knew, and . . .
I am fully convinced that he was the greatest of social reformers and the real
founder of modern Socialism" (Wallace [1905] 1969, 1:89–105, 234–236).

MESMERISM AND PHRENOLOGY

Mesmerism and phrenology were two other areas Wallace pursued keenly
during these early years. He had been introduced to those topics during the
year he taught in Leicester. But it was two lecturers whom he heard in Neath
who confirmed his growing conviction that mesmerism and phrenology were
not only genuine but susceptible to decisive empirical proof as well. The
two lecturers were Edwin Thomas Hicks (who called himself "Professor of
Phrenology") and James Quilter Rumball (a member of the Royal College of
Surgeons and author of several medical treatises). Wallace was amazed by the
accurate depiction of his character by both phrenologists' rapid examination
of the form of his head, made in full public view. Wallace preserved both

these documents and reprinted significant portions of them in his autobiography. He had also used them as some of the evidence given in support of phrenology in *The Wonderful Century: Its Successes and Its Failures* ([1898] 1970). Among the thirty cranial regions examined by Hicks and Rumball, it was their assessment of Wallace's "organ of Veneration" that impressed him most. They both stressed its small size. Wallace was astonished. "My character," he declared, was indeed marked by "disregard for mere authority or rank." In striking contrast, his organs of "Ideality and Wonder [were] both marked as well developed." Here was an explanation for his "intense delight in the grand, the beautiful, or the mysterious in nature or in art" (Wallace [1905] 1969, 1:257–262). Contextualist historiography now offers tools by which attitudes such as those expressed by Wallace can be addressed with increasing sophistication. The scholarly literature on mesmerism, phrenology, and spiritualism is growing rapidly. The current reassessment of the so-called demarcation battles in those three fields puts into question deeply ingrained notions of what constitutes orthodoxy versus heterodoxy in the life sciences (Winter 1997, 28–32). Wallace's phrenological beliefs may have provided some motivation for those career decisions and character traits that puzzled contemporaries and intrigue historians: his independence in matters scientific and cultural and his apparent indifference to professional, financial, and social status.

EPISTEMOLOGY AND THE 1855 ESSAY

Wallace's commitment to clarifying the philosophical bases of his theoretical and practical investigations is apparent in his first major publication. Pledged to some form of evolutionary theory since 1845, Wallace made his first public statement of this position in an essay entitled "On the Law Which Has Regulated the Introduction of New Species." Written in February 1855 at Sarawak and published later that year in the *Annals and Magazine of Natural History,* the essay skillfully weaves together facts from geology and the geographic distribution of animals and plants to construct a hypothesis that explains those facts as a consequence of evolutionary change. Wallace began by arguing that most previous explanations of the present—and often curious—distribution of animal and plants were unsatisfactory because they failed to take into account the past history of the earth and its inhabitants. The influence of recent theories in geology, particularly the doctrine of Lyell and the uniformitarians positing an endless but gradual repetition of geological changes throughout time, is explicit. Wallace considered it incontestable that during the earth's immense history its surface had undergone successive gradual transformations, with a corresponding gradual modification in the forms of organic life as they adapted to new environmental conditions. The present distribution patterns, therefore, must be the result of all previous changes, organic and inorganic. Wallace was particularly concerned with analyzing more closely the spatial and temporal relationships between species. The essay noted that the larger groups, such as classes and orders, are generally spread over the whole earth. In contrast, the smaller ones, such as families and genera, are frequently confined to more limited districts. Further, when genera themselves are widely spread, it is well-marked groups of species that are peculiar to each limited district. Wallace stressed that when "a group is confined to one district, and is rich in species, it is almost invariably the case that the most closely allied species are found in the same locality or in closely adjoining localities, and that therefore the natural sequence of the species by affinity is also geographical" (Wallace 1855, 184–185).

Wallace next argued that the distribution of animals and plants in time, as evidenced by the fossil record, revealed marked similarities to their present geographical distribution. Whereas many of the larger groups (and some smaller ones) extend through several geological periods, there are peculiar groups found in a particular geological period (or formation) and nowhere else. Moreover—just as closely related species in the Amazon Basin occupied adjacent regions—species or genera are more closely related to those occurring in the same geological epoch than they are to species or genera separated from them by longer periods of geological time. Finally, just as the same (or similar) species generally are never found in widely separated regions without also being found in intermediate locations, the geological record does not show any abrupt disjunctions in the fossil remains of a given species. "In other words," Wallace asserted, "no group or species has come into existence twice." He concluded the essay with the famous declaration that "every species has come into existence coincident both in space and time with a pre-existing closely allied species" (Wallace 1855, 186).

Wallace's law drew together a large body of hitherto unrelated facts. It provided a compelling explanation for "the natural system of arrangement of organic beings, their geographical distribution, their geological sequence, the phenomena of representative and substituted groups in all their modifications, and the most singular peculiarities of anatomical structure" (Wallace 1855, 196). The 1855 essay, despite its brevity, is among the most forceful statements of evolution prior to the reading in 1858 at the Linnean Society of the Darwin-Wallace papers announcing the principle of natural selection (Beddall 1972). Darwin was among the first to recognize the significance of, and potential competition represented by, Wallace's 1855 essay (Burkhardt and Smith 1985–, 7:107n. 27). The concept of evolution itself was not novel. By the mid-eighteenth century, authors such as Diderot, Buffon, and Maupertuis were advocating explicit versions of transformist doctrine. At the beginning of the nineteenth century, sufficient "evidence from the fields of biogeography, systematics, palaeontology, comparative anatomy, and animal and plant breeding was already available . . . to have made it possible to develop" convincing arguments for evolution (Mayr 1976, 278). Yet, resistance to the concept was entrenched. Lyell (initially), Richard Owen (the English comparative anatomist and paleontologist), and Georges Cuvier (the brilliant French zoologist and scientific administrator) were among the most formidable opponents of transformism. Moreover, the work of two of the most widely known proponents of evolution, Chambers and Lamarck, was the object of intensive criticism (Hodge 1972; Desmond 1989, 176–180).

The resistance to evolutionism arose from the challenges it posed to biological orthodoxy as well as from concerns about what many perceived to be its radical political, religious, and philosophical implications for British society and culture. The vivid images of the French Revolution were potent forces across the Channel. The English upper classes blamed, rightly or wrongly, the revolution on the corrosive teaching of the Enlightenment philosophes. Lamarck was reviled as a figure who "vomited" his "abominable trash" over a Paris run riot. Lamarck's evolutionary hypotheses were "damned as scientific excrement, fouling the wellsprings of society and subverting Church authority" (Desmond 1998, 89). Chambers, in *Vestiges,* toned down the unsettling social and political visions of radical evolutionism, with his suggestion that the developmental hypothesis represented a case of natural process preordained by God. By making evolutionism more respectable than Lamarck's version, Chambers tamed it sufficiently so that the growing audience of middle-class readers bought his book in vast numbers. But even this slightly sanitized exposition of evolution caused problems for Britain's paternalistic and elitist society. Chambers peppered his discussion with references to phrenology and other "indelicate" topics such as pregnancy and abortion. Women (and not only emancipated socialist females) as well as men were devouring *Vestiges.* Clearly aimed at a broad and popular audience, Chambers was spreading evolutionary and other speculations among larger segments of the British populace than was deemed prudent by the reigning monarchs of science and philosophy. Sedgwick—who warned "our glorious maidens and matrons" against soiling their thoughts with such abominable ideas— voiced the fears of the Anglican establishment. He bitterly characterized the *Vestiges'* mix of transmutation, spontaneous generation, and phrenology as an "unlawful marriage . . . breeding a deformed progeny" (Desmond 1989, 7– 8, 176–178; Secord 2000, 223–226, 240–247). Given Wallace's fascination with evolutionary speculation and phrenology, it was natural for him to be taken with *Vestiges.* His keen interest in the bold visions, if not the detailed arguments, of Lamarck and Chambers sets him apart from many older biologists of the period. But Wallace's reaction to Chambers's *Vestiges* was nuanced. He saw the polemical merit of the book in advancing the general public case for evolutionism but was fully aware of the major scientific inadequacies of Chambers's work.

It has been suggested that Wallace's initial receptivity to Chambers's views derived from his belonging, in the 1840s, more "to the non-biologically-educated public than he did to the world of the professional scientist" (Ruse 1974, 54). The actual situation is more complex. Though the reasons for Wallace's early comfort with evolutionary speculation have still not been established completely, there is no question that he was from the first aware of the conspicuous scientific errors that marred Chambers's *Vestiges* and of the inadequacies of Lamarck's theory. He did not yet belong to the professional scientific community, and his relationship to that community would remain ambivalent throughout his career. But his extensive immersion in the literature and practices of the British naturalist tradition separated him profoundly from the "non-biologically-educated public." Despite its weaknesses, he declared (in his autobiography) *Vestiges* to be an undervalued contribution to the development of evolutionary ideas (Wallace [1905] 1969, 1:255). Bates did not share Wallace's opinion of the worth of *Vestiges*, but their discussions over the issues raised in Chambers's book were one motivating factor in their fateful decision to travel to the Amazon. Although Bates and Wallace differed sharply on a number of Chambers's speculations, their mutual interest in resolving those differences contributed to their desire to see what observational evidence the distant tropics might hold.

For Wallace, the publication of *Vestiges* functioned as one impetus to focus more deeply on possible mechanisms by which species may have been transformed (Schwartz 1990, 140–143). As Bates later recalled, the trip was undertaken to "gather facts, as Wallace expressed it in one of his letters, 'toward solving the problem of the origin of species,' a subject on which we had conversed and corresponded much together" (1863, 1:iii). From as early as 1845, then, Wallace was preoccupied with the "species question." When a decade later, he published his famous "On the Law Which Has Regulated the Introduction of New Species," he had traveled an enormous distance, both geographically and conceptually. Wallace's 1855 law, in contrast to Lamarck's and (especially) Chambers's ideas, was impeccable from a scientific standpoint. Precisely because it was derived from well-established data and sophisticated reasoning, Wallace's law raised the evolutionary debate to a new level of rigor. Lyell, himself, testifies to the forcefulness of this impact. The very first entry in the first of that series of seven notebooks—his "Scientific Journals" written between 1855 and 1861—in which Lyell recorded his thoughts on the possibility of the transmutation of species, was a detailed abstract of Wallace's 1855 essay. These notebooks manifest Lyell's profound, prolonged, and painful reassessment of many of his long-held convictions, including those concerning the origin and development of the human species (1970, xv, xli, 1–8; references to Wallace recur throughout the series of notebooks).

The immediate stimulus for Wallace's 1855 essay had been the 1854 publication of the polarity theory by Edward Forbes, the renowned British naturalist. Forbes contended that paleontological evidence—the abundance of fossils from both the earliest and most recent geological periods, coupled with a relative scarcity of fossils from intermediate periods—was consistent with a divinely ordained scheme of creation necessitating a maximum development of generic types at the opposite poles (in time) of the system of nature (Forbes 1851–1854). Wallace was "annoyed to see such an ideal absurdity put forth" when the facts could be explained simply on the basis of known geological and biological processes. He intended his essay both as a refutation of Forbes and as the occasion for a preliminary statement of his own ideas on evolution (Wallace [1905] 1969, 1:355; Marchant [1916] 1975, 54). Arguing against Forbes, Wallace claimed that during periods of geological stability conditions would be favorable for the appearance and continued existence of new forms of life. Conversely, periods of geological activity and changes of climate in a given region "would be highly unfavourable to the existence of individuals, might cause the extinction of many species, and would probably be equally unfavourable to the creation of new ones." The increase of the number of species during certain epochs and the decrease during others were thus explicable "without recourse to any causes but those we know to have existed, and to effects fairly deducible from them" (Wallace 1855, 192–193).

Wallace considered Forbes's assumption that both the fossil record and human knowledge of it were tolerably complete epistemologically unwarranted. Wallace never tired of stressing that the fossil record was incomplete. Whole geological formations, with their fossil remains from vast periods of time, are buried beneath the oceans and therefore largely inaccessible to human inquiry. And because knowledge of the entire series of the former inhabitants of the earth is necessarily fragmentary, all hypotheses that proceed from the contrary assumption were to Wallace scientifically inadmissible. Quite apart from Forbes's explicit rejection of the doctrine of evolution, his work repelled Wallace by its aprioristic speculation. The 1855 essay was clearly directed against such tendencies in biological thought. "The hypothesis put forward in [my] paper," Wallace emphasized, "depends in no degree upon the completeness of our knowledge of the former condition of the organic world, but takes what facts we have as fragments of a vast whole, and deduces from them something of the nature and proportions of that whole which we can never know in detail. It is founded upon isolated groups of facts, recognizes their isolation, and endeavours to deduce from them the nature of the intervening portions" (Wallace 1855, 185). Wallace's arguments were, of course, not merely "deductions" from facts. He was working within a specific theoretical framework, so much so "that several naturalists had expressed regret that he was 'theorising,' when what 'was wanted was to collect more facts'" (Marchant [1916] 1975, 83).

RATIONALISM AND EMPIRICISM

Wallace was attempting to find his own via media between the competing claims of rationalism and empiricism. Of ancient vintage, this debate became acute in the nineteenth century because of the growing power and prestige of scientific methodology. Rationalism is the thesis that the ultimate source of knowledge is to be found in human reason. What constitutes reason was, and remains, a notoriously tricky question. Thinkers as diverse as Locke, Hume, and Kant—three names that were cited repeatedly in the early Victorian period—propounded different answers to that question. Wallace would have understood reason as that feature of the human mind that differs not just in degree but in kind from bodily sensations, feelings, and psychological states. Rationalists maintain that reason has a unique power for grasping reality. The flux of our experience is comprehensible only through the exercise of reason, which then enables human beings to understand the world in which they live.

The empiricist tradition, in contrast, claimed that all philosophizing begins with actual experience. Few philosophers, however, ever literally maintained that all knowledge comes from experience. Empiricism accorded a place for a priori knowledge but denigrated its significance, particularly when it came to matters of fact. The empirical tradition for Wallace, then, would have represented a significant contrast to rationalism. Rationalists hold that human beings have knowledge about matters of fact that is anterior to experience and yet that does tell them something significant about the world and its various features. Empiricists would deny that this is possible. The history of epistemology has to a large extent been a dialectic between rationalism and empiricism in an effort to meet skeptical challenges posed to both positions. Wallace, as so many of his contemporaries, was confronted with this dialectic. His own resolution was worked out gradually and received its ultimate expression in the period from the 1880s onward. But Wallace's earliest philosophical musings reveal an appreciation of the imperative to reconcile his scientific findings and methodology with his beliefs about the nature of knowledge and the diverse array of human beliefs and actions. Wallace came of age in an era when scientific discoveries and technological successes were beginning to confer a potent authority on science and those intellectuals who were associated with it in the public's eye. Wallace was fully aware of the personal significance of his decision to pursue natural history. Like William James, however, he realized that moral and political beliefs were as central to formulating a philosophy appropriate to the nineteenth century as were scientific theories and discoveries (Siegfried 1990, 16, 248–249).

In the 1850s, Wallace's epistemological framework was not only evolutionary but also secular. What is conspicuous in the 1855 essay, as well as in Wallace's private notebooks of the period, is the lack of any explicit references to the concepts of divine intervention and design in nature (McKinney 1972, 45). In contrast to the majority of British scientists, who still adhered to some form of natural theology, Wallace restricted his early scientific pronouncements to the language of physical and biological causality. The early exposure to secular philosophy, reinforced by his prolonged contact with non-European cultures, freed Wallace, at this period, from incorporating those traditional religious glosses that then colored much of Western biological and geological reasoning. At this stage, Wallace opposed not only creationists but also those evolutionists who incorporated explicitly providential elements into their explanatory schema. The usual evolutionist/creationist dichotomy drawn between biologists in the 1850s is less meaningful than a division among them on the basis of whether they sought or did not seek teleological explanations in their theoretical articulations (Ospovat 1978, 35, 49–52). Wallace's first statements of evolutionism illustrate this dichotomy in a particularly cogent manner. He was clearly sympathetic to certain teleological concepts from the outset of his career. But Wallace's first tentative steps toward an evolutionary teleology were implicit rather than explicit. Certain of Wallace's pre-1858 writings, however, indicate that he had indeed adopted teleological elements as part of his emerging evolutionary hypotheses but in characteristically idiosyncratic fashion (Smith 1992, 29).

There are hints in an essay written in 1843, when Wallace was only twenty, that the youthful naturalist was even then viewing his studies within the broader framework of a purposeful cosmology. Portions of this essay, entitled "The Advantages of Varied Knowledge," were—many decades later— reproduced in Wallace's autobiography. The young naturalist asked, "Can we believe that we are fulfilling the purpose of our existence while so many of the wonders and beauties of the creation remain unnoticed around us? . . . While so many of the laws which govern the universe and which influence our lives are, by us, unknown and uncared for? . . . Can we think it right that, with the key to so much that we ought to know, and that we should be the better for knowing, in our possession, we seek not to open the door, but allow this great store of mental wealth to lie unused, producing no return to us, while our higher powers and capacities rust for want of use?" Wallace concluded these youthful musings with the provocative query, "Can any reflecting mind have a doubt that, by improving to the utmost the nobler faculties of our nature in this world, we shall be the better fitted to enter upon and enjoy whatever new state of being the future may have in store for us?" (Wallace [1905] 1969, 1:201–204). Wallace followed his own advice and during the next two decades not only pursued a diverse array of interests but also suggested how all these bits of information might somehow be fitted together within a broader "general design" of nature. His fascination with Chambers's progressivist cosmology as spelled out in the *Vestiges* (despite Wallace's critique of numerous specific scientific errors in Chambers's work) prodded Wallace to explore the issue of final causation during the 1840s and 1850s. His 1856 essay "On the Habits of the Orang-Utan" specifically suggests the existence of "some general design [in nature] which has determined the details, quite independently of individual necessities," of the many species of animals and plants (1856b). These occasional, but consistent, writings and musings of Wallace during the 1840s and 1850s would find their detailed and explicit exposition in his evolutionary teleology that emerged during the 1860s and after. The crucial point, however, is that Wallace's intellectual evolution from the mid-1840s until the late 1860s—when evolutionary biology mingled and merged with anthropology, spiritualism, and sociopolitical interests in his mind and work—was marked by a series of efforts to find the most suitable integration of the laws of nature as a function of final causation (Smith 1992, 22–23). With the appearance of the 1869 review of new editions of Lyell's *Principles of Geology* and *Elements of Geology* and the publication of "The Limits of Natural Selection as Applied to Man" in *Contributions to the Theory of Natural Selection* (1870), Wallace presented to the world the unambivalent evolutionary teleology that he would expound in ever greater detail during the remainder of his life.

But his main concern at this early period was to rescue the concept of evolution from the fanciful speculations that abounded in the 1840s and 1850s. Teleology, while occupying part of Wallace's philosophical excursions, took second place to a more urgent task. Wallace's first epistemological guidelines were deployed to create an evolutionary theory that would accord with the cannons of empiricist philosophy. Wallace aimed to present an evolutionism that would compel assent precisely because that theory was grounded in demonstrable evidentiary claims. Tactically, Wallace focused on those aspects of evolutionary theory that he (rightly) believed would most effectively defuse the skeptical attacks on his and Darwin's hypotheses presented publicly in 1858 at the Linnean Society and in 1859 with the appearance of

Darwin's *Origin*. Wallace's comparative silence—with the crucial exceptions noted above—on the teleological components of his evolutionism comes as no surprise. He recognized that the battle in 1859 turned on defending the *Origin* to the scientific community and the broader public. He concentrated on elaborating the strongest segments of his and Darwin's theory: the connections between biological change and adaptation and geological/geographical distribution. Wallace realized that it would be counterproductive in the first contentious years after 1859 to emphasize the significant differences between his and Darwin's conceptions of evolutionary cosmology (Smith 1992, 32–34, 36–37, 47–49). Similarly, Wallace did not yet feel sufficiently well armed to enunciate his more controversial views on human evolution. Nor was he yet prepared to elaborate on the broader issue of an evolutionary teleology that would best advance his goal of constructing a worldview that encompassed both detailed biological data and the sociocultural concerns that were integral to his philosophy of nature. The lack of overt references to teleology in Wallace's earliest evolutionary writings was part of a strategy for "going public" with respect to his teleological views "when the time was right" (Durant 1985, 283).[2]

LYELL AND WALLACE: A PHILOSOPHIC KINSHIP

Wallace's emerging evolutionary teleology resonated with important aspects of the philosophical and scientific framework of Lyell's developing evolutionism. As noted previously, Lyell's geological uniformitarianism was an indispensable component of Wallace's first evolutionary pronouncements. It permeated the essays written between 1855 and 1858. While Wallace contested certain of Lyell's specific scientific conclusions—notably Lyell's initial reluctance to accept transmutationism—Lyellian methodology remained crucial to Wallace's speculations on numerous issues. Of the influential scientists with whom Wallace associated on his return to London, Lyell exerted the strongest influence. Wallace saw Lyell (fig. 3) frequently, and they spoke and corresponded at length on a wide variety of subjects, notably human evolution. It was not only scientific subjects that Wallace and Lyell discussed. During the ten years of Wallace's London residency (1863–1872), if Lyell "had any special subject on which he wished for information, he would sometimes walk across the park to St. Mark's Crescent [Wallace's apartment] for an hour's conversation; at other times he would ask me to lunch with him, either to meet some interesting visitor or for friendly talk" (Wallace [1905] 1969, 1:434). These informal but probing conversations between Lyell and Wallace constitute a dialogue between two of the key figures in the evolutionary debates of the 1860s.

Figure 3. Sir Charles Lyell
in later life (engraving by
G. Stodart).

Wallace shared Lyell's concerns regarding the bearing of evolutionary
theory on the question of human origins and human nature. We know from
Lyell's notebooks during 1855–1861 that he was preoccupied with the implications of evolution on the origin of the so-called higher faculties. In an
entry of 1 November 1858, Lyell had written that the "moral world is an
addition [to the evolutionary process] to which nothing preexisting can be
compared—the Free Will of Man—Memory, Pain, reasoning, instinct, sight,
feeling, hearing, smelling, touch, taste, anger, rage . . . existed before, but not
responsibility, sentiment, goodness. The intermediate stages are the enigma.
They may link Man with Higher Beings as yet unrevealed to us"(Wilson
1970, 197). Lyell maintained these views on the uniqueness of certain aspects of human evolution. A decade later, Wallace penned the famous 1869
Quarterly Review article (on the tenth edition of *Principles of Geology*). He
proclaimed that some "power" other than natural selection had been necessary in the evolution of man. Lyell expressed his agreement (Wallace [1905]
1969, 1:428). Lyell's journals and notebooks are replete with references to

philosophers ranging from Plato, Pythagoras, and Aristotle to Bacon, Berkeley, Mill, Adam Sedgwick, and Whewell. Lyell's notebooks were completed just a year or two prior to Wallace's first meeting with him. Philosophical and epistemological issues would have been major topics the two discussed (Wallace [1905] 1969, 1:418–420, 422–424, 434).

It was at Lyell's more formal evening receptions that Wallace met many well-known figures, including the physicist John Tyndall, the period's major historian of ideas W. E. H. Lecky, and the duke of Argyll. Wallace later recalled that although he and the duke "criticized each other's theories rather strongly, he was always very friendly, and we generally had some minutes' conversation whenever I met him." Arabella Buckley (later Mrs. Fisher), who had become Lyell's private secretary in 1863, befriended Wallace and "would point out to [him] the various celebrities who happened to be present" at Lyell's soirées. Buckley and Wallace thus began that "cordial friendship" that would be a continued source of intellectual and social camaraderie throughout his life. Buckley shared Wallace's deep interest in spiritualism, and they investigated the subject together. At one séance they attended, Samuel Butler was present. Butler was skeptical of séance phenomena. He did admit to Wallace, however, that the arguments put forth by Buckley (one of the "clearest-headed people" Butler knew) and by Wallace forced him to concede that there "must be something" of value in the pursuit of such inquiries and inquirers.

Wallace trusted Buckley perhaps more than anyone else. He cherished her integrity, to the point of confiding in her alone of all his friends his constant financial anxieties and woes. She later was instrumental in persuading Huxley and Darwin to petition, successfully as it turns out, Prime Minister Gladstone to grant Wallace a civil service pension of £200 per annum in 1881. It was Buckley who informed Wallace that Lady Lyell regarded him, in the 1860s, as "shy, awkward, and quite unused to good society." Lyell was less concerned than his wife about Wallace's social graces—or lack thereof on his return from the Malay Archipelago—and far more interested in his intellectual prowess. Wallace, in turn, delighted in Lyell's "great liberality of thought and wide general interests." The two held similar reservations about the efficacy of natural selection to account for all aspects of human evolution. Wallace regarded his friendship with Lyell "with unalloyed satisfaction as one of the most instructive and enjoyable episodes in my life-experience." He particularly admired Lyell's philosophical openness. His assessment of Lyell as one of the ablest and most authoritative figures on the Victorian scene derived not only from their shared scientific interests but from their kindred approaches to philosophy. Wallace's description of Lyell could serve as a description of himself. Lyell, Wallace remarked, was a rigorous but flexible thinker: "Although when he had once arrived at a definite conclusion he held by it very tenaciously until a considerable body of well-ascertained facts could be adduced against it, yet he was always willing to listen to the arguments of his opponents, and to give them careful and repeated consideration" (Wallace [1905] 1969, 1:417, 433–435; 2:296–297, 378). Lyell, in one of his notebook entries in 1859, expressed that credo when he quoted a passage from Mill's essay "Dr. Whewell on Moral Philosophy": "The person who has to think more of what an opinion leads to than of what is the evidence of it, cannot be a philosopher or a teacher of philosophers" (Wilson 1970, 289).

NATURAL HISTORY AND METAPHYSICS

Since the 1855 essay was "only the announcement of the theory, not its development" (Marchant [1916] 1975, 54), Wallace dealt only with certain applications of the law that "every species has come into existence coincident both in space and time with a pre-existing closely allied species." The observed affinities among animals (and plants) were an obvious consequence of the law, and Wallace indicated how a combination of two modes of evolutionary development would account for past and present relationships. A new species, having for its immediate "antitype" (or parent stock) a closely allied species existing at the time of its origin, might, in turn, give rise to a third species. If this process continued, with each new species giving rise to but one further species on its model, the resulting system of affinities would be represented by a simple and direct line of succession in time. If, however, one species gave rise, at different times, to two or more new species, the series of affinities would be represented by a forked or many-branched line. Both patterns were evident in the fossil record, and Wallace described the resulting evolutionary network in the now familiar imagery of a "complicated branching of the lines of affinity, as intricate as the twigs of a gnarled oak or the vascular system of the human body" (Wallace 1855, 187). The evolutionary system of natural affinities was not only complex; it was also incomplete. Contra Forbes and other idealist theorists, Wallace asserted the scientific superiority of evolutionary classification to arbitrary systems that assigned a definite number for the divisions of each group. Evolution's trump card was that it proceeded from the more realistic premise that the fossil record was incomplete. Since many species may have become extinct without leaving any trace, Wallace argued that it was difficult, perhaps impossible, to arrive at a precise picture of species history. He also stressed that the historical sequence of fossils was not always ascertainable by the paleontological techniques then available. Despite these reservations, the evolutionary hypothesis did offer a naturalistic explanation of affinities and suggested fruitful avenues for future research. Wallace's evolutionary hypothesis, however, was never envisioned so as to preclude its extension to domains beyond the strictly biological.

Data drawn from the geographical distribution of animals and plants were (and remain) a cornerstone of evolutionary theory, and Wallace demonstrated how his law readily accounted for those facts. The more isolated a region is from other landmasses and the longer its geological isolation, the greater will be the number of species, genera, and families peculiar to it. Conversely, adjacent regions will be populated by identical or closely allied species and genera, as Wallace had observed with the butterflies, monkeys, and fishes of the Amazon. But it was the more singular phenomena of biogeography that provided Wallace with his most striking evidence. He seized on the distributional anomalies of the Galapagos Islands. The fact that each of the islands contained groups of animals and plants peculiar to itself but closely related to those of the other islands, as well as to those of the nearest mainland portions of South America, was inexplicable on the theory of special creation. The contrary would have been expected, since that theory presumed that regions with identical environments, such as the Galapagos Islands, should be populated with identical forms. Conversely, regions with markedly different environments, such as the Galapagos and the nearest South American mainland, should be inhabited by dissimilar forms. To Wallace, the "question forces itself on every thinking mind—why are these things so?" And the solution was clear:

> The Galapagos are a volcanic group of high antiquity, and have probably never been more closely connected with the continent than they are at present. They must have been first peopled, like other newly formed islands, by the action of winds and currents, and at a period sufficiently remote to have had the original species [from South America] die out, and the modified prototypes only remain. In the same way we can account for the separate islands having each their peculiar species, either on the supposition that the same original emigration peopled the whole of the islands with the same species from which differently modified prototypes were created, or that the islands were successively peopled from each other, but that new species have been created in each on the plan of the pre-existing ones.
>
> (Wallace 1855, 188, 190)

In like fashion, the distributions in regions separated by mountain ranges (according to their time of formation) or oceans (according to their depth) become readily understandable.

On the question of whether the succession of species in time had been from a lower, less specialized, to a higher, more complex, degree of organization, Wallace argued that "the admitted facts seem to show that there has been a general, but not a detailed progression. Mollusca and Radiata existed before Vertebrata, and the progression from Fishes to Reptiles and Mammalia, and also from the lower mammals to the higher, is indisputable." His law accounted not only for this development of higher from lower forms of life but also for apparent cases of retrogression in the fossil record. Thus, it is possible for a certain group—such as an order of the phylum Mollusca—to have reached a high level of specialization and complexity at an early epoch. Geological changes would then have caused the extinction of the more specialized (and hence more vulnerable) representatives of the order, while leaving as the sole members of a once rich and varied group only some lower, less-specialized species. These latter would then have served as the antitypes for future species, which might never attain to the high degree of development of the earlier Mollusca. The retrogression in the fossil record is only apparent. In actuality, there had been a progression—although interrupted—of Mollusca and the theory of organic evolution is not contradicted (Wallace 1855, 190–196). Wallace is here grappling with one of the most contentious concepts in nineteenth- (and twentieth-) century evolutionary theory: "progress" (Bowler 1996). Of all terms in the evolutionists' lexicon, "progress" is one of the most problematic because of the multiplicity of meanings that have been attached to it. Wallace's changing conceptions of progress are reflective of his, and his contemporaries', attempts to provide provisional definitions of the relation between so-called higher and lower categories of animals and plants. Indeed, the very adjectives "higher" and "lower" have been rejected by some modern biologists, such as Richard Dawkins, as so "mischievous" as to merit deletion from the discourse of evolutionary biology. In the Victorian period, mischievous as the term "progress" may have been because of its ambiguity and uncertainty in the hands of different thinkers, the concept itself was at the very heart—semantically and culturally—of Victorians' conception of evolutionism in both taxonomic and ideological dimensions (Keller and Lloyd 1992, 6, 263–272). Wallace's early analysis of the concept of progress prefigured his later confident assertions regarding the applicability of that term to describe the course of human and social evolution.

The 1855 essay is a remarkable if somewhat flawed document. It constructed a powerful argument in support of the thesis that new species evolve (though Wallace did not yet employ the word) from closely related, preexisting species but suggested no mechanism for such change. Yet Wallace had produced, from his own observations and insights as well as from the work of Lyell, Chambers, Darwin, Lamarck, and others, a major attack on creationism. Wallace's complex relationship with Lyell surfaced clearly on this point. *Principles of Geology*, with its suggestive remarks on biogeography and the struggle for existence in nature and its convincing demonstration of how geological changes could cause the extinction of certain species, was a fundamental source for the 1855 essay. But on the crucial question of the origin of new species—"the most difficult, and at the same time the most interesting problem in the natural history of the earth" (Wallace 1855, 190)—Lyell had at first explicitly rejected Lamarck's theory of transformism and invoked special creation (Lyell 1830–1833, 2:18–35). Thus Wallace's "hope" that his efforts to deduce a law that determined, "to a certain degree, what species could and did appear at a given epoch, [would] be considered as one step in the right direction towards a complete solution" of the species question was a direct challenge to the major opponents of the theory of organic evolution in the mid-1850s (Wallace 1855, 190).

Wallace was, therefore, surprised at the lack of public response to the appearance of the essay. The death of Forbes the year before had removed the one naturalist who would have been most likely to initiate a critical discussion of Wallace's ideas among British scientists. It was from Bates that he first received some notion of the impact his work was destined to have. "I was startled at first to see you already ripe for the enunciation of the theory," Bates wrote on 19 November 1856. "The idea is like truth itself, so simple and obvious that those who read and understand it will be struck by its simplicity; and yet it is perfectly original. The reasoning is close and clear, and although so brief an essay, it is quite complete, embraces the whole difficulty, and anticipates and annihilates all objections." Bates was prescient in his belief that, although few naturalists would then be "in a condition to comprehend and appreciate the paper," Wallace was assured, ultimately, of a "high and sound reputation" (Marchant [1916] 1975, 52–53). The two men who were best prepared to appreciate the contents and implications of the essay, Lyell and Darwin, read it shortly after it appeared. They were to be deeply influenced by Wallace. Despite the dearth of immediate public recognition, Wallace's essay had brought him to the center of the Victorian evolutionary maelstrom.

BUILDING ON THE 1855 ESSAY

From 1855 to 1858, Wallace sent to England several articles that are mainly descriptive accounts of the fauna and flora of the islands he visited in the Malay Archipelago. Three papers did deal explicitly with the epistemological implications of the 1855 law and reflect the increasing certainty of his evolutionary convictions (Wallace 1856a, 1857, 1858). In "Attempts at a Natural

Arrangement of Birds" (1856a), Wallace developed his contention that a natural system of affinities based on evolutionary relationships provides the only basis for a valid classification schema. Drawing on his extensive knowledge of the birds of South America and of the Malay Archipelago—knowledge gained not only from field observations but also from the "constant habit of skinning" and preparing recently killed specimens—Wallace proposed an arrangement of the Passerine (Perching) order, based on the concept of a forked, or many-branched, line of descent from common ancestors. He held that previous classifications were inadequate or false because they imposed arbitrary divisions that forced "every bird . . . into one of them, [resulting in] the most incongruous and unnatural combinations" of genera and families. Wallace specifically attacked a modification of Cuvier's system, then current in England, which divided the Passerines (which included finches, tanagers, hummingbirds, kingfishers, parrots, and woodpeckers) into five groups according to outward resemblance of beak formation (Wallace 1856a, 194–196). This version of Cuvier's system, Wallace argued, was based on the similarity (or analogy) of superficial traits. As such, it was misleading for the purposes of systematics because those traits often represented independent adaptation (of unrelated organisms) to similar habits and food supply rather than any genetic affinity.

In contrast, Wallace proposed a classification based on a complex of structural traits, internal as well as external, which would reveal the actual (or natural) relationships among different species, genera, and families. He insisted that features such as the texture and arrangement of feathers, the form of nostrils, and the form and strength of the skull afforded more significant taxonomic criteria for assessing affinities because they were less easily adaptable to external conditions. Such an ensemble of characteristics, particularly if it appeared universally throughout a given group, was the strongest evidence for natural as opposed to superficial kinship. Although Wallace was later to amend details of his classification slightly, the 1856 essay exemplified the explanatory potential of evolutionary theory. Not only had he successfully applied the developmental hypothesis to a troublesome ornithological problem, but he had also clarified the question of what constituted an evolutionary transition. The most highly developed members of each group, Wallace asserted, must be most distinctly separated from all the species of any other group and could not possibly be transitional forms. By implication, it was clear that transitions between groups must be sought among the least-developed forms: the common ancestor in the network of branching affinities (Wallace 1856a, 196–199, 204, 207–214).

Wallace continued to refine his evolutionary approach in an 1858 "Note on the Theory of Permanent and Geographical Varieties," which focused on the vexing question of the difference between species and varieties. The conventional view, both biological and theological, held that species were "absolute independent creations, which during their whole existence never vary from one to another, while varieties are not independent creations, but are or have been produced by ordinary generation from a parent species." This definition, though apparently unambiguous, breaks down in practice, and Wallace exploited the dilemma by showing the logical inconsistency of the doctrine of "permanent varieties." Species could be distinguished from varieties on two grounds. Using a quantitative criterion, any form whose characteristics differed from those of a given species, but within a specified limit, would be classed as a variety; any form whose differences exceeded the stated range of variation would be classed as a separate species. Alternatively, the difference between species and varieties could be regarded as qualitative "by considering the permanence, not the amount, of the variation from its nearest allies, to constitute the specific character." Thus, a species would be defined by the permanence of its distinguishing characteristics, whereas a variety would be unstable and might revert back to its parent form. Wallace declared that neither definition was satisfactory. If species differed from varieties in degree only, the line that separates the two would be entirely arbitrary and "so fine that it will be exceedingly difficult to prove its existence." If the only difference between species and varieties was quantitative, Wallace branded "that fact [as] one of the strongest arguments against the independent creation of species." Why, he asked, "should a special act of creation be required to call into existence an organism differing only in degree from another which has been produced by existing laws?" The criterion of permanence fared little better. Certain forms, the so-called geographical varieties, were regarded as possessing characteristics that, though permanent, were not sufficiently distinct to allow their being classed as separate species. Conventional opinion allowed that such varieties shared the character of permanence with true species though, by definition, they were not special creations. Wallace remarked that it was indeed "strange that such widely different origins should produce such identical results." The conclusion was obvious: "The two doctrines, of 'permanent varieties' and of 'specially created unvarying species,' are inconsistent with each other" (Wallace 1858).

Wallace had begun the "Note" somewhat disingenuously by stating that he was not "advocating either side of the question." His actual position is evident from an examination of one of the notebooks he kept during the course of his travels in the Malay Archipelago. This notebook, containing entries from 1855 to 1859, is indispensable for a full understanding of the development of Wallace's ideas. It was probably intended as the draft of an extensive book on evolution, about which he wrote to Darwin late in 1857 and to Bates early in 1858 (Marchant [1916] 1975, 54; McKinney 1972, 30). The entries dealing with the difference between species and varieties are of the utmost interest. They indicate that at least as early as 1855 Wallace had concluded that there was no difference in kind between the two. His comments on the orthodox view that species can vary only within fixed, narrow limits are blunt:

> Lyell says that varieties of some species may differ more than other
> species do from each other without shaking our confidence in the
> reality of species—But why should we have that confidence? Is it not
> a nice prepossession or prejudice like that in favour of the stability of
> the earth which he has so ably argued against? In fact, what positive
> evidence have we that species only vary within certain limits? . . . We
> have no proof how the varieties of dogs were produced. All varieties
> we know of are produced at birth, the offspring differing from the
> parent. This offspring propagates its kind. Who can declare that it
> shall not produce a variety, which process continued at intervals will
> account for all the facts?
>
> (Wallace 1855–1859)

The point of the 1858 "Note" now becomes clear. Convinced that there was no difference in nature between the origin of species and of varieties, Wallace sought to discredit the concept of species as fixed, special creations by showing the inconsistencies that followed from such a definition. He was not able to offer an entirely satisfactory definition of his own. The definition of species was and remains a refractory problem, as the history of, and present debates on, taxonomic classification in evolutionary biology demonstrate (Keller and Lloyd 1992, 302–323). Furthermore, Wallace was not absolutely clear in the 1858 "Note" as to whether he intended the term "variety" to refer to a variant individual or a variant population. He used the term, interchangeably, to denote both concepts. Wallace himself later recognized this conceptual ambiguity and clarified the distinction between the two usages (individual variations vs. the resultant variant population— e.g., subspecies) when the paper was reprinted (in 1870) with the addition of a few key subheadings and footnotes (Kottler 1985, 375–379). But he did demonstrate effectively that there was no essential difference between the origin of species and varieties. Along with Darwin, Wallace thus provided the basis for removing one of the major impediments to evolutionary theory: the philosophical doctrine of essentialism.

Essentialism had retarded the acceptance of evolutionary hypotheses during the first sixty years of the nineteenth century. In England, and to an even greater degree in continental Europe, the typological thinking engendered by essentialism exerted a powerful hold on many naturalists wrestling with the "species question." This philosophical paradigm presumes that the changeable world of appearances is based on underlying immutable essences and that all members of a given class represent the same essence. With a hallowed pedigree going back to Plato, essentialism did not ignore the enormous variability in nature but relegated it to an inferior philosophical and scientific status. Discontinuity and fixity, accordingly, assumed pride of place in the naturalists' essentialist lexicon when tackling the question of species and varieties (Mayr 1976, 282–283; Sober 1980). It was this aspect of Lyell's concept of species that Wallace was targeting in the 1858 "Note." Lyell, of course, was gradually becoming plagued by doubts with the validity of essentialism and publicly confessed his conversion to evolutionism in 1862 (Mayr 1976, 284). But in 1858, Wallace could legitimately count Lyell as a powerful voice in favor of essentialist, as opposed to evolutionary, models for species definition.

Despite their (often significant) differences on the details of the evolutionary process, in this crucial period of the late 1850s Darwin and Wallace wrote in strikingly similar terms in attacking essentialism. In the *Origin*, Darwin made full use of the fact that naturalists had "no golden rule by which to distinguish species and varieties." This absence of any universally agreed on definitions of these two key concepts allowed Darwin to declare: "It must be admitted that many forms, considered by highly-competent judges as varieties, have so perfectly the character of species that they are ranked by other highly competent judges as good and true species. But to discuss whether they are rightly called species and varieties, before any definition of these terms has been generally accepted, is vainly to beat the air" (Darwin [1859] 1964, 49, 296–297). Like Wallace, Darwin's theory of the gradual modification and divergence of species insisted that the difference between species is one of degree, not of kind (Winsor 1991, 103). In the concluding pages of the *Origin,* Darwin boldly predicted that when the views he had meticulously detailed were generally admitted,

> there will be a considerable revolution in natural history. Systematists
> will be able to pursue their labours as at present; but they will not be
> incessantly haunted by the shadowy doubt whether this or that form
> be in essence a species. This, I feel sure, and I speak after experience,
> will be no slight relief. . . . We shall at least be freed from the vain
> search for the undiscovered and undiscoverable essence of the term
> species. The other and more general departments of natural history
> will rise greatly in interest. The terms used by naturalists of affinity,
> relationship, community of type, paternity, morphology, adaptive

characters, rudimentary and aborted organs, &c., will cease to be
metaphorical, and will have a plain signification.

(Darwin [1859] 1964, 484–485)

Though Darwin's prophecies were ultimately—in large measure—fulfilled,
that path is a history filled with complexities and ironies (Winsor 1991, 81–
118). But Wallace's and Darwin's frontal assault on essentialism in biology
was philosophically cogent and polemically effective.[3]

Charles Sanders Peirce, in his 1877 essay "The Fixation of Belief," as-
tutely recognized that the nineteenth-century controversy over evolutionism
was, "in large part, a question of logic" (Hull 1973, 68). Wallace's essays
during the crucial period from 1855 to 1858 show how central his strategy of
poking holes in the logic of essentialist thinking was to his discovery of natural
selection. It was geography—that is, his detailed observations on the distribution of organisms in the Amazon and the Malay Archipelago—which provided one vital clue. Like Darwin, Wallace was emphasizing a new approach
to evolutionism, geographical evolutionism (Mayr 1982, 419). Essentialism
was a major casualty of this new approach. The attack on so deeply rooted a
concept in Western philosophy naturally provoked an outpouring of criticism
from many Victorian biologists and philosophers—as well as defenses of the
new approach (Hull 1973, 67–77). Since essentialism, historically, had often
been linked to the concept of teleology in biological thinking, it would be
surprising if Wallace—as well as Darwin—did not have to come face to face
with teleology and its framework of causality and explanation (Ruse 2000,
222–224; Short 2002). Though the issue is still controversial, there is evidence that Darwin—contrary to what many of his supporters and critics
believed—did not throw out teleology along with essentialism. He attempted
to put certain elements of traditional Aristotelian teleology on a more solid
biological footing (Keller and Lloyd 1992, 324–333; Gotthelf 1999). Wallace, for his part, never thought that the rejection of essentialist philosophy
entailed a rejection of teleology. Indeed, Wallace's rejection of essentialism
but embracement of an updated teleology constitutes a basic element of his
developing evolutionary philosophy.

THE ARU ISLANDS

Despite the lack of any adequate mechanism for evolution, Wallace was interpreting the data from the Malay Archipelago with increasing mastery. His
collecting in the Aru Islands (situated to the southwest of New Guinea and
never before visited by an English naturalist)—which he regarded as the most
successful of his entire travels (Wallace [1905] 1969, 1:357)—provided the

material for the essay "On the Natural History of the Aru Islands," published in 1857. The most striking characteristic of these islands was the absence of many widely distributed species of the western half of the Malay Archipelago (including Borneo, Sumatra, and Java). Equally notable was the similarity— in many cases identity—between Aru species of birds, insects, and mammals (the groups Wallace collected most extensively) and those of New Guinea and, to a lesser degree, Australia. Using that combination of biological and geological reasoning, which he had fashioned into a potent methodological tool, Wallace explained the anomalous distribution patterns of the Aru fauna on the basis of the evolutionary hypothesis and further eroded the special creationist position. Wallace's Aru paper combined meticulous observation with a sophisticated analysis predicated on explicit evolutionary hypotheses. Although relatively brief, it—along with several other papers he wrote at the same period—stands as a seminal contribution to the origins of modern biogeography.

Given the wide interval of sea (averaging 150 miles) separating the Aru Islands from the coast of New Guinea, the close resemblance of species was puzzling. Wallace cited the example of the island of Ceylon. Closer to the mainland of India than Aru is to New Guinea, Ceylon presents a fauna clearly distinct from India. It housed many unique species and, even, unique genera. Sardinia, about as far from Italy as Aru is from New Guinea, also presents a distinct fauna. In contrast, the only major islands that did possess a rich fauna, nearly identical to their adjacent mainland, were Great Britain and Sicily. The relatively singular biogeographical status of Great Britain and Sicily was crucial for Wallace's course of reasoning. He noted pointedly that it "is held to prove that they have been once a portion of such continents, and geological evidence shows that the separation had taken place at no distant period." Arguing by analogy, Wallace declared that Aru must once have formed part of New Guinea. He corroborated this by the fact that the Molucca Sea, which bordered Aru to the west, was of great depth. The sea eastward from Aru to New Guinea and southward to Australia was, in contrast, comparatively shallow. The shallow sea indicated a (geologically) recent land connection that would have provided a common set of ancestors for the present-day faunas of the now-separate landmasses (Wallace 1857, 478–479).

The distributional anomalies of the Aru Islands were of more than merely local significance. They reflected the broader historical changes of the entire Malay Archipelago and afforded Wallace new evidence against special creation. Most naturalists held that as "ancient species became extinct, new ones were created in each country or district, adapted to the physical conditions of that district." Wallace emphasized that, according to Lyell, because extinction generally implied a change in physical conditions (to which existing species were ill-adapted), the new species would be dissimilar to those species they replaced. This theory implied that regions possessing similar climate and topography would house similar fauna. Regions differing markedly in those respects should display unrelated animal populations. If special creation was the law that governed the introduction of species, there could be no contradictions to it, or at the very least, no striking exceptions. But the Malay Archipelago yielded Wallace the precise contradiction he had been seeking:

> Now we have seen how totally the productions of New Guinea [and Aru] differ from those of the Western Islands of the Archipelago, say Borneo, as the type of the rest, and as almost exactly equal in area to New Guinea. This difference, it must well be remarked, is not one of species, but of genera, families, and whole orders. Yet it would be difficult to point out two countries more exactly resembling each other in climate and physical features. . . . If, on the other hand, we compare Australia with New Guinea, we can scarcely find a stronger contrast than in their physical conditions: the one near the equator, the other near and beyond the tropics; the one enjoying perpetual moisture, the other with alternations of excessive drought; the one a vast ever-verdant forest, the other dry open woods, downs, or deserts. Yet the faunas of the two, though mostly distinct in species, are strikingly similar in character.
>
> (Wallace 1857, 481)

Every family of birds (except one) found in Australia also is found in New Guinea. More important, many of the Australian genera are also found in New Guinea. Similar distribution characterizes mammalian and insect groups. Wallace cited the presence of the kangaroo, perfectly adapted to the dry plains and open woods of Australia, in the dense and damp forests of New Guinea (but not of Borneo) as inexplicable on the creationist hypothesis. Similarly, the abundance of monkeys in Borneo—suited to its physical environment—was in direct contradiction to their total absence in New Guinea, whose physical conditions were practically identical. Some law other than special creation, Wallace announced, "has regulated the distribution of existing species . . . or we should not see countries the most opposite in character with similar productions, while others almost exactly alike as respects climate and general aspect, yet differ totally in their forms of organic life" (Wallace 1857, 480–481).

The "other" law is Wallace's own, presented in 1855. Applied to the present case, the creationist contradictions disappear and the apparent distributional anomalies are resolved. At that period in the past when New Guinea and Australia were united, they shared a similar climate and physical geography and housed related or identical species. When the landmasses separated, the climate of both regions would likely have been modified significantly, resulting in the extinction of many species. Subsequently, "new species have been gradually introduced into each [region], but in each closely allied to the pre-existing species, many of which were at first common to the two countries." This process would account for the present similarity (but not identity) between the fauna of New Guinea and Australia. Further, those groups absent from one—such as the monkeys from Australia—would "necessarily be so from the other also, for however much they might be adapted to the country [New Guinea], the law of close affinity would not allow of their appearance, except by a long succession of steps occupying an immense geological interval." Wallace continued the argument with respect to Aru to demonstrate the universal applicability of the 1855 law. Had the Aru Islands been separated from New Guinea for a longer period than was actually the case, the two faunas would be more distinct, though still related. The longer the hypothesized separation, the greater would have been the process of organic change. Some species would have become "extinct in the one country, and unreplaced, while in the other a numerous series of modified species may have been introduced. Then the faunas will come to differ not in species only, but in generic groups. There would then be the resemblance between them that there is between the West India Islands and Mexico." If, finally, the separation of Aru from New Guinea had taken place at a period as remote as that when Madagascar separated from Africa, the Aru fauna would show "an exact counterpart of what we see now in Madagascar." There, although a general resemblance to African forms persists, the long continued divergence of Malagasy species from the ancestral stock has resulted in many peculiar genera and even entire families (Wallace 1857, 482–483).

Wallace had vindicated his theoretical propositions of 1855 by a cogent explanation of distributional data collected in the Aru Islands. He had shown the special creationist argument to be both redundant and false. "Centres of creation," which had been advocated by certain naturalists, were unnecessary unless one literally invoked a "center" in every island or district that possessed a unique species. Wallace also had shown the special creationist argument to be invalid: new species had never been created "perfectly dissimilar in forms, habits, and organization" from those that had preceded them (Wallace 1857, 483). Most significant, he had indicated that anthropological data were as crucial to evolutionary theory as those drawn from the distribution of animals and plants. As mentioned in chapter 2, Wallace had studied the physical and moral traits of the Papuans (natives of New Guinea, Aru, and the Kei Islands) and "noted the very striking differences that exist between them and the Malays, not only in outward features, but in their character and habits" (Wallace 1857, 474, 483). When Wallace later proposed a boundary, "Wallace's Line" (the term was coined by Huxley in 1868), dividing the flora and fauna of the eastern ("Australian") half of the Malay Archipelago from that of the western ("Indian") half, he proposed a similar (but not identical) boundary between the Malayans and Papuans (Wallace 1863).

The roots of Wallace's fascination with tracing ethnological boundaries—geographical as well as socioeconomic and cultural ones—date back to his trade as a land surveyor in the early 1840s. Wallace's firsthand experiences in observing the boundaries that separated the poorer Welsh farmers from their comparatively well-off English neighbors to the east of the south Wales border country left an indelible imprint on the nascent social reformer. Henceforth, Wallace would carry the visual and ideological force of the stark sociopolitical reality and geographic precision of ethnological divides as he moved beyond the confines of Britain to the more exotic arenas of South America and the Malay Archipelago (Moore 1997, 300–307). Ethnological data had, however, at this time a more profound and catalytic effect on Wallace's thinking. It was the question of human evolution and the argument of Thomas Malthus's *Essay on the Principle of Population* (1798) that provided a direct clue to his discovery of natural selection.

DISCOVERY OF NATURAL SELECTION: THE SPECTER OF MALTHUS

The joint discovery of the principle of natural selection by Wallace and Darwin is among the most celebrated episodes in the history of science. Although their paths to discovery displayed common features, there is no doubt that the two naturalists arrived independently at strikingly similar hypotheses on the origin of species. Most historians have analyzed the simultaneous discovery as a matter of scientific priority. Such a focus tends to reconstruct the actual history of the joint discovery in terms of the garland of victory being awarded to either Darwin (more frequently) or to Wallace (less so). Barbara G. Beddall has offered a more realistic and less partisan analytical framework. Instead of invoking the analogy of a zero-sum game with winners and losers, Beddall employs the non-zero-sum game metaphor. Rather than pitting Darwin against Wallace for the spoils of victory—that is, priority—she suggests historians should appreciate and record the enormous contributions made by both Wallace and Darwin. By focusing on the interaction rather than competition between the two formulators of natural selection theory, the manner in which Wallace and Darwin benefited from each other's work becomes the central object of historical reconstruction. This approach has the additional virtue of dispensing with the highly problematic scenarios of "editorial manipulation" and "delicate arrangements" so frequently invoked to "solve" the priority dispute quandary. Such scenarios, while superficially seductive, obscure the complexities of the Darwin-Wallace connection and pose significant problems of interpretation that are not adequately supported by the historical record itself (Beddall 1988a). Wallace's travels provided him with a vast body of observational data by means of which he was able to translate his evolutionary speculations (first suggested in a letter to Bates in 1845) into the rigorous theory announced in 1858. And, like Darwin, Wallace's empirical data were refracted through an epistemological lens. A key component of his "philosophical biology" was Malthus's *Essay on the Principle of Population.*

Although the exact nature of Malthus's influence continues to be the subject of debate, his *Essay* provided Wallace with a critical insight that enabled him to solve the question of how species originate (McKinney 1972, 149–150; Young 1985a, 610, 633–635; Bohlin 1991). Wallace's autobiographical rendition of his moment of discovery provides a dramatic, if remote (it was written nearly half a century after the event), statement of scientific creativity (Wallace [1905] 1969, 1:360–363). It was during an illness on the island of Ternate, in late February 1858, that Wallace, pondering those subjects that had most engaged him during his Malaysian travels, recalled the work of Malthus that he had read some twelve years before. Wallace's illness proved a blessing in disguise. "Suffering from a sharp attack of intermittent fever, and every day during the cold and succeeding hot fits [having] to lie down for several hours," Wallace had "nothing to do but think over any subjects then particularly interesting me." Malthus's vivid demonstration of "the positive checks to increase"—disease, war, accidents, and famine—which keep the population of savage races down to a much lower average than civilized races sparked Wallace's chain of reasoning:

> It then occurred to me that these causes or their equivalents are continually acting in the case of animals also; and as animals usually breed much more rapidly than does mankind, the destruction every year from these causes must be enormous in order to keep down the numbers of each species, since they evidently do not increase regularly from year to year, as otherwise the world would long ago have been densely crowded with those that breed most quickly. Vaguely thinking over the enormous and constant destruction which this implied, it occurred to me to ask the question, Why do some die and some live? And the answer was clearly, that on the whole the best fitted live. From the effects of disease the most healthy escaped, from enemies, the strongest, the swiftest, or the most cunning. . . . Then it

suddenly flashed upon me that this self-acting process would necessarily *improve the race,* because in every generation the inferior would
inevitably be killed off and the superior would remain—that is, *the
fittest would survive.*

(Wallace [1905] 1969, 1:361–362, emphasis in original)

It at once became clear to Wallace that natural selection (though he did not
yet use that term) was the mechanism he had been seeking. Combining Lyell's
description of the gradual fluctuations of land and sea, climate, food supply,
and predators with his own field experience of organic variation in nature,
Wallace realized that—given sufficient time—new species would evolve in response to altered environmental conditions. The exquisite and often complex
adaptations of animals were now explicable not as the product of design but
as the outcome of evolutionary change. Wallace's autobiographical account
of his historic discovery concluded with a fittingly histrionic flourish:

> The more I had thought over it the more I became convinced that
> I had at length found the long-sought-for law of nature that solved
> the problem of the origin of species. For the next hour I thought over
> the deficiencies in the theories of Lamarck and of the author of the
> "Vestiges," and I saw that my new theory supplemented these views
> and obviated every important difficulty. I waited anxiously for the
> termination of my fit so that I might at once make notes for a paper
> on the subject. The same evening I did this pretty fully, and on the
> two succeeding evenings wrote it out carefully in order to send it to
> Darwin by the next post, which would leave in a day or so.
>   I wrote a letter to him in which I said that I hoped the idea would
> be as new to him as it was to me, and that it would supply the miss-
> ing factor to explain the origin of species. I asked him if he thought
> it sufficiently important to show it to Sir Charles Lyell, who had
> thought so highly of my former paper.
>
> (Wallace [1905] 1969, 1:362–363)

Upon recovery, Wallace wrote out his theory as an essay entitled "On the
Tendency of Varieties to Depart Indefinitely from the Original Type" and
mailed it to Darwin with the request that he show it to Lyell, "should he
think it sufficiently novel and interesting" (Beddall 1968, 299).

Wallace's terse account of his moment of discovery of natural selection
and his effort to so inform Darwin has generated a historical industry of its
own. There is no reason not to accept Wallace's account in *My Life* as his
genuine recollection of the events leading to the joint announcement at the

Linnean Society. The magnitude of this event, however, has caused controversy over the precise details of many of Wallace's, and Darwin's, assertions as to the sequence of events leading to 1 July 1858. Unfortunately, Wallace's original 1858 manuscript about varieties has never been found. The only known version of this crucial text is the one published in the August 1858 *Journal of the Proceedings of the Linnean Society* (Wallace [1858] 1969). The original manuscript was sent to Darwin as an enclosure in a letter (itself missing), referred to by Wallace in his account in *My Life* quoted above, and was subsequently sent by Darwin to Lyell in a letter dated 18 June 1858. The date of this particular letter from Darwin to Lyell has also been questioned. Since Darwin told Lyell that he received Wallace's enclosure and letter "today," that is, 18 June 1858, the actual date of Darwin's letter to Lyell is enveloped in a mystery of its own (Burkhardt and Smith, 1985–, 7:xvii–xviii, 107–108, 512).

The absence of Wallace's letter and original manuscript, coupled with the lack of any conclusive evidence for dating Darwin's letter to Lyell, renders moot the question of whether Darwin held up Wallace's packet for any reason (Burkhardt and Smith 1985–, 7:xviii). Certain points concerning the precise steps leading to the joint publication, as well as to the nature of the debt owed by Darwin to Wallace and/or Wallace to Darwin in the final articulation of the theory of natural selection, remain—for now—irresolvable (Beddall 1988a, 2). The documentation that does exist in the form of notebooks, journals, and letters, in addition to the published articles and books of Wallace and Darwin, have, nonetheless, permitted scholars to establish fairly rigorously the historical reconstruction of the Wallace-Darwin relationship both before and after 1858. Thus, it is clear that Wallace, although aware that Darwin was preparing for publication his great work on species and varieties, did not know that the latter, too, had discovered natural selection but had not yet published on it. Darwin, in contrast, had likely discerned Wallace's progress on the species question from his letters as well as the pre-1858 articles. Lyell's and Edward Blyth's alerting Darwin to the existence and significance of Wallace's 1855 text should certainly have shaken, or at least roused, Darwin. Blyth's letter to Darwin was particularly fulsome in its praise of Wallace's hypotheses. Darwin could not fail to see that Wallace's 1855 essay propounded a powerful statement of the relationships between closely allied species and that it tried to show how geographical distribution would provide the key to understanding these. But Darwin quickly made a note to himself that the essay contained "nothing very new." A recent biographer of Darwin suggests that, although Darwin was obviously aware of the evolutionary overtones in Wallace's essay, he "blindly stared straight past the implications in Wallace's words. . . . He was not prepared to see the possibility that someone else might be hesitantly circling around before arriving at the same theory. His own work, not Wallace's, was primary." But Darwin could not minimize the significance of Wallace's discovery (Browne 1995, 537–538).

As Lyell had warned, Darwin was forestalled. He was in a quandary over the proper course of action to follow. Wallace had not specifically instructed Darwin to publish the essay. To his credit, Darwin realized that publication was the only honorable step (Burkhardt and Smith 1985–, 7:107). Lyell and Hooker arranged a compromise by which both Wallace and Darwin were accorded priority. On 1 July 1858, Wallace's essay was read before the Linnean Society, preceded by extracts from an unpublished essay on natural selection written by Darwin in 1844 and from a copy of a letter dated 5 September 1857 from Darwin to Asa Gray (fig. 4) that discusses the "principle of divergence"—an important part of the theory not discussed in the 1844 manuscript. Neither of the two principals was present at that meeting. Wallace was still in the Dutch East Indies. Darwin was in Down, grieving over the death of his infant son Charles Waring Darwin on June 28 from scarlet fever and worrying about his other children (Porter 1993, 29). The jointly published papers, along with the letter to Gray, appeared on pages 45–62 of volume 3 (1858) of the *Journal of the Linnean Society of London (Zoology)* as Charles Darwin and Alfred Russel Wallace, "On the Tendency of Species to Form Varieties; and on the Perpetuation of Varieties and Species by Natural Means of Selection," with Wallace's essay covering pages 53–62.[4]

Wallace, in distant Malaysia, was ignorant of the distress his essay had caused Darwin and of the skillful manner in which Lyell and Hooker had extricated Darwin from his dilemma. Wallace never publicly questioned the propriety of the joint publication. He wrote home that Darwin had shown his essay to "Dr. Hooker and Sir C. Lyell, who thought so highly of it that they immediately read it before the Linnean Society," thus ensuring Wallace "the acquaintance and assistance of these eminent men on [his] return home" (Marchant [1916] 1975, 57, 131). Wallace's persistent deference to Darwin was generous but curious in the extreme. He later issued statements establishing the independence of his discovery and emphasized that his essay had been printed without his knowledge, "and of course without any correction of proofs" (Beddall 1968, 313). Yet it is primarily by Wallace's own efforts that the theory of evolution by natural selection is usually known as Darwinism. There is a double irony here. There were significant differences between Wallace's and Darwin's formulations of the theory—differences that intensified through the years (Kottler 1985). Some of those differences were apparent as early as 1860, when Wallace annotated the copy of the first edition of the *Origin* that Darwin had his publisher Murray send to him in the East Indies. This annotated copy is now part of the Keynes Collection, housed in the

Figure 4. Photograph of Asa
Gray at age fifty-seven.

library at the University of Cambridge (Beddall 1988b). Nonetheless, when
Wallace published his masterly textbook exposition of the theory of natural
selection in 1889, he titled it, simply, *Darwinism*—this, despite the fact that
the final chapter of the book spelled out in detail Wallace's conviction that
natural selection did not account for important aspects of human evolution.
The second irony is that Wallace's deference has encouraged many twentieth-
century historians to relegate his own contributions to evolutionary science
to a lesser rank than Darwin's. But in 1858, the relatively obscure Wallace
could well be satisfied with having his name indelibly associated with that of
a member of Britain's scientific elite.

THE 1858 ESSAY: BREVITY AND BRILLIANCE

The object of Wallace's 1858 essay was to show "that there is a general principle in nature which will cause many *varieties* to survive the parent species,
and to give rise to successive variations departing further and further from the
original type." His argument proceeded from the premise that the "struggle

for existence" among animals in the wild arises from the disparity between the immense number of animals born and the limited resources necessary to sustain life. Ineluctably, Wallace wrote, this led to the survival of those individuals (within a given species) which are best equipped to meet and overcome the checks imposed by the precariousness of the food supply, the constant predations of enemies, and the vicissitudes of the seasons. Analogously, among the several allied species of a group, those which are best adapted to surrounding conditions will increase at the expense of other species, which themselves diminish in population and, in extreme cases, become extinct (Wallace [1858] 1969, 23–26). Turning to the central issue of the relation between varieties and species, Wallace noted that variations from the typical form of a species must have some definite effect, however slight, on the habits or capacities of the individuals possessing them. Changes such as difference in color (by rendering the animal more or less conspicuous and thus affecting its safety) or alteration in the strength or dimension of limbs or other external organs (by rendering the animal more or less capable of procuring food), for example, would affect the survival power of the variant. Those varieties possessing useful variations will tend to increase in numbers and keep their numerical superiority. Those possessing useless or harmful variations will tend to diminish.

Wallace asked what would happen in a district populated by a parent species plus varieties if some alteration of environmental conditions (such as drought or invasion of new predators) occurred that rendered existence more difficult. His answer was clear. Those individuals that formed the least numerous and feeblest variety would suffer first and, under continued environmental pressure, become extinct. If the altered conditions persisted, the same fate might meet the parent species, leaving only the superior variety. Wallace pointed out that this variety would not revert to the parent form because it would constitute a better-adapted population with altered heredity traits. But this new, improved, and populous race might itself, in time, give rise to new varieties that, by the same general law, become predominant and replace their own parent forms completely. If the process of "progression and continued divergence" continued through a sufficiently vast period of time, Wallace concluded, the ultimate variety will have departed far enough from the original type to be classed as a separate species. The origin of new species, therefore, is (in part) the result of the struggle for existence between closely related members of a population and the fact that variations among those members do frequently occur in nature (Wallace [1858] 1969, 27–29).

One of the strongest traditional arguments to demonstrate the fixity of species was that varieties produced under domestication are unstable. Left to themselves, they generally revert to the normal form of the parent species.

This instability was also thought to characterize varieties occurring in the wild. Naturalists inferred that wild varieties would either revert to the parent form or, at most, vary within strictly defined limits. Wallace rejected that analogy as invalid. His essay proves just the opposite with respect to wild varieties. Wallace also showed how the frequent reversion of domestic varieties followed directly from natural selection. Domestic animals are artificial in that they are protected by man and thereby removed from the rigors of the struggle for existence. Variations that arise among them are selected and bred according to human requirements. Often, those that would render a wild animal unable to compete with its fellows are no disadvantage whatever in a state of domesticity. Short-legged sheep, pouter pigeons, and poodle dogs "could never have come into existence in a state of nature because the very first steps towards such inferior forms would have led to the rapid extinction of the race; still less could they now exist in competition with their wild allies" (Wallace [1858] 1969, 31). Wallace's point was evident: if domestic animals were turned wild they would either become extinct or vary in a direction that would again adapt them to existence in the wild. They would of necessity return to something approximating the original species.

Wallace's terse conclusion to the 1858 essay encapsulated his achievements at this stage:

> There is a tendency in nature to the continued progression of certain classes of *varieties* further and further from the original type—a progression to which there appears no reason to assign any definite limits. . . . This progression, by minute steps, in various directions, but always checked and balanced by the necessary conditions, subject to which alone existence can be preserved, may, it is believed, be followed out so as to agree with all the phenomena presented by organised beings, their extinction and succession in past ages, and all the extraordinary modifications of form, instinct, and habits which they exhibit.
>
> (Wallace [1858] 1969, 38)

The Darwin-Wallace joint publication marked a turning point in the history of biology. Remarkably, it seems to have generated minimal response at the time from the relevant scientific community (England 1997). Wallace's essay was the more impressive contribution, as Darwin himself noted (Marchant [1916] 1975, 112), despite the fact that he had not intended it for publication in that form. And though Darwin assured Wallace that his "share in the theory will [not] be overlooked by the real judges, as Hooker, Lyell, Asa Gray, etc.," it was to Darwin that full public recognition came with the publication of *On the Origin of Species* ([1859] 1964) the following year (Marchant [1916] 1975, 112, 115). Wallace recognized that Darwin s book, brilliantly written with a wealth of illustrative examples, would advance the evolutionary cause among the general public as well as the scientific community. In recommending it to his friend George Silk, he declared that "Mr. Darwin has given the world a *new science,* and his name should, in my opinion, stand above that of every philosopher of ancient or modern times. The force of admiration can no further go!!!" (Wallace [1905] 1969, 1:372–73).

Actually, there were limits to Wallace's admiration. Darwin relied heavily on the analogy between human selection and natural selection in presenting the case for evolution, whereas Wallace considered that analogy suspect and misleading (Wallace [1858] 1969, 31). He held it to be a major weakness that Darwin utilized so extensively the evidence of variation and selection among domestic animals and plants and devoted his own career to demonstrating that the theory of evolution could be supported solely by the evidence of variation in the wild. Darwin's use of data from artificial selection, however, did clarify the concept of variation by showing that natural selection acted both on individual differences (to produce varieties) and, secondarily, on differences between varieties (to produce species)—a distinction that was not entirely clear in Wallace's essay (Bowler 1976). In 1858, however, the differences between Darwin and Wallace were far less significant than the fact of their joint discovery. Catapulted to the forefront of Victorian science, Wallace could devote the remainder of his Malaysian travels to gathering that additional evidence necessary to support the theory against the anticipated hostility of its critics.

This oft-told tale of the joint discovery of natural selection is germane to our understanding of Wallace's rise to eminence as a Victorian social critic. It was Wallace's scientific achievements in the 1850s that lent greater authority to his later pronouncements on social, political, and religious issues. But Wallace himself always resisted a compartmentalization between his scientific work and his increasingly visible presence in the arena of public affairs. The union of science and social activism already present in the Malay travels would become increasingly overt in the next few decades. Wallace's lecture tour in North America during 1886–1887 epitomized his philosophical maxim that science and culture were inseparable.

TRANSATLANTIC CONTACTS

In 1885, Wallace received an invitation from the Lowell Institute in Boston, Massachusetts, to deliver its prestigious annual course of lectures the following year. The opportunity to proselytize on behalf of evolutionary theory and the financial rewards an extended American tour would be expected to yield were attractive. Wallace left England on 9 October 1886, arriving in New York on 23 October, "after a cold and disagreeable passage." Five days later, he left for Boston to deliver his eight Lowell lectures, which were highly successful. Wallace spent the next year traveling across North America, repeating the Lowell lectures with equal success in major American cities and in Toronto and Kingston in Canada. He met many of the United States' most distinguished scientists and leading political, social, and intellectual figures. In addition to speaking on scientific subjects and observing at firsthand the flora and fauna of North America, Wallace spoke publicly on his political and theistic views. Wallace's North American sojourn thus situated him squarely within the broader context of late Victorian cultural controversies on both sides of the Atlantic (Wallace 1886–1887).

The tour constituted a high point of Wallace's career both as statesman of evolutionary science and as ardent proponent of the controversial doctrines of (increasingly) socialist politics and of spiritualism (Wallace [1905] 1969, 2:105–106, 129, 160, 200). One of the most intellectually significant aspects of the tour was his exposure to new currents in American philosophy. The emerging school of pragmatism, as it was then being formulated by William James and Charles Peirce, intrigued him. Wallace's relationship with James and Peirce was crucial to the elaboration and clarification of his own philosophical stance. These transatlantic contacts proved more decisive in the domain of evolutionary philosophy than of evolutionary science—although the two were never separable in Wallace's mind.

The parallels in the lives and careers of Wallace and James are striking. Both men pursued, through their studies of science, social issues, and epistemology, a worldview that would render intellectual analysis a potent tool for cultural reform. They both championed new ideas and sociopolitical causes as a means to effecting a transition to a more just social order. "Philosophic study," James wrote, "means the habit of always seeking an alternative, of not taking the usual for granted, of making conventionalities fluid again, of imagining foreign states of mind." Both men wrote books in the last decade of their lives (James died in 1910) that gave their particular philosophical messages to the world, messages that incorporated significant theistic elements. For James, the "final" statement of his convictions appeared in the trio of books by which he most wanted to be remembered: *Pragmatism* (1907), *The Meaning of Truth: A Sequel to "Pragmatism"* (1909), and *A Pluralistic Universe: Hibbert Lectures . . . on the Present Situation in Philosophy* (also 1909; Simon 1998, xvi–xxiii, 204, 276–279, 301–305). Wallace's trio of books were *Man's Place in the Universe* ([1903] 1907), *The World of Life* (1910a), and *Social Environment and Moral Progress* (1913b). Wallace and

James ultimately arrived at a similar philosophical position. They rejected both Anglo-German idealism (particularly in its neo-Hegelian incarnation) and scientific materialism as wholly adequate responses to the dilemmas facing late Victorian culture. They sought, instead, a comprehensive worldview that integrated scientific, ethical, spiritual, and theistic elements in a cosmos that had both a visible as well as an unseen order (Myers 1986, 447–449). Both Wallace and James were fully aware of the audacity of their quests. Of the two, it was James who most keenly experienced—and wrote of—the personal anxieties that such audacity entailed. He was convinced that "his public personality contradicted a hidden, more authentic, self" (Simon 1998, xvi). Wallace, in contrast, seemed to be at one with himself. Though open about what others perceived as paradoxes in his life and writings, Wallace nonetheless came increasingly to believe that there was no schism between his public and personal "selves." His evolving philosophy of nature—with its amalgam of science, politics, and theism—reflected a growing assurance about the coherence of his life's work.

Already famous for his lucid expository writings, Wallace was less well known for his gifts as a public lecturer. According to Sir William Barrett, few could approach him for the clearness and vigor of his oratory, "which commanded the attention of every one of his hearers" (Marchant [1916] 1975, 430). As both his admirers and critics in North America would attest, Wallace's declarations on matters scientific and sociocultural always aroused heated debate. His tour afforded Wallace's American and Canadian hosts and audiences a firsthand opportunity to witness one of the late Victorians' most brilliant and contentious public figures in action. Wallace's evolutionism had by 1886 come to include significant doses of spiritualism, theism, and social reformism. The elder statesman of evolutionary biology would both dazzle and infuriate his listeners in lectures that merged the theory of natural selection with forays into the incendiary arena of evolution's implications for the pressing questions of Victorian industrial society.

AN ENGLISHMAN IN BOSTON

Wallace arrived in Boston on 28 October 1886. His first Lowell lecture (1 November), "The Darwinian Theory," was delivered before a crowded audience. The major newspapers had all sent reporters. According to the *Boston Evening Transcript* on 2 November 1886, "The first Darwinian, Wallace, did not leave a leg for anti-Darwinism to stand on when he got through his first Lowell lecture last evening. It was a masterpiece of condensed statement—as clear and simple as compact—a most beautiful specimen of scientific work.

Mr. Wallace, though not an orator, is likely to become a favourite as a lecturer, his manner is so genuinely modest and straightforward." The account concluded with the key statement that "Dr. Wallace . . . then defined his own position upon the relation between man and the lower animals. Physically they are connected, but mentally there are powers which never could have been developed from lower animals." Thus, from the outset of his North American tour, Wallace's explication of "Darwinism" was, for all those who cared to notice, an explication of "Wallaceism."

Wallace stayed in Boston, with excursions to Poughkeepsie (home to the prestigious women's college Vassar), New York, and Baltimore, for the next two months. At Vassar, Wallace's lecture on evolution was part of the process by which the new biology was introduced into the curriculum. In 1889, Marcella O'Grady was appointed instructor in biology and by 1893 was named full professor, a status a female could rarely attain except at a women's college. O'Grady's 1894 course "Higher Biology" was the first at Vassar to deal explicitly with evolutionary theory (Wright 1997, 631–635). During Wallace's stay in Boston, he had the opportunity to meet most of Boston's (and Cambridge's) intellectual elite: the preeminent American botanist Asa Gray (who dined on several occasions with Wallace in order to introduce him to most of the important biology professors at Harvard), Oliver Wendell Holmes, the geologist James Dana, James Russell Lowell, William James, and Alexander Agassiz. Wallace was given an extensive personal tour of the Harvard Museum of Comparative Zoology by Agassiz. The museum was a bastion of antievolutionism under Alexander's father, Louis Agassiz, until the 1870s. The situation moderated substantially when Alexander succeeded his father as curator in 1874 (Winsor 1991). By the time of Wallace's trip to America, museums had ceased to be static organizations designed to preserve the status quo. There was an active transatlantic debate as to how best to ensure that natural history museums could serve their scientific functions. The debates extended as far as the colonial outposts of the European empires (Sheets-Pyenson 1989). Wallace was sufficiently impressed with certain of the institutional aspects of the Harvard Museum to provide a detailed account of it in the second volume of his *Studies Scientific and Social* (1900c).

At a meeting of the National Academy of Science, Gray asked Wallace to answer a question concerning geographical distribution. He replied by describing the phenomena of seed dispersal by wind. Wallace explained how this ostensibly innocuous mechanism accounted for the varying proportion of endemic species in oceanic islands. He cited "the total absence in the Azores of all those genera, whose seeds could not be air-borne (either by winds or birds), [as] throwing light upon some of the most curious facts in plant-distribution." Wallace was surprised to find that these topics—which he, Darwin, and Hooker, had made well known in Great Britain—were novel ones to his American audience. At another meeting of the National Academy, Wallace witnessed a confrontation between two of America's foremost paleontologists, Edward Drinker Cope and Othniel Marsh. Their clash reminded him of the acrimonious debates in England between Owen and Huxley. Wallace's conversation with Holmes, "the Autocrat of the breakfast table," touched not on evolution but on spiritualism. Holmes admitted that, although he had as yet little personal knowledge of the phenomena of spiritualism, he was inclined to accept its doctrines. Invited to a meeting of the New England Women's Club, Wallace heard a talk on "what socialists want." His own public declaration as a socialist was still a few years away, but he told the club that he found the views expressed "very vague and unpractical." Challenged to clarify his reaction, Wallace gave a half-hour impromptu talk in which he argued that a true "social economy" need, at minimum, be founded on the twin pillars of land nationalization and equality of opportunity. He afterward wrote this out more systematically and published it as the chapter "Economic and Social Justice" in *Studies Scientific and Social* (Wallace [1905] 1969, 2:110–111). Wallace utilized his final weeks in Boston to good advantage. At various dinners, he met more members of Boston's intellectual elite, including Russell Lowell and Edward Waldo Emerson (son of Ralph). Wallace also made the acquaintance of Boston's community of spiritualists, many of whom he found quite equal to the former in terms of "general intelligence." He attended some "very remarkable seances" (Wallace [1905] 1969, 2:115–116).

WILLIAM JAMES

Of all the Bostonians he met, Wallace's relationship with William James (fig. 5) was among the most significant. Wallace met with James frequently in Boston and corresponded with him during the remainder of the North American tour (Wallace 1886–1887). Like Wallace, James's religious/theistic thinking permeated his scientific and other studies. This is most evident in *Varieties of Religious Experience* (1902) and his work in experimental psychology. James's abundant forays into psychical research demonstrate that passion for empirical observation and pursuit of "concrete facts" that Wallace maintained in his own investigations of spiritualism. One of James's earliest published writings was a review of Wallace's "The Origin of Human Races and the Antiquity of Man Deduced from the Theory of Natural Selection" ([James] 1865). James, as did many others, regarded Wallace's 1864 essay as a brilliant resolution of the bitter controversy in England and North America between the monogenists and polygenists. The monogenists (or "unity" theorists) held that man is essentially a single species and that the various human races are merely local and temporary variations produced by different environmental conditions. The polygenists argued, in contrast, that the different races of man constitute, in effect, separate species, each of which had always been as distinct as it was at present. Wallace, by an ingenious application of the principle of natural selection, effected a compromise between the opposing anthropologists. He argued that though racial differences do antedate the recent historical period of human evolution, the several races derived from a single species. Early man, Wallace postulated, would have been subject to the harsh selective pressures of the environment. As humans spread to different regions of the globe in prehistory, the extremes of food, climate, disease, and predators would have produced, through the mechanism of natural selection acting on physical variations, those marked racial differences in appearance that remain so striking. But, Wallace suggested, these early humans were also subject to natural selection of intellectual variations. And once the human brain had evolved to a sufficient degree of power and flexibility, developing man would respond to the selection pressure of the environment by social, cultural, and intellectual adaptations. Wallace concluded that the development of man's superior brain essentially suspended the action of natural selection on bodily structures. The different physical racial characteristics would henceforth remain fixed, and further human evolution would proceed by the selection of mental and cultural variants (Wallace 1864b).

In his review, James applauded Wallace's theory and asserted that "Natural Selection, then, in its action on man, singles out for preservation those communities whose social qualities are the most complete, those whose intellectual superiority enables them to be most independent of the external world. The physical part of him is left immutable, and his mental and moral advance is secured." So reasonable and irrefutable did Wallace's essay appear to James that he declared it astonishing that Wallace's solution to the monogenist-polygenist debate was "made so late." James felt that Wallace's insight, like the theory of natural selection itself, underwent a gestation period prolonged by the obstinacy of less forward-looking thinkers. He concluded the review by asking why "may there not now be lying on the surface of things, and only waiting for an eye to see it, some principle as fertile as Natural Selection, or more so, to make up for its insufficiency (if insufficiency there be) in accounting for all organic change?" (James [1865], 263). James is here picking up on Wallace's concluding remarks in the 1864 essay. Wallace had intimated that evolution "neither requires us to depreciate the intellectual chasm which separates man from the apes, nor refuses full recognition of the striking resemblances to them which exist in other parts of his structure" (Wallace 1864b). Wallace did not explicitly state in 1864 that natural selec-

Figure 5. Photograph of William James in later life.

tion alone could not account for the origin and development of mankind. Nor did he predict a utopian future of socioeconomic and ethical harmony. But a reader as astute and "prepared" as James could clearly read a great deal more into Wallace's 1864 essay than the theoretical resolution of the anthropologists' racial controversy.

James's piece on Wallace, along with another review James wrote on Huxley's *Lectures on the Elements of Comparative Anatomy,* establishes that he was an active participant in the evolutionary debates in America as early as the mid-1860s. James saw evolutionary theory as crucial to the emerging field of experimental psychology. The review of Wallace points to further similarities between Wallace's and James's emerging worldviews. Like Wallace, James regarded the problem of human consciousness as residing at the epicenter of evolutionary speculation. The review also reveals the influence of Charles Sanders Peirce and Chauncey Wright on James's intellectual development and hints at his debt to Swedenborgian and transcendentalist thought. Peirce and Swedenborg, as will be shown, were important sources for Wallace's epistemological outlook. Another intellectual link between Wallace and James was that both viewed psychical research as an adjunct to experimental psychology (Taylor 1996, 10–14). Finally, both James and Wallace exhibited a willingness to leave certain problems short of definitive resolution. They were able to live with that tension between skepticism and belief that tormented so many of their contemporaries by using such tension creatively in fashioning holistic theories of psychic phenomena (Still 1995, 5). James concluded his *Varieties* with the assertion that "the conscious person is continuous with a wider self through which saving experiences come; he neither affirms nor denies salvation (as defined in orthodox terms), but holds to the chance of it."

JAMES AND SPIRITUALISM

In his essays on psychical research, James admitted that he was at times baffled by the phenomena of spiritualism. He felt, nonetheless, that each limited, tentative observation would help to point in a definite direction and would add up to the beginning of a solid "science of the psychic" (James 1986, xxx). James's keen interest in investigations of spiritualist phenomena from a scientific perspective made him extremely sympathetic to the efforts of the group of prominent British intellectuals pursuing that goal. In addition to Wallace, they included Frederick Myers, Henry Sidgwick, Arthur Balfour (who later became England's prime minister), and William Fletcher Barrett (physicist and member of the Royal College of Science in Dublin). Barrett founded the London Society for Psychical Research (SPR) in 1882. It was Barrett's visit to America, with a contingent of Britain's leading psychical researchers, that led ultimately to the formation of the American Society for Psychical Research in December 1884. Barrett, then forty years old, was a vigorous speaker who possessed impeccable scientific credentials. He had come at the invitation of several members of the American Association for the Advancement of Science, including the Harvard astronomer Edward Charles Pickering, Samuel Scudder (an entomologist who headed the Boston Museum of Natural History), and Alpheus Hyatt, editor of *Science*. Barrett persuaded a group of America's most eminent intellectual figures—including Pickering, Simon Newcomb (head of Johns Hopkins' astronomy department and perhaps the nation's most famous scientist), Alexander Graham Bell, Henry Bowditch (Dean of Harvard's Medical School), and James—that psychical phenomena deserved as much scientific attention as electricity, magnetism, light, and gravitation. Barrett emphasized that none of these central fixtures in the nineteenth-century scientific firmament were themselves visible or tangible except through their effects. He challenged his American hosts with assertions that if electricity could travel unseen, why not thoughts? If gravitation pulled objects toward earth, why not a psychic force that would cause objects to rise? Barrett's personal as well as intellectual gifts had a profound impact on James and like-minded associates. His rhetoric proved decisive in emboldening them to pursue experiments that would yield evidence of psychic phenomena that could compel scientific assent (Simon 1998, 191–193).

THE SWEDENBORGIAN CONNECTION

Wallace's conviction that the investigation of psychic phenomena was a legitimate field of scientific inquiry placed him at the center of the transatlantic enterprise championed by James and Barrett in the 1880s and 1890s. That Barrett was a close personal, as well as scientific, friend of Wallace's, renders the Wallace/James nexus even more significant (Marchant [1916] 1975, 426–428, 433–440). In 1885, Wallace had written an article on the harmony of spiritualism and science for the Boston *Sunday Herald*. He argued that many scientific critics of spiritualism misused the term "science" in taking it as synonymous with "a limited branch of science, namely—physics. There are," he countered, "whole regions of science in which there is no such regular sequence of cause and effect and no power of prediction." Wallace cited meteorology and biology as two such subjects. He noted that no one maintained that these were not sciences. Wallace asserted that such critics' dismissal as nonscientific any field "where *will* intervenes" were wholly mistaken. "We have," he pointed out, "the human will as a constant factor in sociology, in anthropology, in ethical science, in history, in psychology, yet no one maintains that all these studies are opposed to science even if they have, as yet, no claim to rank among established or exact sciences" (Wallace 1885b). Wallace's article was criticized in a note appearing in the *Journal of Science*. The critic singled out Emanuel Swedenborg as a victim "of delusion and imposture." In his reply to the critic, Wallace rejected the denigration of Swedenborg (Wallace 1885a). Wallace's explicit references to Swedenborg—in both his article and the reply to his critic—points to yet another similarity between Wallace and James. Swedenborgian thought figured prominently in James's philosophical preparation (Varila 1977). Swedenborg influenced Wallace somewhat later in his career. He owned copies of two of Swedenborg's books—*God, Creation, Man* (1905) and *The Earths in Our Solar System* (1894). Wallace's annotations indicate that he saw certain parallels between his own philosophy and that of the admittedly far more mystical Swedenborg. The annotated copies are now in the Alfred Russel Wallace Library in Special Collections at the Edinburgh University Library. Wallace also possessed copies of two studies of Swedenborg: J. S. Bogg, *An Illustrated Life of Swedenborg* (1911) and J. V. Hultkrantz, *The Mortal Remains of Emanuel Swedenborg* (1910).

For Swedenborg, teleology was the key concept linking the study of animal and human nature. In his two-volume *The Animal Kingdom, Considered*

Anatomically, Physically and Philosophically (English trans., 1843–1844), Swedenborg declared that "the world with its forces and forms, [is] but a complex of means to an universal end. . . . For what purpose are sensations given, but to produce intellectual ideas in human minds? In themselves they are but mediate or instrumental causes, and aspirations to intellectual ideas, but as soon as they enter the higher sphere of human minds, they begin to live more sublimely, or to understand. For what purpose, again, are intellectual ideas, unless to subserve the supreme life, or wisdom" (quoted in Varila 1977, 132). Wallace's philosophy of nature, as expressed most fully in *The World of Life,* mirrors aspects of Swedenborgian thought. In that book, a distillation of his life-long study of natural history, Wallace concluded that abundant evidence "indicated a prevision and definite preparation of the earth for Man." Wallace allowed that such teleology was "an old doctrine, supposed to have been exploded." The central thrust of *World of Life* was to show the hollowness of teleology's critics. "To all who accept the view that the universe is not a chance product," he argued, "will . . . no longer seems to be outside the realm of scientific inquiry." Wallace regarded his own mature evolutionary teleology as rendering Spencer's doctrine of the unknown "Reality which underlies both spirit and matter" both more concrete and more intelligible (Wallace 1910a, 399–400).

Like James and Peirce, Wallace felt he had refashioned Swedenborgian and other earlier transcendental epistemologies into a comprehensive philosophy of nature more consonant with the findings of nineteenth-century science. In 1903, Peirce recalled the "immense sensation" that the Darwin-Wallace theory of evolution created among his circle of friends (especially its impact on Chauncey Wright, the mentor of several Harvard intellectuals, including Peirce and James). Peirce considered evolutionary theory a new paradigm for scientific thinking, which extended beyond biological disciplines. He saw in evolution ideas of development that compelled philosophers as well as scientists to recognize "that there is a mode of influence upon external facts which cannot be resolved into mere mechanical action." Significantly, considering Wallace's own ambivalent relation to Spencer's concept of evolution, Peirce deemed Spencer's interpretation a purely mechanistic account. Since for Peirce "all matter is really mind," he emphasized the teleological implications of evolutionary theory in his own extensive writings on the subject (Parker 1998, 15–16, 200).[5]

Wallace regarded as misguided the characterization of Swedenborg as a mere speculative visionary whose worldview had been rendered obsolete by modern science. He dismissed the allegation that Swedenborg was the victim "of delusion or imposture." Turning the tables against Swedenborg's (and his own) critics, Wallace ridiculed the hubris of those "people who deny that we have any evidence whatever of the existence of spirits, [yet nonetheless] claim to know *a priori* exactly what spirits ought to know and ought to tell us, if they do exist!" He insisted that the proponents of dogmatic scientific naturalism often were guilty of overzealous, and illegitimate, proclamations on behalf of science. Wallace argued that "science itself does not yet know the 'origin of the energy' of gravitation, yet the theory of gravitation is its proudest boast." He declared provocatively that "science only guesses at the 'origin of energy' of the magnet; and in tracing all terrestrial energy to the sun it only removes the difficulty one step, and cannot do more than make more or less guesses as to where the energy of the sun comes from" (Wallace 1885a). In an interview with Albert Dawson in 1903, Wallace discussed why he had come to write *Man's Place in the Universe* and clarified certain aspects of his evolutionary teleology. Dawson concluded that "Dr. Wallace inclines to the view—and thus do revelation and Science clasp hands—[that] man is made in the image of his Maker, and I do not think he sees anything inherently absurd in the Swedenborgian idea that the whole universe may total up into the shape of a huge Man-God." These are Dawson's words, but they may be taken to reflect closely what Wallace himself stated (Dawson 1903).

Since Wallace wrote no systematic treatise on the philosophy of science, it has usually been assumed that his interest was minimal. However, numerous sections of his published books and articles—as well as many of his published and unpublished letters and notebooks—show Wallace's abiding concern with the epistemological bases of late-nineteenth-century science. Wallace was deeply involved in the debates not only on the status of evolutionary theory but on the broader questions of what defined science as a unique avenue of human inquiry. In an article published in 1900 in the *New York Sun,* he expanded on his philosophical views:

> So many of the objections which are still made to the theory of evo-
> lution . . . rest upon a misconception of what it professes to explain,
> and even of what any theory can possibly explain, that a few words
> on its nature and limits seem to be necessary. . . . Evolution, even if it
> is essentially a true and complete theory of the universe, can only ex-
> plain the existing conditions of nature by showing that it has been de-
> rived from some pre-existing condition through the action of known
> forces and laws. It may also show the high probability of a similar
> derivation from a still earlier condition; but the farther back we go
> the more uncertain must be our conclusions, while we can never
> make any real approach to the absolute beginnings of things. Her-
> bert Spencer, and many other thinkers before him, have shown that

if we try to realize the absolute nature of the simplest phenomena, we are inevitably landed either in a contradiction or in some unthinkable proposition. . . . It follows that all explanations of phenomena can only be partial explanations. They can inform us of the last change or the last series of changes which brought about the actual conditions now existing, and they can often enable us to predict future changes to a limited extent; but both the infinite past and the remote future are alike beyond our powers. Yet the explanations that the theory of evolution gives us are none the less real and none the less important, especially when we compare its teachings with the wild guesses or the total ignorance of the thinkers of earlier ages.

(Wallace 1900a)

Wallace's epistemological preoccupation is evident. The passage also reveals why and how he conceived additional explanatory factors—including theism and spiritualism—to be cognitively as well as ethically germane to evolutionary philosophizing. The parallel to American pragmatism is overt. Like that of Peirce and James, Wallace's evolutionary philosophy emphasized the fundamental connection between thought and action, between theory and practice. His conviction that evolutionary theory afforded a powerful tool by which social and political change could be effected mirrors the pragmatic criteria of truth and meaning. As did James and Peirce, Wallace believed deeply in the possibility of a convergence of human opinions on a stable body of scientific propositions. In both the near term and in the future, he expected evolutionary philosophy to provide viable and salutary guides for individual and collective action (Potter 1996, 74 and, on Peirce's "British Connection," see 17–36).

Like Wallace, James wrote no complete philosophy of science. Yet there exist, scattered throughout his enormous body of work, important references to scientific definition and procedure. These can be assembled to construct a "Jamesian outlook" on the philosophy and practice of science from both an "internal" and "external" perspective. The internal dimension involves James's views on contentious questions such as the ontological status of scientific laws, the relationship between conceptual laws and sensory experience, and the "language of reality." The external focus of James's deliberations is the need to connect scientific procedures with ethical and religious dimensions of human discourse. Jamesian metaphysics demands an element of "commitment"—the famous "will to believe." Peirce, too—although he was accorded far higher standing as a philosopher than either James or Wallace—refused to divorce logic and epistemology from metaphysics, cosmology, and ethical and social commitment (Peirce [1903] 1997, 36, 65–75, 205–220).[6] Despite differences in their particular enunciations of pragmatism, Peirce and James shared a fundamental philosophical outlook. Both believed that no theory of science or of truth could be elaborated without reference to a larger worldview. Any attempt to approach reality, they contended, required knowledge of the motives of the inquirer. James and Peirce espoused a holistic methodology that entailed that scientific statements could not be left in their immediacy as "fact." Rather, raw facts had to be understood as manifestations of patterns of development underlying the evolution of the entire universe (Gavin 1992, 56–95, 99–109, 202, 209). This holistic approach incorporated a religious component, however broadly defined. Peirce saw the universe as a "vast representamen, a great symbol of God's purpose, working out its conclusions in living realities" (Hartshorne and Weiss 1958–1965, 5:119). Wallace's evolutionary teleology and epistemology emerged from a kindred vision of nature and reality.

The Swedenborgian influence on the development of American pragmatism is well documented (Taylor 1996, 12–13, 182–183). Fundamental aspects of James's and Peirce's basic epistemologies were derived from their affinity to certain Swedenborgian precepts (Varila 1977). For Wallace, the two most suggestive components of Swedenborg's philosophical system were those of teleology and the "spirituous fluid." Swedenborg's *The Economy of the Animal Kingdom* is predicated on the assumption that the fundamental "substance of the animal kingdom is the spirituous fluid." In his chapter "The Human Soul," Swedenborg declared that the basis of both biology and philosophy is the maxim that the spirituous fluid can be considered to be, also, man's mental life. He asserted that from "the anatomy of the animal body we clearly perceive, that a certain pure fluid glances through the subtlest fibres . . . and nourishes [and] actuates . . . everything therein." If the human soul resides in the body, Swedenborg argued that its physical embodiment was the spirituous fluid. Anatomy and moral philosophy thus become one. Swedenborg further maintained that "if this fluid be regarded as the purest of the organs of its body, and the most exquisitely adapted for the reception of life, then it lives not from itself, but from Him who is self living, that is from the God of the universe, without Whom nothing whatever in nature could live, much less be wise" (Swedenborg 1843–1844, 2: 35, 211, 216, 233).

Swedenborg's main purpose in his elaborate conjectures on the structure of mental life and the relationship of the soul and body was to emphasize the centrality of the spirituous fluid. He deemed its primary function as having "intuition of ends." The spirituous fluid was "conscious of all things," with the power "to determine" events. For Swedenborg, teleology provides the basic explanatory model for both science and philosophy. The spirituous fluid served as his nodal point for integrating the diverse phenomena of nature, including that of human cognition, activity, and, ultimately, wisdom. Swedenborg's cosmological vision, when shorn of its more overwhelming mystical symbolism, became an element in James's pragmatic philosophy. These Swedenborgian elements in James's thought contributed to both the praise and scorn with which James's "new" notions in his theory of knowledge were greeted (Varila 1977, 99, 125–135). When Wallace met James in 1886, the two discussed Swedenborg. At a farewell dinner for Wallace at Parker's Hotel, hosted by the wealthy Boston merchant John M. Forbes, the guests included (in addition to James) Holmes, Lowell, Gray, and Edward Waldo Emerson. Wallace noted that during the dinner, "luxurious in the extreme," the conversation included "some pleasant interchange of ideas" on topics ranging from politics to travel to spiritualism (Wallace [1905] 1969, 2:115–116). Ralph Waldo Emerson, Edward's father, had been instrumental in bringing Swedenborgian and transcendentalist connections to James's attention (Taylor 1996, 182). The parallels between Wallace's and James's philosophical outlooks provided ammunition for critics who emphasized peculiarities or paradoxes in their respective writings and careers.

ENTER FREDERIC MYERS

Another link between Wallace and James was their close friendship with, and high admiration for, Frederic William Henry Myers (1843–1901). Myers, a well-known British essayist and psychical researcher, was a founding member of the British Society for Psychical Research (in February 1882). James soon joined the British SPR. In 1885, he began an active correspondence with Myers that lasted until the latter's death in 1901. He visited Myers when in London (Skrupskelis and Berkeley 1992–, 6:511). Other prominent members of the SPR were Henry Sidgwick, Barrett, Oliver Lodge, Arthur Balfour, and Hensleigh Wedgwood. Wallace and James considered Myers instrumental in the effort to subject spiritualism to the canons of scientific empirical research (Wallace [1905] 1969, 2:334–337; James 1986, 89–106). In a letter to James, Myers listed Wallace and James as among those persons in Europe and America "both earnestly and intelligently interested in psychical research" (Skrupskelis and Berkeley 1992–, 7:134–135). James wrote a flattering review of Myers's *Science and a Future Life* (1893 [James 1986, 107–110]). James was also active in the American Society for Psychical Research. He was one of the nine members of the committee that organized the first official meeting of the ASPR in January 1885 at the American Academy of Arts and Sciences. The ASPR counted among its membership such luminaries of American science and medicine as Benjamin Peirce, Gray, George Fullerton, and Charles

Loomis Dana. Simon Newcomb, then head of the Smithsonian Institution, was the society's first president. James was extremely active in the work of the ASPR in the early years of its existence. He sat on the committees on mediumship, thought transference, and hypnotism (which he chaired). The committee on hypnotism was crucial in introducing to the United States the research of the French exponents of the experimental psychology of the subconscious. James sought to replicate the experimental findings of Alfred Binet and Pierre Janet, whose work had been studied by Myers, Frank Podmore, and Edmund Gurney at the SPR in London. The activities of the British group, as transmitted by James and his Boston colleagues, served as the vehicle by which news of the work of Josef Breuer and Sigmund Freud on hysteria first entered the American psychological literature. James's reinterpretation of this new European research within the framework of pragmatism helped define American psychology as the scientific study of consciousness conducted in empirically testable laboratory settings (Taylor 1996, 22–24, 146).

James's involvement with Myers was part of the broader development of his thought during the 1880s and 1890s. As one of America's foremost psychologists and philosophers, James was positioned to insist that both disciplines included an inescapable subjective element. He dove into the controversial fields of abnormal psychology, psychical research (including experiments into telepathy, clairvoyance, and automatic writing), and the mind-cure movements that were then arising in the United States as alternatives to orthodox medicine. Like Wallace, he repeatedly declared puzzlement as to why most scientists seemed simply to ignore the data on which claims in these areas were made. Also, like Wallace, James attracted crowds to his lectures because he apparently inspired a sense of optimism in his listeners. He told his audiences that they had untapped reserves of energy that could be utilized to live more vigorous and healthy lives. James's religious and metaphysical speculations fueled the attacks of some of his skeptical colleagues. But few could deny his stature as one of the most innovative North American psychologists at the turn of the century. James's opposition to the imperialism he saw behind the Spanish-American and other wars, his concerns that minority rights were being trampled in the rush toward economic and industrial advance, and his socialistic proclamations for a more equitable distribution of societal wealth were positions all shared by Wallace. Both men saw these as ingredients of a worldview that integrated science with moral, social, political, religious, and cultural beliefs and activities. Refusing to be pigeonholed into one field, James was a giant of late Victorian North American intellectual and social history (Myers 1986, 10–12). James, like Wallace, was elusive not because of evasion or secrecy; they were elusive because their wide-ranging thoughts and activities defied neat categorization.

Wallace's relationship with Myers was as powerful as James's. Wallace counted Myers as "among the eminent men whose first acquaintance and valued friendship I owe to our common interest in spiritualism." The two met at some séances in London in 1878, and Myers invited Wallace to his lodging in Bolton Row, Mayfair. They discussed the experimental researches into psychic phenomena being conducted by Myers, as well as Gurney and Sidgwick. Wallace regarded Myers as the "first English writer to attempt to educe order out of the vast chaos of psychic phenomena, to connect them with admitted physical and physiological laws, and to formulate certain hypotheses that would serve to connect and explain a considerable portion of them" (Wallace [1905] 1969, 2:334–336). There were, however, subtle differences between Wallace's and Myers's conceptions as to the ultimate implications of psychic phenomena.

Myers is famous for his theory of the "subliminal self," which he set forth at a meeting of the SPR in 1892. The subliminal self provided a conceptual framework in which the diverse array of psychical phenomena could be interpreted as manifestations of a stream of subliminal consciousness intruding on the "normal" state of consciousness. Hypnotism, telepathy, automatic writing, and mediumistic messages could be unified by Myers's hypothesis. It was just this unifying function of the subliminal self theory that many professional psychologists found attractive. Ironically, it was the posited unifying power of the subliminal self that caused Wallace to reject, or at least keep his distance, from that speculation. The irony resides in the fact that Wallace was, almost congenitally, disposed to seek for unifying principles in nature. Like many other Victorian natural philosophers—including figures such as Lodge and Crookes—Wallace was uncomfortable with the very notion of gaps in the fabric of the universe. Lodge's deployment of the ether was motivated, in part, by his conviction that the continuity of nature was a fundamental assumption of scientific inquiry. The continuity principle was also central to Balfour Stewart's *The Unseen Universe; or, Physical Speculations on a Future State* (1875). Crookes declared that "all the phenomena of the universe are presumably in some way continuous" in both his presidential address to the sixty-eighth meeting of the BAAS and in his 1897 presidential address to the SPR (Oppenheim 1985, 382).

It was not Myers's speculative unity, however, to which Wallace objected but the specific motive that underlay it. Myers's attempt to extend the boundaries of what was then considered the normal personality necessitated, in his view, joining the baser with the elevated planes of psychic being. Wallace— who fully shared Myers's conviction that psychic phenomena were empirically conclusive evidence of the presence of spirits—was not keen to emphasize any petty antics of the souls of the departed. To him, they represented inferior personality traits individual spiritual beings possessed prior to their physical deaths (Wallace 1896a, 262–264). Wallace wanted to emphasize only those nobler aspects of the human personality, as they were manifested in spiritual beings. His overarching purpose in embracing the existence of spiritual beings was that they guided humans to a higher moral development. Myers, in contrast, was mainly concerned with the spirit world to assure himself, personally, that the soul was not annihilated by bodily death. In Turner's apt phrase, "Immortality rather than morality was Myers's foremost concern" (Turner 1974, 122–130). Wallace sought to demonstrate that human beings and spiritual entities were linked in an enterprise of moral amelioration of the human species. He devoted the second half of his life to articulating an evolutionary cosmology that gave pride of place to social and political activism. Wallace, nonetheless, regarded Myers as one of the dominant forces in the rigorous study of psychic phenomena in the late Victorian period. James had an equally high assessment of Myers's role as one of the major figures responsible for bringing the scientific temper to bear on the humanists' interest in psychical phenomena and their bearing on the problems of human life and destiny (James 1893).

One of Myers's abiding interests in psychical research related to the question of the possibility of life after death. His final thoughts on the subject were published posthumously in 1903, by his wife and son, as *Human Personality and the Survival of Bodily Death* (Taylor 1996, 147). Wallace considered it a "great work" (Wallace [1905] 1969, 2:334). James was similarly intrigued by Myers's view that psychical research might shed light on the question of life after death (Myers 1986, 613). James's esteem for Myers testifies to his profound intellectual and personal commitments to the reality of an unseen universe. These commitments were shared by Wallace and Peirce. Though the dynamics of the relationship between James and Peirce were not always calm, both championed the conviction that the aggressive scientific naturalism that emerged in the 1860s and beyond, on both sides of the Atlantic, was at once misguided and dangerous. For James and Peirce, scientific naturalism's gravest defect was the tendency to minimize the role of theistic beliefs in knowledge formation (Raposa 1989; Croce 1995). Like James (and his father Henry James, Sr.), Peirce ultimately adopted an evolutionary theistic cosmology (Raposa 1989, 77). Peirce, moreover, had the highest regard for Wallace's philosophical acumen as manifested in many aspects of his thoughts and writings, including Wallace's evolutionary cosmology (Peirce 1906, 161). Throughout their respective careers, both James and Peirce wrote extensively on the methodological and metaphysical implications of evolutionary theory and the debates arising therefrom in England, France, Germany, and North America (Kuklick 1977; Taylor 1990).

WALLACE AND PEIRCE

Like James—and like Wallace—Peirce's scientific and philosophical writings came to be increasingly permeated by theism. Unlike the more dogmatic scientific naturalists, Peirce (fig. 6) did not regard science and religion, or reason and faith, as opposing domains. He maintained that faith plays an important role in the process of scientific inquiry, particularly in the enunciation of hypotheses. He saw a "sublime manifestation of the Deity" in human attempts to analyze and solve problems through the utilization of scientific method. Peirce considered this a pragmatic argument for God's reality (Croce 1995, 16; Potter 1996, 169–194). There is an affinity between the epistemological framework of Wallace's evolutionary theism and Peirce's conception of scientific inquiry. Peirce's confidence in the scientific method stemmed from a conviction that the capacity to discover ultimate truths was limited and fallible. But "when we practice our best, scientific method of inquiry," he believed, "we can catch glimpses, always growing in clarity, of divinely created certainty" (Croce 1995, 16). Wallace's and Peirce's commitment to scientific theism is evident in their shared view that a scientific outlook was fully compatible with faith. Despite his (and Wallace's) frequent condemnation of the traditional institutional aspects of religion, Peirce believed that a "community of faith" was necessary for the development of a better society in the late Victorian period (Raposa 1989, 11–13). In words that resonate with Wallace's conception of a benevolent social order, Peirce wrote that "man's highest developments are social; and religion, though it begins in a seminal, individual inspiration, only comes to full flower in a great church coextensive with a civilization. This is true . . . supereminently so of the religion of love. Its ideal is that the whole world shall be united in the bond of a common love of God accomplished by each man's loving his neighbor" (Hartshorne and Weiss 1958–1965, vol. 6, par. 443). Peirce endorsed neither all of Wallace's political views nor his relentless defense of the veracity of most spiritualist mediums. But he was wholly sympathetic to the conception of universal justice expounded in Wallace's socialism (Peirce 1906, 161).

Peirce considered Wallace one of the major "scientific notables" of the late Victorian period. He regarded Wallace's lucid powers of argumentation, on a bewildering range of issues, to be sufficiently potent "and so surprisingly strong that some one of his works, say his 'Studies, Scientific and Social,' ought to be made the basis of a course of lectures on logic." Peirce reinforced his laudatory opinion of Wallace's logical and argumentative sophistication by citing similar assessments by Darwin, Mill, Spencer, Huxley, Norman Lockyer, and Chauncey Wright. To Peirce, "almost everybody whose judgment concerning the logic of science had any particular value . . . have

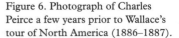

Figure 6. Photograph of Charles
Peirce a few years prior to Wallace's
tour of North America (1886–1887).

ranked Wallace among the past masters in scientific argumentation." What
others perceived as paradoxes or eccentricities in Wallace's life and work,
Peirce regarded as the necessary attributes of a bold and original thinker.
Declaring that Wallace's concept of natural selection, as it began to emerge
more explicitly in the 1880s, was "superior to Darwin's," Peirce considered
Wallace's evolutionary teleology and cultural vision a notable achievement
(Peirce 1906, 160–161). Peirce's enthusiastic judgment is unsurprising, considering the significant place accorded theism in Peirce's epistemology. Both
men elaborated worldviews that were compatible with theism, though both
articulated their specific outlooks in an idiosyncratic manner (Raposa 1989,
70). Moreover, both Peirce and Wallace shared the conviction that all knowledge was, and ought to be, social in character. Reason, itself, they believed,
was communal. Peirce was blunt on this point, declaring that even "logic is
rooted in the social principle" (Menand 2001, 229–230).

One of Peirce's clearest expositions of evolutionary theism is the series
of essays he wrote for the *Monist,* during 1891–1893. Like James, Peirce
had been deeply influenced by the theory of evolution by natural selection.
Peirce's relationship to Darwin is complex. He regarded the evolutionism of
Darwin—as well as that of Spencer and other prominent exponents of evolutionary science—as critically important but limited. Fairly or not, Peirce

characterized natural selection in the phrase: "Every individual for himself, and the Devil take the hindmost" (Hartshorne and Weiss 1958–1965, vol. 6, par. 293). "Evolutionary Love," the last of his *Monist* essays (1893), sets out Peirce's conception of evolutionary theism most succinctly (Raposa 1989, 72–87). Its objective was to demonstrate that wholly naturalistic interpretations of evolution were incomplete. Peirce endorsed the scientific significance of natural selection. But, like Wallace, he held it to be inadequate to account for the whole of the evolutionary process. For Peirce, strictly naturalistic evolutionary theories were "pseudo-evolutionism." He deemed them "scientifically unsatisfactory" because they gave "no possible hint of how the universe came about" and were "hostile to all hopes of personal relations to God" (Hartshorne and Weiss 1958–1965, vol. 6, par. 157). Peirce's cosmology is permeated by the central role played by both divine and human love in affecting, and effecting, the course of evolution. His scientific theism ultimately rests on the criterion of purposefulness. For Peirce, evolution is inextricably bound to a social ethic that is both radically communitarian and utopian (Raposa 1989, 76, 82). Notwithstanding the differences between Peirce's and Wallace's specific versions of evolutionary theism—Peirce's "Supreme Being" accords more closely to traditional Christian conceptions of the deity than does Wallace's—the parallels between their philosophical frameworks are striking and warrant more detailed scrutiny.

"IS NATURE CRUEL?"

Peirce's perception of the problem of evil accords with Wallace's treatment of that conundrum. Wallace devoted all of chapter 19 in *World of Life* to the query that had hounded philosophers and theologians for centuries: "Is Nature Cruel? The Purpose and Limitations of Pain" (Wallace 1910a, 369–384). Evolutionists now had to tackle it. In Peirce's analysis of the perennial philosophical and practical ruminations on evil, he softens the dilemma of the classical theists who believed in an omnipotent and benevolent Deity by invoking evolutionism. In place of Leibniz's "whatever is, is best" argument, Peirce contends that whatever is will prove to work out for the best in the (evolutionary) long run. He is able to assert that, "thus, the love that God is, is not a love of which hatred is the contrary; otherwise Satan would be a coordinate power; but it is a love which embraces hatred as an imperfect stage of it." Peirce's solution to the problem of evil is that particular manifestations of evil, such as human suffering or pain, are warning devices. They serve God's purposes as a means for "detaching us from false dependence on Him" (Hartshorne and Weiss 1958–1965, vol. 6, pars. 287, 507; Raposa 1989, 87–92). In the *World of Life*, Wallace attempted to answer the numerous charges that the theory of evolution by natural selection necessitated, even glorified, the widespread destruction of many living forms as the price for evolutionary advance. In language analogous to Peirce's, Wallace admits that evolutionary struggle is often "so utterly abhorrent to us that we cannot reconcile it with an author of the universe who is at once all-wise, all-powerful, and all-good. The consideration of these facts has been a mystery to the religious and has undoubtedly aided in the production of that widespread pessimism which exists to-day; while it has confirmed the materialist, and great numbers of students of science, in the rejection of any supreme intelligence as having created or designed a universe which, being founded on cruelty and destruction, they believe to be immoral" (Wallace 1910a, 369–370).

Wallace maintained that such sentiments arose from a misunderstanding of natural selection. Citing J. Arthur Thomson (who reviewed Wallace's *Darwinism*) and Huxley's article in *Nineteenth Century* (1888), Wallace asserted that leading evolutionists themselves contributed to the broader public confusion regarding the role of pain in the universe. He targeted Huxley's contention that since myriads of generations of herbivorous animals "have been tormented and devoured by carnivores" and that carnivores and herbivores alike were "subject to all the miseries incidental to old age, disease, and over-multiplication," one could not maintain that the world was governed "by what we call benevolence." Wallace rejected Huxley's caustic opinion that were human ears sharp enough, we would hear sighs and groans of pain like those heard by Dante at the gate of hell. Like Peirce and James, Wallace felt it was not his or Darwin's own words but the constant assertions of prominent Darwinians, and their disciples, "repeated and exaggerated in newspaper articles and reviews . . . that [have] led so many persons to fall back on the teaching of Haeckel—that the universe had no designer or creator, but has always existed; and that the life-pageant, with all its pain and horror, has been repeated cycle after cycle from eternity in the past, and will be repeated in cycles for ever." Wallace averred that "we have here presented to us one of the strangest phenomena of the human mind—that numbers of intelligent men are more attracted by a belief which makes the amount of pain which they think does exist on the earth last for all eternity in successive worlds without any permanent and good result whatever, than by another belief, which admits the same amount of pain into one world only, and for a limited period."

Wallace's teleology afforded a quite different conclusion. "Whatever pain there is," he declared, "only exists for the grand purpose of developing a race of spiritual beings, who may thereafter live without physical pain—also for all eternity!" He rejected all conceptions "of a universe in which pain exists *perpetually* and *uselessly*, [for] one in which the pain is strictly *limited*, while its beneficial results are *eternal!*" Wallace's extended analysis of the existence, purpose, and limitations of pain provides one of the most cogent examples of his scientific theism. By embedding the principle of utility within a teleological framework, Wallace believed he had answered the question of nature's cruelty. Pain was an essential factor in an evolutionary process "that has succeeded marvellously, even gloriously, inasmuch as it has produced, as its final outcome, MAN, the one being who can appreciate the infinite variety and beauty of the life-world, the one being who can utilise in any adequate manner the myriad products of its mechanics and its chemistry" (Wallace 1910a, 370–373).

Wallace characterized the human tendency "to transfer *our* sensations of pain to all other animals [as] grossly misleading." He considered it far more probable that "there is as great a gap between man and the lower animals in sensitiveness to pain as there is in their intellectual and moral faculties; and as a concomitant of those higher faculties." Wallace denied the sensation of pain to the vast majority of the animal kingdom. Even for those more complex animals that he had to admit feel pain, he claimed their consciousness of their suffering was fleeting. Death by powerful predators such as the lion or puma is generally so swift that the momentary pain inflicted by the deep wounds of claws and teeth of carnivores is masked by the rapid death of the prey. For Wallace, natural selection leaves scant place for pain except for one species, humans. In this chapter of *The World of Life*, Wallace elaborated on arguments he discussed briefly in *Darwinism*. "*We* require to be more sensitive to pain," he noted, "because of our bare skin with no protective armour or thick pads of hair to ward off blows, or to guard against scratches and wounds from the many spiny or prickly plants that abound in every part of the world; and especially on account of our long infancy and childhood."

Wallace felt that his analysis of the purpose and limitations of pain provided the solution to a problem that had long puzzled him. Why had humans lost their hairy covering, especially from their backs where it would have been so useful in carrying off rain, at an early stage in evolution? The answer was now clear. In his essays of the 1860s, Wallace had invoked only utilitarian arguments to explain this striking difference between humans and the higher primates. His developed teleology allowed a complementary agency, divine provision, to account more fully for the anomaly. The loss of hair most likely occurred, Wallace surmised, at precisely that point when ancestral forms "first became Man—the spiritual being, the "living soul" in a corporeal body, in order to render him *more sensitive.*" Mankind's earliest advance toward civilization, the use of fire, immediately became a daily and hourly danger to the species. It was the advent of sudden and acute sensations of pain that rendered this momentous event in human history benevolent rather than hazardous. As life became more complex, dwellings, clothing, and cooking increasingly exposed early humans to loss, injury, and death by fire. Natural selection gradually eliminated those individuals less sensitive to fire and more careless in the use of it. This juxtaposition of guiding providence and the rigors of natural selection, which had come to constitute the very core of Wallace's evolutionary theism, dictated the emergence of pain and suffering as necessary stages in human development toward a higher end.

Wallace's views parallel Peirce's in "Evolutionary Love." For both men, pain and suffering are necessary ingredients in human evolution because they inevitably result in a higher destiny for mankind. Wallace asserted that no "other animal *needs* the pain-sensation that we need; [and] it is therefore absolutely certain that no other possesses such sensations in more than a fractional degree of ours." But Wallace is no Dr. Moreau. He warned that his evolutionary explanation of pain gave no support to vivisection. Wallace asserted that the moral arguments against vivisection retain their full force, whether "the animals suffer as much as we do or only half as much." Vivisection, in his view, breeds "a callousness and a passion for experiment" on living organisms that is at once brutal and immoral. For Wallace, nature's cruelty must be seen within the broader context of the overarching plan of the "Ruler of the Universe." Any glorification of pain and suffering, whether for experimental purposes or for militaristic or economic gain, are doctrines not "worthy of an evolutionist, or of a believer in God" (Wallace 1910a, 373–381, 383). The parallels between Wallace's and Peirce's scientific theism is indicative of a broader transatlantic harmonization of evolutionary teleology.

Peirce's faith shaped his philosophical pronouncements on the presence of both certainty and uncertainty in scientific inquiry. His conviction that the scientific method incorporated theistic elements was endorsed by James and an influential group of New England intellectuals. Peirce had been one of the leading members of the Metaphysical Club, a Cambridge intellectual discussion group of Harvard students and graduates active in the 1860s and 1870s. In addition to Peirce and James, the Metaphysical Club's members and visitors included the future Supreme Court Justice Holmes, the popularizer of science John Fiske, and Frances Ellingwood Abbot (Fisch 1964; Kuklick 1977; Menand 2001, 201). Holmes and Wallace, as noted above, were proponents of spiritualism. Abbot was well known for his critique of agnosticism and his arguments on behalf of scientific theism. Wallace owned copies of Abbot's *The Way out of Agnosticism or the Philosophy of Free Religion* and *Scientific Theism.*[7] Clearly, Wallace found in the New England Cambridge setting an extremely hospitable environment for the doctrines of evolution and evolutionary theism that he was expounding in his tour.

WALLACE AND T. A. BLAND'S *IN THE WORLD CELESTIAL*

Wallace continued his spiritualist investigations—and his integration of them into his evolutionary philosophy—after leaving Boston and arriving in Washington, D.C., on 31 December 1886. (See chap. 5 for the detailed analysis of Wallace's three-month sojourn in Washington.) The relationship he formed in the nation's capital with T. A. Bland is particularly germane to an understanding of Wallace's philosophical and epistemological framework. Bland shared both Wallace's spiritualist and political convictions. Editor of the magazine *Council of Fire*, Bland was appalled at the rampant land speculation in America. He was particularly ardent in his denunciation of "the robbery of land granted [originally] as Indian reserves." He invited Wallace to give a talk to a group of his friends explaining the principles of land nationalization. Wallace noted in his journal that he "preached on 'Land Nationalisation' " but that his ideas seemed too "wild" even to as sympathetic a group as that gathered by Bland and that he didn't "think [he] made a convert." In his autobiography, Wallace later contextualized this failure by noting that in the mid-1880s "there was hardly a professed socialist in America" (Wallace [1905] 1969, 2:129–130). But the climate for spiritualism in Washington was decidedly more salubrious.

Bland considered Wallace one of the great "headlights of humanity" in the nineteenth century, an avatar of cultural as well as ethical advance (Bland 1906). Wallace, in turn, was impressed with Bland. He possessed copies of two of Bland's books: *Pioneers of Progress* (1906) and *In the World Celestial* (1905). It is the latter work, as Wallace's annotations in his copy (in Edinburgh University Library) demonstrate, that affords insights into Wallace's philosophical worldview. *In the World Celestial* is ostensibly a work of fiction, depicting the relationship between two characters, Paul and Pearl. Bland in his introduction asserted that the experiences recounted actually occurred to his friend, "a well-known and popular author." Bland explicitly vouched "for the integrity of Paul and assure[d] the reader the story is true in its essential facts. Pearl is a real character, and . . . the story of the love which budded on earth and blossomed in Heaven is not a fiction, but a genuine romance of two worlds" (Bland 1905, author's introduction). Whatever the literary merit of this book, Wallace's copious markings and annotations indicate that he fully endorsed Bland's views as expressed in this "romance of two worlds."

In the World Celestial was a spiritualist tract for its times. Paul and Pearl, lovers on earth, have continued their relationship in the spiritual realm. Bland is specific about the physical structure of spiritual space, which is a nestlike set of seven spheres surrounding earth. The earth—as are the other planets—is "surrounded by seven belts of ether, and a still finer fluid which your scientists call argon. These belts are spheres of spirit life. . . . Thus, all space is occupied by the planets and their spirit spheres." In Bland's topography, spirits must make the progressive journey from the first to the seventh sphere as their intellectual and moral qualities become more and more refined. Although spirits from inferior spheres cannot pass to higher spheres until sufficiently developed ethically, spirits in the "higher spheres can visit the lower at will, and mingle with their inhabitants freely. [Moreover,] spirits from all the spheres can visit the planet from which they came, but they cannot make their presence known to those in their physical bodies, except through the agency of persons endowed with some sort of medial gift" (Bland 1905, 87). Wallace marked this passage as a "good account of spheres." Wallace's conviction of both the reality of the spirit realm, and the interaction between spirits and earthly inhabitants, is unambivalent. He was often viewed as a bit too keen on defending the reality of spirit manifestations rendered visible by persons "endowed with some sort of medial gift." A brief examination of the context of Bland's work provides insight into the depth of Wallace's spiritualist convictions and their function in his broader epistemological perspectives.

First, those convictions reinforced his arguments from the principle of utility that the "higher faculties of man" required more than the operation of natural selection for their development. Wallace marked the following passage in Bland's tale, in which Pearl informs Paul that "you doubtless know that man has, as a natural and necessary inheritance from his savage ancestors, all the faculties possessed by the lower animals, and that in addition to these he has, not only the reasoning faculties of much higher order than any brute possesses, but moral consciousness, ideality, and other endowments which are purely human, and that it is through the development of these higher faculties alone that he is enabled to control those faculties which are common to brute and man, and live a human, humane, and harmonious life" (Bland 1905, 121). The significance of this passage for Wallace is enhanced by the fact that it occurs within a lengthy discourse by Pearl (120–128) during which she expounds on a number of historical and religious topics. The "lurid theology" of Calvin, with its harsh doctrine of predestination, is declared barbarous compared to the teachings of modern spiritualism. Yet even Calvin has been "reformed" under spiritualist guidance. Pearl informs Paul that Calvin "has long since revised his creed and he deeply deplores having given to the world a system of doctrines which so greatly misrepresent the character of the all-loving Father. He is an unseen member of every great assembly of Calvinist churches, doing all that he can do to bring about a radical change in the creed which he gave to the world" (Bland 1905, 123).

The image of Calvin as a fifth-columnist would surely have appealed to the radical reformer in Wallace.

Pearl mentions Swedenborg explicitly as "one kindred writer who spoke of such spiritual spheres both actually and metaphorically." Paul is further informed by his partner in two worlds that "good deeds, prompted by pure motives, done on earth, are the treasures in heaven referred to by Jesus." Wallace's markings indicate that he read these pages carefully. His annotations substantiate the claim that there was no abrupt break or inherent contradiction between his earlier secular attitudes and his mature evolutionary theism. Wallace objected not to religion but to those traditional institutions and creeds that placed too great an emphasis on fear, hatred, and punishment. Wallace's scientific and spiritualist principles enabled him to refashion theism into a modern evolutionary teleology. This framework enabled him to integrate religious sentiments and sociopolitical reforms with scientific advances into a strategy for the renovation of late Victorian society. Without such integration, Wallace was certain that the positive achievements of the "wonderful century" would be obliterated. When Paul and Pearl visited Washington, as spirit tourists, they attended many of the numerous séances popular in the city. More dramatic were their visits—spirits apparently passed easily through security points—to a presidential cabinet meeting. There, as Bland recorded, Paul found the current president surrounded by a host of spirits. Among these were such mighty figures from American history as Benjamin Franklin and Alexander Hamilton. Each of these, quite literally, spiritual advisers proffered advice to the president on the critical decisions that lay before him. There was also a "spirit lobby" in Congress. Its "wise and good" efforts at persuasion in influencing votes, as Pearl wryly noted, ensured that "this government would [not] be far more corrupt, unjust, and despotic than it is" (Bland 1905, 125–126, 130–133).

Wallace's laudatory opinion of Bland's *In the World Celestial* is indicative of his conviction that political, social, and theistic forces were interdependent. It was such interdependence that gave Wallace, as it did so many others in Britain and North America, hope that the ravages of late Victorian commercialism and capitalist industrialization could be moderated and a just society effected. The Reverend H. W. Thomas, president of the World's Liberal Congress of Religions, had written the introduction to Bland's book. One of Thomas's main objectives was to counter what he perceived as the materialist thrust of the times. Modern science and modern theism, he believed, could together forge a basis for a new outlook on humanity's place in nature. Thomas, like Bland and Wallace, had little use for either dogmatic Catholicism or the Protestant Reformers. He condemned their emphasis on hell as eternal damnation for "unintentionally [doing] a great wrong to mankind; they closed the doors between the two worlds. Nothing could be done for lost souls; their doom was forever hopelessly sealed; and even the saved were far removed from the concern of those on earth. The result has been to chill the emotions, to lessen interest in the life to come and to weaken faith in immortality." Modern spiritualism and refashioned theism, according to Thomas, afforded a far more reasonable, "and certainly not unscriptural," framework for religious and ethical precepts and practices. "God," he declared, "is life, 'the living God,' 'the God of the living, and not of the dead.' The change that we call death is an incident in the evolution of life. . . . The universe at center is mind—spirit; man at center is divine; is the child of God." Thomas considered "the fact of form"—as it was being increasingly revealed in scientific findings in all fields—fundamental to late Victorian philosophies of nature. Suns, stars, and satellites as well as their various orbits are "forms." The universe itself is form, and there are the myriad forms of life that cellular biology is continually demonstrating. For Thomas, the "prototypes of all these are thought forms—mind, spirit forms; the universe, it is the objectivized thought of the Infinite mind, reason, beauty, justice, and love." He noted that this conceptual scheme was akin to the old "Platonic doctrine of Divine Ideas." H. W. Thomas asserted that his views were in complete accord with "what we know as the ideal philosophy" (Thomas 1905). The tradition of idealist philosophy was as significant at the close of the nineteenth century in North America as it was in Great Britain and much of continental Europe (Otter 1996).

Wallace was attracted to certain strands of idealist philosophy. He endorsed Thomas's description of "ministering spirits" as "often walking unseen by our side . . . [and whose affections] Death does not lessen, but intensifies." Spirit love, Thomas averred, "is, of all love, the tenderest, the most forgiving, the greatest, for it has risen above the discords and hatreds of time." Bland concluded his own introduction on a triumphant note:

> The old theology is losing its hold upon the real beliefs of many
> thoughtful minds; the larger and better faith and hope of the new is
> taking the place of the old. And not only this, with the larger knowl-
> edge of his mighty surroundings and the mastery of material forces,
> man is coming to see, to feel, and to fill his larger place in the uni-
> verse. The power of man has been augmented a hundred, a thousand
> fold; steam, electricity and telegraphy have made the once distant
> and unknown parts of the earth seem near and common. And one
> now knows that the universe is one—"the one Being"; . . . In the
> occult world we are finding that hypnotism, clairvoyance and telepa-

thy are facts. The unseen world is nearer and more real; not only is
wireless telegraphy a fact, but we may send thoughts to those far way;
and all this is making more common and real the idea of the possible
communion of earth with heaven.

(Bland 1905, author's introduction)

In 1885, Wallace had written in strikingly similar terms of the impact of
science and spiritualism on theology:

> We who have satisfied ourselves of the reality of the phenomena of
> modern Spiritualism in all their wide-reaching extent and endless
> variety, are enabled to look upon the records of the past with new
> interest and fuller appreciation. It is surely something to be relieved
> from the necessity of classing Socrates and St. Augustine, Luther and
> Swedenborg, as the credulous victims of delusion or imposture. The
> so-called miracles and supernatural events that pervade the sacred
> books and historical records of all nations find their place among nat-
> ural phenomena, and need no longer be laboriously explained away.
> The witchcraft mania of Europe and America affords the materials
> for an important study, since we are now able to detect the basis of
> fact on which it rested, and to separate from it the Satanic interpreta-
> tion which invested it with horror, and appeared to justify the cruel
> punishments by which it was attempted to be suppressed. . . . In
> these and many other ways history and anthropology are illuminated
> by Spiritualism.
>
> To the teacher of religion it is of vital importance, since it enables
> him to meet the skeptic on his own ground, to adduce facts and ev-
> idence for the faith that he professes, and to avoid that attitude of
> apology and doubt which renders him altogether helpless against the
> vigorous assaults of Agnosticism and materialistic science. Theology,
> when vivified and strengthened by Spiritualism, may regain some of
> the influence and power of its earlier years.
>
> Science will equally benefit, since it will have opened to it a new
> domain of surpassing interest. Just as there is behind the visible world
> of nature an "unseen universe" of forces, the study of which con-
> tinually opens up fresh worlds of knowledge often intimately con-
> nected with the true comprehension of the most familiar phenomena
> of nature, so the world of mind will be illuminated by the new facts
> and principles which the study of Spiritualism makes known to us.
> Modern science utterly fails to realize the nature of mind or to ac-
> count for its presence in the universe, except by the mere verbal and

unthinkable dogma that it is "the product of organization." Spiritualism, on the other hand, recognizes in Mind the cause of organization,
and, perhaps, even of matter itself; and it has added greatly to our
knowledge of man's nature, by demonstrating the existence of individual minds indistinguishable from those of human beings, yet separate from any human body. It has made us acquainted with forms
of matter of which materialistic science has no cognizance, and with
an ethereal chemistry whose transformations are far more marvellous
than any of those with which science deals. It thus gives us proof that
there are possibilities of organized existence beyond those of our material world, and in doing so removes the greatest stumbling-block
in the way of belief in a future state of existence—the impossibility
so often felt by the student of material science of separating the conscious mind from its partnership with the brain and nervous system.

(Wallace 1885b)

Wallace's close and enthusiastic reading of Bland's book is now contextualized. His linkage of spiritualism, theism, and science reflected a pervasive
ideology among important segments of late Victorian society and culture.
The Wonderful Century and *The World of Life* should be recognized for what
they were. These later works of Wallace were not the eccentric musings of
a declining mind but powerful syntheses of late-nineteenth/early-twentieth-
century intellectual currents. They incorporated and influenced the thoughts
and activities of members of elite and popular cultures on both sides of the
Atlantic. Figures as eminent as Peirce and James recognized the significance
of Wallace's efforts to formulate a proactive evolutionary theism. His attempt
to establish a viable framework for harnessing the forces of science and technology to a humane, and environmentally conscientious, social and political
order was central to his life.

WALLACE'S EPISTEMOLOGY: IDEALIST AND REALIST?

Wallace's epistemology resists any neat label. His philosophical position bears
the imprint of both idealism and realism. He sought to integrate elements of
idealist and realist metaphysics to construct a scaffold for his own study and
interpretation of nature. Such an integrative attempt was far from unique
in the later Victorian period. Only when realism and idealism are viewed as
antithetical is a synthetic approach deemed odd, if not impossible. There are
many definitions of these two dominant philosophical traditions. It is sufficient for the purpose of situating Wallace's position on the epistemological
spectrum to indicate the most basic characteristics of each tradition. Idealism,

whatever form it takes, "requires that what is thought of as real be constructed by mental conditions, whether those conditions be limited to human agents or a transhuman agency, understood as divine or as the Absolute. Realism, on the other hand, in the minimal sense requires that what is real be to some extent or in some way independent of mental construction—indeed, what is real determines, contributes to, or at least resists the [agency] of mental activity, human or transhuman" (Hausman 1993, 144).

Traditional historiography has tended to view the latter half of the nineteenth century, especially in Britain, as a period in which philosophical realism, and its more specific formulation in scientific naturalism, was triumphant. The propagandists for the primacy of scientific naturalism in explicating "truth" were vocal and influential. But they were decidedly not the only voices heard. Recent scholarship has identified British idealism as a crucial component of late Victorian culture. David George Ritchie, T. H. Green, F. H. Bradley, Edward Caird, Bernard Bosanquet, and their followers explored compelling alternatives to materialist and atomist/reductionist models in the study of nature and of society (Otter 1996, 6–7, 143–148). The interplay between realism and idealism was equally pronounced in North America (Hausman 1993, chap. 4). Thus, Wallace's admixture of realist and idealist concepts in his own philosophizing is reflective of the broader context of transatlantic thought.

There are two main reasons why Wallace found it appropriate to attempt a synthesis of realism and idealism. First, as noted above, he considered scientific naturalism inadequate to account for all aspects of evolution. But second, and equally crucial, Wallace was drawn to the idealists' conviction that philosophy and sociopolitical reform should be inextricably related activities (Nicholson 1990). Although the attempts to establish a "biological sociology" were as diverse as they were inconclusive, the lure of that particular Holy Grail was irresistible. Realists and idealists alike were drawn to the quest to utilize evolutionary biology to illuminate the broader questions of consciousness, thought, and purpose in social evolution (Otter 1996, 101–119). In this case, politics often made curious bedfellows of disparate thinkers.

John A. Hobson summed up the rich and often contradictory world of Victorian evolutionary philosophy when he remarked that it "was not the least of Spencer's victories that he has forced evolution on the idealists—those who approach Unity from the other side" (Hobson [1904] 1988, 63). Similarly, Wallace's alliance with Ritchie was complicated. Ritchie was unsympathetic to Wallace's incorporation of spiritualism into the evolutionary process (Ritchie 1889, 115). But both shared a commitment to the concept of design in nature. More tellingly, for Wallace, Ritchie wedded an evolutionary teleology to an explicitly socialist and collectivist solution to the "social question." Ritchie appealed to socialists to have patience and "reverence for the long toil of the human spirit." Although he noted that much work was required before socialism could bring the "good elements" in regressive sentiments and institutions to the fore, he asserted that the "Divine purpose . . . is gradually revealing itself in the education of the human race" (Robbins 1982, 86). As did many other Victorians, Wallace enunciated a holistic philosophy in which the indisputable findings of science could be shown to be compatible with a broader metaphysical and ethical framework. As cofounder of the theory of natural selection, Wallace's particular philosophical synthesis was bound to generate vigorous reactions—positive and negative—among his contemporaries.

WALLACE AND SPENCER

Wallace's long and complex relationship with Spencer reveals much about Wallace's attempt to reconcile idealism and realism. He frequently expressed his admiration for Spencer, declaring himself "an enthusiastic admirer of Mr. Spencer's writings, and a follower of his philosophy" (Wallace 1869b). Spencer exerted on Wallace, as on many others coming of age in the mid-nineteenth century, a powerful intellectual as well as social influence. The two were close friends. They exchanged views on a wide variety of issues ranging from animal and human fertility to land nationalization. Spencer was one of the first Londoners Wallace had sought out on his return from the Malay Archipelago. Accompanied by Bates—who, like Wallace, had read and "been immensely impressed" with *First Principles*—Wallace hoped for enlightenment of the great unsolved problem of the origin of life. He realized only too well that the *Origin of Species* had (deliberately) left that problem "in as much obscurity as ever." Both he and Bates eagerly looked to "Spencer as the one man living who could give us some clue to it." Wallace's account of that first meeting says much about his desire to probe deeply the fundamental questions posed by evolutionary theory. It also captures Spencer's tendency to retreat into the comforts of accepting the barriers posed by the "unknowable":

> [Spencer's] wonderful exposition of the fundamental laws and conditions, actions and interactions of the material universe seemed to penetrate so deeply into that "nature of things" after which the early philosophers searched in vain . . . that we both hoped he could throw some light on that great problem of problems. . . . As young students of nature we wished to have the honour of his acquaintance. He was very pleasant, spoke appreciatively of what we had both done for the

practical exposition of evolution, and hoped we would continue to work at the subject. But when we ventured to touch on the great problem, and whether he had arrived at even one of the first steps towards its solution, our hopes were dashed at once. That, he said, was too fundamental a problem to even think of solving at present. We did not yet know enough of matter in its essential constitution nor of the various forces of nature; and all he could say was that everything pointed to its having been a development out of matter—a phase of that continuous process of evolution by which the whole universe had been brought to its present condition. So we had to wait and work contentedly at minor problems.

Wallace was disappointed but not discouraged. His own developing evolutionary philosophy permitted him to approach questions about the origin of life and the course of human evolution by epistemological avenues "which Spencer and Darwin neglected or ignored" (Wallace [1905] 1969, 2:23–24).

Wallace digested the volumes of Spencer's *Synthetic Philosophy*. Although he thought highly of Spencer, his early reverence was soon tempered with a more critical and qualified respect. Their divergent positions on key issues in the natural and social sciences became apparent. Wallace and Spencer remained united more by a passionate dedication to evolution and to social reform than by shared philosophical outlooks. Wallace derived much amusement "from the often unexpected way in which [Spencer] would apply the principles of evolution to the commonest topics of conversation" (Wallace [1905] 1969, 2: 33). It was Spencer's phrase "survival of the fittest" that Wallace preferred to his and Darwin's term "natural selection." Wallace thought "survival of the fittest" less subject to misleading personification and a direct, not metaphorical, expression of the process of evolution (Marchant [1916] 1975, 141).[8] Their mutual respect, despite their differing epistemologies, endured. In 1874, Spencer asked Wallace to look over the proofs of the first six chapters of *The Principles of Sociology* "and give him the benefit of my criticisms, 'alike as naturalist, anthropologist, and traveller.' " Wallace found little requiring emendation, but "sent him a couple of pages of notes with suggestions on points of detail, which, I believe, were of some use to him" (Wallace [1905] 1969, 2:27).

But Wallace was wary of Spencer's habit of dismissing many books if a scant perusal of their contents did not accord with his own preconceived notions. He was particularly annoyed by Spencer's refusal to read Henry George's *Progress and Poverty* and Henry Buckle's *History of Civilization* (Wallace [1905] 1969, 2:27, 30–31; Marchant [1916] 1975, 391–392). Wallace rejected Spencer's analysis of the origin of the moral sense in humans. "If mind," he wrote, "with all its powers is simply a function of organized matter, then Mr. Spencer's theory of the origin of morals is the only one which can be held by a student of science. If, however, there is any thing in man more than his physical organization, then it becomes a subject of strict scientific and philosophical inquiry to determine from a study of the phenomena of his mind in various stages of growth and under various conditions, what is the mental substratum required to account for the development of the faculties we actually find in him" (Wallace 1869b). Wallace's teleology rendered Spencer's utilitarian explanation for the development of the moral sense unacceptable. Wallace's reasoning on this score paralleled Peirce's dictum that theism, like science, can be true to its own essence, only by subjecting its doctrines to the test of experience and criticism.

Peirce's epistemology was first elaborated in a series of "Illustrations of the Logic of Science" in the *Popular Science Monthly* in 1877–1878. The scientific method, he argued, is but one of several ways of fixing beliefs. Beliefs are essentially habits of action. For Peirce, the scientific method began with an experimental investigation of the observable effects of forces on objects. That accomplished, science provided guidelines by which the human intellect would recognize those old, inadequate habits that were no longer appropriate. New, flexible habits of action could be established that allowed for better adaptation to the never-ending novelties revealed by empirical research. Peirce's metaphysical writings, with their emphasis on chance and continuity, were further illustrations of his logic of science. When pragmatism became a popular movement in the late 1890s and early 1900s, Peirce was dissatisfied with all of the forms of pragmatism then current, including his own original exposition. His last productive years were devoted to a radical revision of pragmatism. He labored on articulating the principle of what he had come to call "pragmaticism." Peirce's later epistemology incorporated evolutionism more directly than previously. Knowledge, including science, he now emphasized was developmental and open to its own correction. Peirce's pragmaticism explicitly rejected the ideology of the ethical, metaphysical, and theistic neutrality of science. The ideological neutrality of science had been expressed most forcefully in America by Wright. Peirce worked toward effecting a viable rapprochement between what he termed the "religious metaphysics" and the "physical metaphysics" of the late nineteenth century (Anderson 1995, 61–63).

Wallace pursued a similar approach in his attempt to construct an epistemological synthesis of diverse elements in late Victorian thought. This synthetic approach did not preclude opting for a specific position if the circumstances warranted. When forced to choose between the utilitarian hypothesis and the intuitional theory, which explained the origin of morality "by the supposition that there is a feeling—a sense of right and wrong—in our nature antecedent to and independent of experience of utility," Wallace endorsed the intuitional hypothesis. This put him on a collision course with certain aspects of Spencerian epistemology. Wallace declared that he was decidedly of the opinion "that there is a limit to the sphere which [Spencer's] philosophy embraces, and that the limit is to be found in the doctrine of the origin of morals" (Wallace 1869b). In the end, Wallace sided with the idealist philosophers. He endorsed the views of Ritchie, Bosanquet, and Henry Sidgwick. They all objected to what they perceived as Spencer's overly positivist and reductionist epistemology. Spencer's strident evolutionary defense of individualist and laissez-faire sociology in such works as *The Man versus the State* (1884) was a further irritant to the idealist camp (Wallace [1905] 1969, 2:28–29 [esp. 29n. 1], 33, 272; Otter 1996, 94–98, 127–133).

Wallace's relationship to Spencer illuminates the "branching-tree" course of Wallace's evolution as a philosopher and social critic. Spencer's *Social Statics* (1851) provided Wallace with a direct stimulus to his writing the 1864 essay on the origin of human races (Wallace 1864b, clxx). *Social Statics* also forcibly demonstrated to Wallace "the immorality and impolicy of private property in land." It was instrumental in focusing him onto the question of land nationalization (Wallace 1880, 735). Wallace's mature views on political economy, specifically socialism, retained an individualist accent. He fervently believed that a socialist state must incorporate a voluntaristic, not coercive, ideology. But he rejected the central axioms of laissez-faire political economy that had become so closely associated with Spencer (Coleman 1999, 11–15). Wallace's siding with the idealist philosophers in their critique of Spencer did not prevent him from remaining his own man. Wallace's life was informed by an eclecticism that permeated his approach to all issues. This eclectic imperative manifested itself in Wallace's philosophical evolution. It was an indelible feature of his personality and intellect throughout his life.

———— • ◆ • ————

NOTES

1. I have quoted from the original letter as reprinted in McKinney 1969, rather than the very minimally emended version Wallace printed in *My Life* (Wallace [1905] 1969, 1:254–255).

2. I have borrowed Durant's phrase "going public" because of its forcefulness and accuracy. Durant's analysis, however, is restricted to an account of Darwin's career. Wallace's strategy for "going public," of course, is markedly different from Darwin's.

3. Currently, there is renewed interest among historians and philosophers of biology with respect to whether essentialism was the dominant paradigm for taxonomy in the first half of the nineteenth century or, instead, only one of several concepts influencing naturalists in the decades prior to the publication of *Origin*. Mary Winsor and others argue that essentialism had little impact on the actual practice of working naturalists in the eighteenth and nineteenth centuries (Winsor 2003). In contrast, others argue that essentialism at the level of species concepts (or, rather, a family of such essentialisms) was the main paradigm (e.g., Stamos 1999). For Wallace and Darwin, however, the evidence is strong that they both regarded essentialism as a powerful paradigm that had to be purged from biological theory.

4. For a discussion of the complex and still ambiguous events leading to the joint publication, see Beddall (1968, 299–318); McKinney (1972, 142–146); Desmond and Moore (1991, 467–472); and Porter (1993). Beddall's account, emphasizing the disappearance of the manuscript of Wallace's essay as well as the disappearance of critical letters from Wallace, Lyell, and Hooker sent to Darwin during this crucial period, is the most temperate yet forceful indictment of Darwin's, Lyell's, and Hooker's handling of the joint publication and suggests, once again, that Darwin may have owed a debt (unacknowledged) to Wallace in the genesis of the *Origin*.

5. On the evidence for Swedenborgian influences on Peirce, see Varila (1977, 20–21).

6. Peirce also delivered a series of eight Lowell Lectures between 23 November and 17 December 1903 (Peirce [1903] 1997, 18, n. 2).

7. Wallace's annotated copies are in ARWL. Abbot sent Wallace a letter with the gift copy of *The Way out of Agnosticism*, which is enclosed in that volume. Dated 27 April 1890, Abbot wrote Wallace to please

> do me the honor to accept the accompanying little book of mine . . . and kindly to express your critical opinions of the new argument, grounded solely on science and philosophy, which it presents in support of theism. Somehow I entertain a stronger hope of sympathy in this endeavor from you than I do from most scientific men of the day; for you have shown what seems to me a deeper insight than they into the indestructible nature of our great religious convictions. I hope to receive a number of such critical judgments from leading men of our time, with a view to publication in a future edition of my book; and that must be my excuse for venturing to make this request. Ignoring wholly the traditional grounds, I make my appeal solely to the modern intelligence; and I hope my patiently worked-out results will be found of lasting value.

8. These two terms have been among the most contentious in the history of evolutionary biology. Two recent treatments of the their semantic and ideological ambiguities are: M. J. S. Hodge (1992, 212–219) and Diane B. Paul (1988).

CHAPTER 4

The Making of a Victorian Spiritualist:
Multiple Directions and Inevitable Tensions

VICTORIAN SPIRITUALISM IN CULTURAL CONTEXT

Victorian spiritualism has been a problematic area of inquiry. At its most basic level of definition, spiritualism can be characterized as the belief that departed souls, and other nonmaterial conscious entities, could influence and communicate with humans—usually through a medium (by means of physical phenomena) or during unusual mental states such as trances (Wallace 1869b, 1874, 1875b, 1885b). Spiritualism's claims and experimental procedures were embraced by a significant number of people, scientists as well as laypersons. But equally large numbers rejected those claims and the experimental proofs that were adduced in support of them. This ambivalent contemporary reception of nineteenth-century spiritualism has both attracted and repelled later scholars. The reasons for its mixed reception in the Victorian period are complex but comprehensible. The reasons for the generally marginalized status it has been accorded by cultural historians and historians of science during the twentieth century are clearer but more disturbing. Spiritualism and what may be termed the other "occult sciences" (including phrenology, mesmerism, and psychical research) seemed less than completely reputable to early- and mid-twentieth-century historians trained to emphasize the triumph of positivism. In the 1980s, however, a newer breed of scholars began to reconsider the significance of spiritualism in the Victorian era.[1] Their analyses have uncovered the reasons for the widespread appeal of spiritualist and related beliefs. The intellectual relevance and pervasive cultural importance of spiritualism, mesmerism, and phrenology, and their varied intersections, can no longer be construed as marginalized phenomena.[2] They were too deeply ingrained in the minds and hearts of too many people to be anything other than profound, if contentious, characteristics of Victorian culture.

The first revisionist studies emphasized the intense sense of religious doubt and questing that drew individuals to spiritualism. It, and related avenues into the realm of the nonmaterial universe, were powerful influences because they afforded vehicles for mediating between the often competing claims of traditional religions and modern science. This influence reached thousands of men and women from diverse social, economic, and theological backgrounds. Some of Britain's foremost intellectuals and scientists, such as Henry Sidgwick, Edmund Gurney, and Frederic W. H. Myers, were among those most deeply affected (Mandler et al. 1997, 77–84). Studies in the past decade have broadened the explanatory scope of spiritualism's appeal by addressing issues of gender and sexual politics, theories of narrative and rhetorical strategies, interiority and constructions of subjectivity and the "self," and the epistemological claims of science (Barrow 1986; Owen 1990; Noakes 1998; Winter 1998). It is against this rich background of scholarship that Wallace's own complex relationship to spiritualism—as well as mesmerism and phrenology—may now be more fully assessed. He was part of a broad sociocultural as well as intellectual community for whom investigations into the spiritual realm were epistemologically significant, politically influential, and emotionally rewarding. Wallace's encounters with spiritualism during the 1860s reveal much about the thoughts and plans of the man who returned to London after prolonged travels in the tropics.

A LONDON LIFE: 1862–1871

When Wallace arrived back in London in March 1862, he was a scientist of considerable repute. He associated with most of the intellectual luminaries of the day. Two factors, however, tempered his rise in the British scientific establishment. First, Wallace encountered a series of disappointments in securing a major teaching or research post in the metropolis. This experience prefigured his life-long battle to earn a living commensurate with his talents and achievements. Second, Wallace's controversial stance in matters scientific and social began to emerge more openly. The decade of the 1860s constitutes, in microcosm, the ambivalent pattern his life would follow for the next half century. The distinction between, and integration of, the public and private aspects of any Victorian scientist was complex. Different scientists had their own particular patterns of navigating the torturous geography of public/private and individual/group levels of scientific activities to gain professional status and public legitimacy (Rudwick 1982). Wallace surely sensed but did not accept the split image of him generated in the minds of many of his scientific colleagues owing to his spiritualist pursuits during the 1860s. His life is a record of his efforts to dispel such perceptions by integrating his diverse convictions and activities. These efforts were only partially successful. There were struggles and setbacks, professional as well as personal, on the road to the final articulation of his comprehensive worldview. Through it all, Wallace was sustained by his growing certainty in the scientific validity and ethical superiority of a holistic philosophy of human nature.

Wallace's London years are crucial for understanding the personal and public trajectory of a highly innovative figure in Victorian culture. They were the years that witnessed some of his most brilliant scientific accomplishments. These were also the years in which Wallace's association with spiritualism was most intense. At one level, the link between Wallace's science and his spiritualist activities is obvious. He repeatedly made the point that he approached the phenomena of spiritualism with the critical training and epistemology of the scientist. His goal was to establish a scientific foundation for the phenomena and claims of spiritualism. But this obvious (to him) connection tells us little about the personal and emotional motives that drew Wallace to the study of spiritualism. Nor does it explain why he became one of its foremost champions for more than half a century. The spiritualism-as-science approach fails to penetrate the deeper sources of Wallace's, and many others', persistent commitment in the face of widespread skepticism. One of the strongest lures of spiritualism was the message of self-help and optimism with which its teachings were imbued. Phrenology and mesmerism attracted large followings for the same reason. Wallace had earlier applauded Chambers's *Vestiges* and George Combe's *The Constitution of Man* for bringing the significance of mesmerism and phrenology to his attention (Wallace [1905] 1969, 1:234–235). The broad appeal of these and kindred works lay in their progressive conclusions regarding human development (Winter 1994; Secord 2000, 161–163, 267).

The linkage between spiritualism, mesmerism, and phrenology in Wallace's developing concept of human nature is clear. But there is an additional connection between spiritualism and mesmerism that bears directly on his attraction to those movements. Spiritualism, like mesmerism, was a public activity. The procedures by which advocates (and critics) sought to assess the authenticity and propriety of séance and other phenomena involved "observing and interpreting social characteristics" of the participants and "judging the relationships between people." Allegations and counterallegations concerning fraud were staples of spiritualist controversies. But questions of fraud lead to larger issues. Fraud was a constant question in spiritualist experiments precisely because spiritualist phenomena were inextricable from the social context in which they occurred. These phenomena were made manifest through the mediation of actual persons, with real passions, socioeconomic status, and personal as well as cultural agendas. The very act of deciding what the phenomena "meant required that one assert what one thought social relations were, or ought to be" (Winter 1998, 66).[3] Wallace genuinely believed in the viability of bringing spiritualism within the scope of legitimate scientific inquiry and validation. But he was also a willing and active participant in the cultural construction of the phenomena he studied so intently. Why, and how, do Wallace's spiritualist activities relate to his own personality and to his conception of what social—and societal—relations ought to be?

THE PUBLIC AND PRIVATE WALLACE

Social relations operate at two levels: the personal relationship between individuals and the relationship between individuals and the society in which they live. Wallace wrote extensively on the latter aspect. His literary output concerning social, political, and economic matters is enormous. In contrast, he wrote little about his personal life. Wallace's immersion in spiritualism during the 1860s provides important clues to his character. In addition to its cultural dimension, spiritualism fulfilled private needs and aspirations. Since spiritualism was one of the arenas in which females played an influential, perhaps even dominant, role, Wallace's commitment affords insights into his relationship with women. Lady Lyell thought Wallace "shy, awkward, and quite unused to good society" (Wallace [1905] 1969, 1:433). Lady Lyell was a formidable London hostess and her assessment of Wallace's social graces is as indicative of her own social conventions and expectations as it was of Wallace's. His unassuming demeanor would have been apparent with respect to women, and especially women of elevated social status. The fact that spiritualist circles afforded less intimidating surroundings for male/female interaction would certainly have been a relief to one as shy as he. Wallace was drawn to spiritualism for a number of reasons. But the opportunity for socializing with women must be recognized as one of those reasons. Wallace felt comfortable with the females whom he encountered at séances and similar gatherings because he shared not only their spiritualist beliefs but their egalitarian and progressive sociopolitical views as well. Wallace was frequently criticized for defending mediums (the majority of whom were female) in celebrated cases of alleged fraud. If the interpretation of spiritualist phenomena involved asserting what one thought social relations should be, then Wallace's tenacious defense of spiritualism is indicative of his growing commitment to gender egalitarianism.

For those less prone to judge a person's character by the ability to shine in the glittering salons of London high society, Lady Lyell's dismissal of Wallace as awkward was misguided. At a meeting of the BAAS in Dundee in 1867, during that same period when Wallace was frequenting the Lyell soirees, Wallace stayed at the home of Professor W. A. Knight. Knight's impression of Wallace was the reverse of Lady Lyell's. "I, and everyone else who then met him at my house," Knight commented, "were struck, as no one could fail to be, by his rare urbanity, his social charm, his modesty, his unobtrusive strength, his courtesy in explaining matters with which he was himself familiar but those he conversed with were not; and his abounding interest, not only in every branch of Science, but in human knowledge in all its phases, especially new ones. He was a many-sided scientific man." Knight also noted specifically Wallace's "vivid sense of humor," a trait that many others enjoyed in their encounters with Wallace (Marchant [1916] 1975, 451–452). Were Lady Lyell and Professor Knight talking about the same person? They were. Wallace was shy in those social surroundings in which he felt uncomfortable. But in those settings in which he felt at ease, such as spiritualist circles, shyness gave way to openness. Wallace's zest for life surfaced. It was this passion and humor that sustained him through the vicissitudes in his long journey to self-fulfillment and serenity. His strength of character impressed even those with whom he disagreed on sociopolitical and scientific matters.

In London, Wallace first went to live with his sister and brother-in-law, Frances (Fanny) and Thomas Sims. They had a house and nearby photographic business in Westbourne Grove. Fanny was an ardent spiritualist, and she and Wallace shared their convictions openly with each other (Raby 2001, 186, 190–191). Although gaunt and drawn, with a beard dotted with white, Wallace was still an imposing figure. An inch taller than six feet, with square though not very broad shoulders—he had not yet acquired that scholar's stoop that would become increasingly more noticeable as the years progressed—he was spare and very active. Wallace was passionate about his frequent walks. Until late in life, he remained capable of taking long country hikes, which never failed to afford him great pleasure (Marchant [1916] 1975, 349). Despite the twelve years spent in exotic lands, Wallace's disposition remained basically unaltered. He had become accustomed to servants, bargained toughly with Chinese traders and Malay pirates, and dealt with sea captains, sultans, and rajahs. Yet the cheerful outlook on the world and that touch of humility were still, and would remain, hallmarks of Wallace's personality (Brackman 1980, 236). According to his two children, "He was very independent, and it never seemed to occur to him to ask to have anything done for him if he could do it himself—and he could do many things, such as sewing on buttons and tapes and packing up parcels, with great neatness." Wallace's lack of vanity, rare in the gilded age of late Victorianism, extended to his mode of dress. "He was not very particular about his personal appearance, except that he always kept his hair and beard well brushed and trimmed. . . . His clothes were always loose and easy-fitting, and generally of some quiet-coloured cloth or tweed. . . . He wore no ornaments of any kind, and even the silver watch-chain was worn so as to be invisible" (Marchant [1916] 1975, 350). But Wallace's years abroad, filled with physical and intellectual trials and rigors, had transformed the youthful autodidact into a figure of powerful motivation.

Wallace's determination to persevere and succeed, despite obstacles, is one of his most characteristic traits. One anecdote from this period captures well this facet of his personality. "Soon after my return home in the spring of 1862," Wallace recorded,

> my oldest friend and schoolfellow, Mr. George Silk, introduced me
> to a small circle of his friends, who had formed a private chess club,
> and thereafter, while I lived in the vicinity of Kensington, I was in-
> vited to attend meetings of the club. One of these friends was a Mr.
> L— a widower with two daughters, and a son at Cambridge Uni-
> versity. I sometimes went there with Mr. Silk on Sunday afternoons,
> and after a few months was asked to call on them whenever I liked in
> the evening. . . . On these occasions the young ladies were present,
> and we had tea or supper, and soon became very friendly. The eldest
> Miss L— was, I think, about seven or eight and twenty, very agree-
> able though quiet, pleasant looking, well educated, and fond of art
> and literature, and I soon began to feel an affection for her, and to
> hope that she would become my wife. In about a year after my first
> visit there, thinking I was then sufficiently known, and being too shy
> to make a verbal offer, I wrote to her, describing my feelings and ask-
> ing her if she could in any way respond to my affection. Her reply
> was a negative, but not a very decided one.

Wallace felt that it was his "undemonstrative manner [which] had given her no intimation of my intentions. . . . At first I was inclined not to go again, but on showing the letter to my sister and mother, they thought the young lady was favourably disposed, and that I had better go on as before, and make another offer later on." Wallace misread the "friendly" situation. When he did make a formal offer again to Mr. L——, he was now told that "his daughter wished to break off the engagement. The blow was very severe, and I have never in my life experienced such intensely painful emotion" (Wallace [1905] 1969, 1:409–410). This emotional trauma was one factor predisposing Wallace to react favorably to the comforting doctrines of spiritualism. But he required more than spiritualism to comfort him completely.

Wallace learned his lesson from "Miss L." and proceeded to hone his skills in the Victorian marriage market. In the spring of 1865, he took a small house with his mother in St. Mark's Crescent, Regent's Park, near the Zoological Gardens. It was here that he saw "most of my few scientific friends," including the botanist William Mitten. Scarcely a year later, Wallace, then forty-three, married the botanist's "eldest daughter, then about eighteen years old." His marriage to Annie Mitten proved an enduring and mutually satisfying one. Wallace gave no further details as to the reasons for the success of his second proposal. He noted only "that it never occurs to me at any time to talk about myself; even my own children say that they know nothing about my early life; but if any one asks me and wishes to know, I am willing to tell all that I know or remember" (Wallace [1905] 1969, 1:410–412). That Wallace, so voluminously eloquent in his publications, remained reticent about his inner life renders a detailed analysis of some major personal episodes speculative to a degree.

CONSTRUCTING A SCIENTIFIC CAREER

England in the early 1860s was a nation in ferment. Rapid changes affected the nature and scope of the emerging community of professional scientists. Groups like the X Club (Barton 1998a, 410–444), the BAAS, and numerous other bodies, both formal and informal, all sought to define what should be meant by that new evolutionary specimen: the professional scientist. Wallace had to define a place for himself in this complex institutional battleground. The rubric of professional scientist was itself a contentious term. The 1860s were a time of intense discussion about the character of the Victorian scientific community. Its functions in society, and the values by which it judged the work of its members, were under scrutiny. One of the dominant features of the disputes over evolution, ostensibly carried out in the spirit of objective scientific neutrality, was the key role played by ideology (Moore 1991; Fichman 1997). The tactics of the rising Darwinian camp, in particular, amounted to an organized marginalization of certain prominent (and promising) biologists. Those whose works contained elements deemed problematic for the emerging definition of professional scientists came under fire.

Henry Charlton Bastian was one of many casualties of the battle to promote the image of a naturalistic and value-neutral professionalized science. Bastian has been stigmatized as the loser in his debates with Huxley and Tyndall in the late 1860s/early 1870s over the status of spontaneous generation within the acceptable canon of evolutionary science. He was subjected to a singularly effective, if scientifically dubious, assault by members of the X Club and their allies. Their efforts to discredit Bastian's theoretical and experimental work were elements in the general strategy to create a socially irreproachable community of professional scientists. That community had no room for Bastian and other evolutionists whose ideas held associations with radical politics and amateur science. By the early 1870s, Huxley had succeeded, more by his potent rhetorical skills than by scientific argumentation, in discrediting Bastian and his claim that spontaneous generation was inextricably tied to Darwinian evolution. Huxley's concern was never actually with the validity of spontaneous generation. He privately vacillated on the subject for many years. Huxley was far more anxious to ostracize a rising scientist whose theories and experiments seemed to sully the professional standing and respectability of the Darwinian camp and the X Club. Since Bastian refused to amend his theories on spontaneous generation, so that they no longer carried explicit associations with radical politics and amateur science, Huxley used his formidable professional standing to excommunicate him from the High Church of X Club Darwinians (Strick 1999). Bastian is a prime case of an evolutionist whose theories prevented him from gaining the imprimatur of professionalizing science.

Wallace was sufficiently persuaded by Bastian's evidence and interpretations to write a highly enthusiastic review of the latter's best known book, *The Beginnings of Life* (1872). Wallace's review is but one signal of the ambiguity of his own reputation within the rising Darwinian camp during that critical period of the struggles to define professional science (Wallace 1872b). Wallace was not the only prominent evolutionist to disagree with Huxley's assessment of Bastian. A considerable number of biologists regarded "Bastion's version of scientific naturalism *with* spontaneous generation as an equally valid competitor to the X Club's version without it—perhaps as having an even better claim to be the version most compatible with the doctrine of [evolutionary] continuity." The Bastian controversy is important to the assessment of Wallace's professional standing. It demonstrates that the claim by Huxley and his cohort to be the "official" voice of Darwinism, as well as constituting the London scientific establishment, is open to substantial historiographic critique. One could be a "good Darwinian," as Wallace claimed to be until his dying days, without marching to Huxley's orders. Wallace's defense of Bastian served only to further alienate him from the X Club Darwinians (Strick 1999, 74–80). Anton Dohrn complained to Darwin that Wallace's laudatory two-part review of Bastian was evidence that Wallace was "drift[ing] away, and now most unfortunately associates himself with such a man as Bastian! [Wallace's] two articles in *Nature* are the worst thing he ever did in his life,— and it becomes really difficult for his friends to speak with respect of him" (Groeben 1982, 40–41). Wallace later realized that Bastian's extreme mechanistic materialism was irreconcilable with his own deepening commitment

to spiritualism and evolutionary theism. But it was too late. His belated repudiation of Bastian did nothing to endear him to the dogmatic scientific
naturalists (Strick 2000). Wallace's spiritualism, though it caused him to distance himself from the persona non grata Bastian, provided ammunition for
Huxley and his coterie to augment tensions with respect to Wallace's status
in the professional scientific community.

WALLACE AND THE PROFESSIONAL ELITE

The expansion in the numbers of scientists and the widespread dispersion
of scientific ideas at the popular level and within institutions of education
signaled a shift in social and intellectual authority in England. As competing
voices and factions sought to capture the prized role of becoming recognized
as *the* professionalizing elite, it became clear that what was at stake was not
only the authority of science in Victorian culture. A new public image of science was also being forged. Huxley, Tyndall, and their allies actively engaged
in formulating codes of ethics, strengthening professional organizations, and
establishing professional schools. Educational institutions, public lectures,
popular journals and newspapers, and books of science popularization were
crucial to this goal (Perkin 1989; Turner 1993, 174–176). The process of
professionalization was rife with conflicts among those who were within the
scientific community but who had different ideological goals. Those deemed
outside the emerging scientific community—amateurs, religious authorities,
and the public—posed another set of definitional problems. The notion of a
"working-class science" further complicates an already dense historiographic
situation. Yet another role played by the new "man of science" was that of
public moralist (Collini 1991). Where did Wallace fit in?

By the early 1860s, the sheer magnitude of his theoretical accomplishments rendered him no longer an amateur. But was Wallace a member of the
emerging professional scientific elite? Although he corresponded extensively
with his scientific peers, held posts in several scientific societies, and was
author of a stream of innovative and highly influential papers in the most
prestigious scientific journals, Wallace remained at the fringes of the professionalizing elite. Moreover, if professionalization was as much about social
status as about science, Wallace seemed blithely unconcerned. Whereas Huxley and Tyndall were social climbers—and others like Hooker and Lyell who
did not need to do any climbing since they were already "there"—Wallace
had no such ambitions. He did not want to control the thinking of others in
matters scientific, as did Huxley, Tyndall, and their coterie. Nor did he feel
that the public image of science needed continual polishing (e-mail message
from Ruth Barton to the author, February 28, 1999). Although Wallace was

fervent about his scientific beliefs, he either adopted or was pushed into the stance of outsider-as-insider. Since Wallace was a social outsider, as well as an intellectual maverick, he was likely regarded as not especially crucial to the goals of the architects of emerging professional science. Moreover, the notion of "professional science" in the nineteenth century is currently under scrutiny. Several historians have recently suggested that professional science became exclusive and specialist only in the early twentieth century (Barton 1998a, 444n. 86). Wallace's ambivalent status was, thus, not unique. Given his eminent scientific achievements, however, it was egregious. Wallace's increasingly public involvement with spiritualism, as well as his critiques of the ethics of industrializing Victorian society, were seen as problematic by those advocating an ideologically neutral vision of science. Wallace's sociopolitical, spiritualist, and theistic pronouncements irked Huxley, Tyndall, and their coterie. But once the Victorian scientific community is recognized as far more diverse than the scientific naturalists' model, Wallace ceases to be the anomaly created by much of twentieth-century historiography. Involvements in spiritualism and social reform were seen by Wallace and many others as entirely consistent with a broader conception of science's—and scientists'— role within the wider cultural context.

Wallace established a network of relationships with many of the leading scientists during his first years back in London. He formed close ties with Darwin, Lyell, Huxley, Tyndall, Spencer, Lubbock, W. B. Carpenter, Crookes, Hooker, Galton, Alfred Newton, Philip L. Sclater, St. George Mivart, William Flowers, Norman Lockyer, and the eminent chemist Raphael Meldola. Wallace met frequently with them, both at scientific meetings in the capital and at their homes. His two most important relationships were with Darwin and Spencer. Soon after his return to London, Wallace accepted Darwin's invitation to visit him in Down. In that rural court, Wallace "had the great pleasure of seeing him in his quiet home, and in the midst of his family." During the next several years, Wallace would visit Darwin whenever the latter came to London to stay with his brother, Dr. Erasmus Darwin, in Queen Anne Street—"which he usually did every year when he was well enough. . . . On these occasions I usually lunched with him and his brother. . . . He also sometimes called on me in St. Mark's Crescent for a quiet talk or to see some of my collections" (Wallace [1905] 1969, 2:1–6, 33–34). A major fruit of their friendship was the brilliant scientific correspondence between the two. This correspondence was an intense interchange. Darwin wrote Wallace from Down (22 January 1866) to "thank you for your paper on Pigeons, which interested me, as everything that you write does. Who would have ever dreamed that monkeys influenced the distribution of pigeons and parrots! But I have a still higher satisfaction; for I finished yesterday your paper in the *Linnean*

Transactions. It is admirably done. I cannot conceive that the most firm believer in Species could read it without being staggered. Such papers will make many more converts among naturalists than long-winded books such as I shall write if I have strength" (Marchant [1916] 1975, 137–38). Darwin recognized and admired Wallace's lucid and persuasive writing style as a formidable tool in the arsenal for advancing the evolutionist cause. Despite growing differences on matters both scientific and cultural, Darwin and Wallace remained close and affectionate colleagues for the remainder of Darwin's life.

Wallace also sought out the acquaintance of Spencer. As mentioned in chapter 3, Wallace had been "immensely impressed" with Spencer's *First Principles* and looked to him as "the one man living who could give [him] some clue . . . to the great unsolved problem of the origin of life—a problem which Darwin's 'Origin of Species' left in as much obscurity as ever." Thus began another relationship in which scientific and philosophical concerns formed the core of personal friendship. Although Wallace would later come to differ greatly from Spencer "on certain important matters, both of natural and social science," the two never ceased to value each other's views (Wallace [1905] 1969, 2:23–33). As further testimony to their friendship, Wallace named his son, born in 1867, Herbert Spencer Wallace. The name, as he informed Darwin—admitting that "I quite forget whether I told you that I have a little boy, now three months old"—honored both his brother, Herbert, and Spencer. Darwin, collecting materials for *The Expression of the Emotions in Man and Animals* (1872), congratulated Wallace, adding that he hoped the new child would deserve his namesake but that "he will copy his father's style and not his namesake's. Pray observe, though I fear I am a month too late, when tears are first secreted enough to overflow; and write down date" (Marchant [1916] 1975, 155–158). Wallace's description of his relationship with the remainder of the London scientific community as for the most part only "social" is revealing. He was deeply familiar with the scientific work of all of them, as they were with his own evolutionary (and other) hypotheses and empirical data. Yet Wallace, from the outset of his long career and residence in England, desired to distance himself somewhat from them professionally. That he made an exception in the case of Darwin and Spencer is not surprising. Both of them also maintained a degree of separation from the intense machinations of the emerging professional scientific community. In Darwin's case, this can be attributed in part to his conscious decision to live at some physical distance from London (a decision similar to that made later by Wallace). In Spencer's case, his standing as a professional scientist was always ambiguous. In contrast, Wallace's relationship with other colleagues, especially Huxley, was cordial but professionally restrained. He was often invited to Huxley's house in Marlborough Place. He became a welcome visitor, greatly enjoying the companionship of Huxley's wife and children. Yet Wallace felt, at least during the early 1860s, that he lacked the scientific "authority" of Huxley, already a rising star in the professional community. Huxley was a major figure at the School of Mines, the Zoological Society (of which he was made a vice-president in 1861), and the Geological Society (whose 1862 annual address he was asked to deliver by its infirm president [Desmond 1998, 302–304]).

Wallace's professional path could not have been more different from that of Huxley (fig. 7). He was somewhat in awe of Huxley's specialist knowledge of physiology. During the course of the 1860s, however, Wallace "gradually acquired confidence in my own judgment, so that in dealing with any body of facts bearing upon a question in dispute . . . I would always draw my own inferences from them, even though I had men of far greater and more varied knowledge against me. Thus I have never hesitated to differ from Lyell, Darwin, and even Spencer [and Huxley]" (Wallace [1905] 1969, 2:42). Though never reluctant to engage in public debate, Wallace displayed a relative indifference to professional advancement in terms of positions of power within the scientific community. But he was not without certain career ambitions. Indeed, because Wallace spent the crucial decade of the 1850s away from London, he was able to establish himself more readily as a scientist of repute and not merely a gifted amateur naturalist. The boundaries of class and status, so sharply defined in England, were far less obvious in colonial outposts of the Empire.

WALLACE'S EARLY VIEWS ON HUMAN EVOLUTION

The more fluid society of European (including English) travelers and settlers in the East Indies afforded Wallace an easy entrée into such gentlemanly circles as the clergy, government, industry, and medicine. This would have been impossible for him in England. He became, instantaneously, a member of the colonial establishment in the Malay Archipelago. The contacts open to him and the favors and privileges granted by the European gentry all greatly facilitated his scientific endeavors during 1854–1862 (Camerini 1997, 371–372). When Wallace did return to England in 1862, his professional credentials, though impeccable, did not necessarily carry the same weight in London as they did in the Malay Archipelago. But as codiscoverer of natural selection, Wallace was inevitably thrust to the center of controversy surrounding the question of "man's place in nature." Both he and Darwin had been preoccupied with human evolution from the start. However, the wording of the communication to the Linnean Society in 1858 announcing their joint discovery obscured this fact. The publication of Darwin's *Origin*

Figure 7. A confident
Thomas H. Huxley at age
thirty-two (from an 1857
photograph by Maull and
Fox).

the following year continued the duo's public reticence on the subject of man. It was Wallace who, in 1864, first abandoned that reticence by demonstrating that evolution could provide a comprehensive methodological framework for the scientific study of man. Wallace's writings in the 1860s are crucial for understanding his contentious interpretation of the implications of evolution for human concerns, including spiritualism. His thoughts and activities in this decade vividly illustrate the profound but ultimately ambiguous role evolutionary theory played in Victorian culture.

Wallace extended the biogeographical arguments he used in analyzing zoological distribution in the Malay Archipelago to encompass the islands' human inhabitants. "On the Varieties of Man in the Malay Archipelago," read at the Ethnological Society of London meeting on 26 January 1864 (Wallace 1865b), proposed that the geological history of the archipelago had a significant influence in determining the character and distribution of mankind there. Wallace asserted that a line analogous to that which marks the zoological boundary between the Indo-Malayan and the Austro-Malayan regions divides the archipelago into "two portions, the [human] races of which have strongly marked distinctive peculiarities" (Wallace 1865b, 211). The focus

on mapping accurate human as well as animal and plant boundaries domi-
nates three of his most important scientific essays of the early 1860s (Wallace
1860, 1863, 1864a). Wallace's persistent attention to mapping biogeographical regions precisely—and to demonstrating that these regions reflected the
evolutionary history and migrations of organisms—was part of a broader
reconceptualization of "visualizing" biology that gathered force in the mid-
nineteenth century.[4]

As discussed in chapter 2, Wallace observed that striking contrasts existed between the Malays (inhabiting the western half of the archipelago)
and the Papuans (inhabiting New Guinea and some adjacent islands). These
contrasts, Wallace argued, were the result of their separate evolutionary histories. His protracted intimacy with the Malays and Papuans had enabled
Wallace not only to establish their geographical distribution but to specify
their differing physical as well as behavioral, moral, and intellectual characteristics in detail (Wallace 1865b, 201–205). "On the Varieties of Man,"
taken in conjunction with "The Origin of Human Races and the Antiquity
of Man Deduced from the Theory of 'Natural Selection' " (Wallace 1864b),
were Wallace's first public extensions of evolutionary theory to humans. They
made explicit his conviction that anthropological issues were the legitimate
concern of the evolutionary biologist, a conviction not entirely popular then.
Although the evidence for man's great antiquity was generally accepted by the
mid-1860s, resistance was still strong toward a complete evolutionary explanation of man's nature and history (Burrow 1970, 131). Wallace would later
include factors other than natural selection in his account of human evolution. But these early 1860s essays were powerful contributions to the growing
chorus of geologists, archeologists, physical and cultural anthropologists—as
well as biologists—who by the mid-1860s were radically altering Victorian
conceptions of human prehistory (Van Riper 1993, 159–160, 172, 220).

Wallace admitted that his biogeographical treatment of man was somewhat speculative and, in parts, based on inadequate data. Wallace's "human
line" cannot be taken as proving that the Malays came from the West, originally, and the Papuans from the East. Modern blood group analysis has shown
that line to be an actual boundary between certain human populations, thus
lending support to some aspects of Wallace's theory. But the boundary he
posited may only mark the eastern limit of one particular wave of colonization from the West. Other western emigrants may have spread further east,
giving rise to the Papuan/Polynesian races now found in the archipelago's
eastern half.

Additional factors complicate Wallace's analyses of human distributional
patterns. *Homo sapiens* can cross the seas more readily than other mammals.
Consequently, certain assumptions that are valid in treating the distribution

of other animals may not be applicable in the reconstruction of human evolution (George 1964, 115; for Wallace's own qualifications, see [1905] 1969,
1:421). Despite these caveats, Wallace successfully demonstrated in "Varieties of Man in the Malay Archipelago" that evolution by natural selection
could function as a potent explanatory model in the study of humans. Wallace's paper, a legitimate product of his belief that "true science only begins
when hypotheses are framed to express and combine the facts that have been
accumulated," signaled his entry into the vigorous anthropological debates
of the period (Wallace 1865b, 215). Two months later, he was prepared to
offer his major contribution to those debates and stimulated the evolutionary
bias that soon permeated Victorian anthropology and social theory.[5]

THE ORIGIN OF HUMAN RACES

The question of the origin and relation of the several human races had provoked a controversy in England with profound cultural as well as strictly
biological implications. Wallace's "The Origin of Human Races and the Antiquity of Man Deduced from the Theory of 'Natural Selection'" (1864b)
was a brilliant effort to resolve the dispute between the monogenists and
polygenists. The essay, which was read with great interest by William James,
as well as Spencer, Lyell, Darwin, and most of the scientific community in
Britain and America (Marchant [1916] 1975, 277–278), was not restricted
to the question of the origin of human races. Wallace extended the argument to include the sensitive issue of racial superiority. He suggested that
those races that were exposed to harsher climatic conditions would become
hardier, more provident, and more social than the races that lived in subtrop-
ical and tropical regions, where food was more abundant and "where neither
foresight nor ingenuity are required to prepare for the rigours of winter."
Wallace appealed to history to support biology on this point. He claimed that
all "the great invasions and displacements of races have been from North to
South, rather than the reverse." Wallace cited the successive conquests of the
Indian peninsula by races from the northwest, and the conquest of southern
Europe by the "bold and adventurous tribes of the North," as proof that
the inhabitants of temperate regions are always superior to the races of the
tropics. The "great law of 'the preservation of favoured races in the struggle
for life,'" he declared, operated as inexorably in the human realm as it did
throughout the rest of the natural world (Wallace 1864b, clxiv).

Wallace's 1864 essay is notable for its statement of the racial superiority
of Europeans. In the coming years, he modified this opinion as his political
views became more explicitly reformist and egalitarian. But in 1864, Wallace
echoed the imperialist sentiments of the period. He asserted that natural

selection "leads to the inevitable extinction of all those low and mentally undeveloped populations with which Europeans come in contact." The indigenous populations of North America, Brazil, Australia, Tasmania, and New Zealand succumbed "not from any one special cause, but from the inevitable effects of an unequal mental and physical struggle." As late as 1864, then, Wallace was confident that the European race, and its descendants, would always conquer the savage races with which it comes in contact "in the struggle for existence, and . . . increase at [their] expense, just as the more favourable increase at the expense of the less favourable varieties in the animal and vegetable kingdoms" (Wallace 1864b, clxv). Despite Wallace's personal opposition to overt forms of racial discrimination, his first publications on human development demonstrate how readily evolutionary concepts and vocabulary lent themselves to racist social theory (Haller 1971; Bannister 1979, 180–200; Lorimer 1997). But Wallace's own background and travels predisposed him to question the moral superiority of European to other cultures. He could not deny that Europeans had been empowered by the scientific and technological developments of the eighteenth and nineteenth centuries. They occupied a position of undeniable global economic, industrial, and military supremacy. Spiritualist, theist, and socialist precepts, however, soon emboldened Wallace to articulate unequivocal condemnations of the misuse of European and North American technical prowess (Stack 2000, 693–694).

That Wallace read his paper to the Anthropological Society of London has an added interest. The controversies over race within the rival Anthropological and Ethnological Societies had erupted into full-scale war. The Anthropologicals were a break-away group from the Ethnological Society. Led by the openly racist Robert Hunt, they accused the Ethnologicals of being less scientific because of their attachment to the "rights of man mania" (Stocking 1987, 248–254). The Anthropologicals also advocated polygenesis as a theory and justified slavery as a policy. They differed sharply from their parent society on ideological and political as well as scientific grounds (Richards 1989b). Since Wallace's 1864 paper was intended to ameliorate the tense situation between the proponents of monogenism and polygenism, he did not see the opposition between the two societies as irreconcilable at this point. However, when it became apparent that the Anthropological Society was unrelenting in its doctrinaire racism and antiegalitarian politics, Wallace quit. He joined forces with Huxley, Lubbock, and others over the next several years to limit and control the contributions of the Anthropologicals to the BAAS in order to give what they called "proper direction" to anthropology and to reunite the societies (Barton 1998a, 439). These efforts paid off. Under the leadership of Huxley, the two rival societies were merged in 1871 in the newly formed and professionally more respectable Anthropological Institute. Wallace's views on the "woman question" were also in transition at this point. In 1864, he had spoken against the Ethnological Society's admission of women to its meetings on the grounds that "consequently many important and interesting subjects cannot possibly be discussed there" (Richards 1989a, 264). Like his views on races, Wallace's views on women were to change significantly during the next decade and after. During the mid-1860s, however, Wallace had not yet fully articulated his critique of the assumption that science was an ideologically neutral enterprise. Consequently, it is during this decade that Wallace acted more in accordance with the emerging image of the professional scientist than he did either earlier or from the 1870s onward (Fichman 1997, 100–103, 112–114).

Wallace's 1864 paper further implied that evolution accounted for the preeminent status humankind held within the animal kingdom. At that period when the human mind had become of greater importance than bodily structure, Wallace declared that "a grand revolution was effected in nature—a revolution which in all the previous ages of the earth's history had had no parallel." Since humans could now respond to changing environmental conditions by an advance in mental capabilities, Wallace suggested they were in "some degree superior to nature, inasmuch as [they] knew how to control and regulate her action" (Wallace 1864b, clxviii). One of the most contentious issues to surface in the debates about evolution was the question of man's status in the hierarchy of nature. Evolutionary theory sharpened the perennial concern as to where humans stood in the "great chain of being." This not only involved the diverse opinions advanced by those who held that traditional religious teachings dictated that humans and nonhuman animals were onto-logically different. Scientific opinion was also divided on this point. One need look no further than the ongoing debate between Wallace and Darwin as to the adequacy of natural selection to account for the origin and development of humans to recognize how incendiary this question was in the Victorian period (and remains so to the present day [Kottler 1985, 420–424]). Darwin inclined toward the view that placed humans squarely and completely within the realm of those forces that accounted for the evolution of all species. Wallace, in contrast, believed that human evolution in certain crucial respects was an exception: forces other than natural selection operated in the history and future development of mankind.

Wallace's views on human evolution are complex as well as controversial. He had from his earliest writings and thoughts been convinced of humanity's unique status. Joel Schwartz has argued that sometime after 1858, Wallace first came to decide that humans and nonhuman animals had evolved according to different processes (Schwartz 1984). But Wallace had entertained this idea of humans as an exception to the evolutionary history of other animals from the 1840s (Smith 1992, 2–3, 6–8, 13–14). The evidence for the continuity of Wallace's thought on this matter will be discussed later in the chapter, in the section dealing with the precise role played by spiritualist beliefs in Wallace's theory of human evolution. By 1864, Wallace's maturing evolutionary concepts provided him with additional motivation to assert that those who maintained that human attributes argued for a "position as an order, a class, or a sub-kingdom by [themselves], have some reason on their side." Nor were humans merely at the summit of organic nature. The continued action of natural selection and spiritual agencies destined *Homo sapiens* to an ever higher level of existence. Persuaded that further mental and moral evolution was possible, Wallace described humanity's future ecstatically in the concluding paragraph of the essay:

> Each one will then work out his own happiness in relation to that of his fellows; perfect freedom of action will be maintained, since the well balanced moral faculties will never permit any one to transgress on the equal freedom of others; restrictive laws will not be wanted, for each man will be guided by the best of laws; a thorough appreciation of the rights, and a perfect sympathy with the feelings, of all about him; compulsory government will have died away as unnecessary (for every man will know how to govern himself), and will be replaced by voluntary associations for all beneficial public purposes; the passions and animal propensities will be restrained within those limits which most conduce to happiness; and mankind will have at length discovered that it was only required of them to develop the capacities of their higher nature, in order to convert this earth, which had so long been the theatre of their unbridled passions, and the scene of unimaginable misery, into as bright a paradise as ever haunted the dreams of seer or poet.
>
> (Wallace 1864b, clxviii–clxx)

"Origin of Human Races" testified to Wallace's conviction that the findings of biology, particularly of evolutionary theory, bore directly on social and political questions. During the remainder of the decade, he continued to publish articles, reviews, and commentaries of anthropological importance. He wrote "On the Progress of Civilization in Northern Celebes" (1864), "How to Civilize Savages" (1865), "Phallic Worship in India" (1865), "On Physico-Anthropology: Its Aims and Methods" (1867), "On the Primitive Condition of Man" (1869), "The Origin of Moral Intuitions" (1869), "On Instinct in Man and Animals" (1870), and, finally, a review of Galton's 1869 *Hereditary Genius* for the 17 March 1870 issue of *Nature* (Smith 1991, 482–

488). Clearly, Wallace had thought long and hard before penning the two bombshells that appeared in 1869: his review of Lyell's tenth edition of *Principles of Geology* and "Limits of Natural Selection as Applied to Man." A major, but not exclusive, impetus for going public with his long-held belief that factors additional to natural selection had guided human evolution were Wallace's first sustained encounters with spiritualism in the mid-1860s.

EARLY ENCOUNTERS WITH SPIRITUALISM

During his tropical journeys, Wallace had heard of the strange phenomena associated with spiritualism said to be occurring in America and England. Some of the accounts seemed "too wild and outre to be anything but the ravings of madmen." Other reports appeared to be well confirmed. Wallace determined, therefore, to ascertain on his return to London whether the alleged phenomena were legitimate or merely the results of chicanery or suggestion (Wallace [1905] 1969, 2:276). Wallace's early involvement with mesmerism and phrenology predisposed him to consider that there might be "mysteries connected with the human mind which modern science ignored because it could not explain" (Wallace 1875b, 131–132).

Wallace's first documented séance attendance took place in 1865, at the home "of a friend—a sceptic, a man of science, and a lawyer." Wallace, too, was initially skeptical that "the marvels related by Spiritualists" he had either read or heard about "could literally be true." What changed his attitude was the force of personal evidence. He also declared that it was not any "dread of annihilation . . . or from [an] inordinate longing for eternal existence" that he approached spiritualism. In the jungles of the Amazon and the Malay Archipelago he had at least three times "to face death as imminent or probable within a few hours." Wallace felt only a "gentle melancholy at the thought of quitting this wonderful and beautiful earth to enter on a sleep which might know no waking." Prior to the 1860s, Wallace regarded the question of "conscious existence . . . independent of the organised body" as not yet answerable. He approached his first séance with an open mind, "utterly unbiassed by hopes or fears, because I knew that my belief could not affect the reality." It was the "unrelenting" accumulation of personal evidence at séances following his first that gradually but profoundly made Wallace convinced of the reality of spirit manifestations (Wallace 1875a). He regarded spiritualism as a legitimate field for scientific investigation. Hypotheses could be tested empirically, and spiritualist claims would be verified or refuted according to the canons of nineteenth-century scientific methodology. Wallace set out to overcome the skepticism of many of his professional associates and establish spiritualism as a valid "science of human nature which . . . appeals only to facts and experiment [and which] affords the only sure foundation for . . . the improvement of society and the permanent elevation of human nature" (Wallace 1875b, 228–29). He would later endorse his friend Barrett's application of Ockham's razor to the maze of spiritualist phenomena. "But why," Barrett challenged spiritualism's detractors, "should we think it so extravagant to entertain the simplest explanation that occasionally a channel opens from the unseen world to ours, and that some who have entered that world are able to make their continued existence known to us" (Barrett 1908, 64). Wallace underlined the phrases "the simplest explanation" and "occasionally" and in the margin wrote "YES!" in his copy of Barrett's book, which is in Edinburgh University Library's Special Collections.

Wallace's desire to embark on a detailed study of spiritualism is fully consistent with his philosophy of nature and his temperament. Evolution, in Wallace's view, encompassed diverse explanatory elements. Evidence for his conviction that there were forces other than strictly physical or material ones at work in the evolution of animals as well as humans comes from an early essay on orangutans. In 1856, Wallace argued that the large canines of the male of the species do not seem to be adaptive. "Here," he observed, "we have an animal which lives solely and exclusively on fruits and other soft vegetable food, and yet has huge canine teeth. It never attacks other animals, and is rarely attacked itself; but when it is, it uses, not these powerful teeth, but its arms and legs to defend itself." Wallace preempted the expected "indignant" reaction that he was suggesting that the orangutan, and many other animals including humans, were provided with organs of no use to them. His answer was direct:

> Yes, [I] do mean to assert that many animals are provided with organs and appendages which serve no material or physical purpose. The extraordinary excrescences of many insects, the fantastic and many-coloured plumes which adorn certain birds . . . , the colours and infinitely modified forms of many flower-petals, are all cases, for an explanation of which we must look to some general principle far more recondite than a simple relation to the necessities of the individual. [I] conceive it to be a most erroneous, a most contracted view of the organic world, to believe that every part of an animal or plant exists solely for some material and physical use to the individual—to believe that all the beauty, all the infinite combinations and changes of form and structure should have the sole purpose and end of enabling each animal to support its existence—to believe, in fact, that we know the one sole end and purpose that exists in organic beings, and to refuse to recognize the possibility of there being any other.

Naturalists are to apt to *imagine*, when they cannot *discover*, a use for
everything in nature: they are not even content to let "beauty" be a
sufficient use, but hunt after some purpose to which even *that* can be
applied by the animal itself, as if one of the noblest and most refining
parts of man's nature, the love of beauty for its own sake, would not
be perceptible also in the works of a Supreme Creator.

What this 1856 essay demonstrates is that even as his mind was gearing up for
the articulation of natural selection, Wallace's conceptualization of evolution
was broad. His aim was nothing less than "our complete appreciation of all the
variety, the beauty, and the harmony of the organic world" (Wallace 1856b).
Natural selection would be one of the major keys to understanding how the
harmony of the organic world had come about. But spiritualism was another
key. Wallace considered spiritualism as a fruitful standpoint from which to ex-
plicate the broader meaning of evolution, particularly at the moral/intellectual
level. For him, natural selection and spiritualism were mutually supportive
elements in a grander scheme of things.

Wallace's early exposure to mesmerism provided him with a potent epistemological and experimental analogue for studying the phenomena of spiritualism. Both mesmerism and spiritualism had been subjected to the twin
demons of either hostile (usually uninformed) criticism or enthusiastic (often naive and overzealous) endorsement. Wallace could deploy his own
formidable powers of exploring contentious issues with that rigor and independence that characterized his entire career. One more factor attracted
Wallace to spiritualism. Its teachings required no dogmatic, unbending submission of its broad spectrum of adherents. Spiritualism afforded the serious
investigator a fertile field for investigating experimentally phenomena of an
unusual and hence—to the insatiably curious Wallace—highly alluring character (Smith 1992, 1, 18).

The curiosity to explore new ideas has long been recognized by sociologists of science as an indispensable component of the psychological make-up
of those individuals who choose science as a career. Wallace was no exception.
He spent many long and dreary hours in uncertain experimentation, with
scrupulous attention to detail, pursuing what he and many others thought
was, potentially, an exciting and novel domain of scientific inquiry. Far from
being naive when it came to such experimental researches, Wallace performed
or read about sophisticated attempts to validate spiritualist claims experimentally. The fraud and conjuring that later came to characterize much of popular
spiritualism, as paid mediums and accomplished magicians became standard
fixtures on the spiritualist circuit, were only one aspect of a broad movement.
Wallace, in fact, was obsessive about ferreting out such legerdemain in his

spiritualist enterprises in the 1860s. It has been suggested that Wallace often failed to detect imposture not because his critical approach or observational skills were deficient—his entire career proves the contrary—"but because he was a naturalist, not an accomplished magician. We must deal with Wallace under his given circumstances and experiences; if he had access" to the techniques of a Houdini some of his more adamant defenses of séance phenomena may have been toned down or abandoned (Malinchak 1987, 77). Not surprisingly, Wallace's "conviction threshold was lowered whenever he came across some apparent fact or reference involving a member of his family" at certain séances. Alfred and his sister Fanny shared a certainty that spiritualism put them in touch with their dead brothers (Raby 2001, 185–186). The strongly personal dimension of Wallace's spiritualist beliefs is not incompatible with the overall stance of objectivity he maintained in his investigations of the scientific status of spiritualism and other psychic phenomena.

SPIRITUALISM AND WALLACE'S WORLDVIEW

Even had Wallace moderated the tenacity of his defense of spiritualist phenomena, he would still have adhered to the movement. Spiritualism powerfully reinforced his political and theistic convictions. The ethical teachings of spiritualism were as important, if not more important, to him than the scientific legitimacy of its claims. His encounters with spiritualism, from the beginning, were in accord both with his epistemological framework concerning nonmaterial reality as well as his cultural vision. Two works that Wallace deemed as "forming the best-reasoned and the most logically arranged body of evidence for psychical phenomena in existence" were written by Robert Owen's son, Robert Dale Owen: *Footfalls on the Boundary of Another World* (1861) and its sequel *The Debatable Land between This World and the Next* (1871 [Wallace [1905] 1969, 2:294–295]). Wallace reviewed the latter work in 1872 (Wallace 1872c). His enthusiastic review elicited "very interesting" letters from Dale Owen and from Eugene Crowell, a New York medical doctor. Crowell's letter came with a copy "of his exceedingly valuable work, 'Primitive Christianity and Modern Spiritualism' (2 vols.), in which almost every miraculous occurrence narrated in the Old or New Testaments is paralleled by well-authenticated phenomena from the records of modern spiritualism, many of them having been witnessed and carefully examined by Dr. Crowell himself" (Wallace [1905] 1969, 2:295). Wallace's personal copies of both of Dale Owen's books are marked throughout with enthusiastic triple vertical lines in the margins. Dale Owen's combination of spiritualism and political reformism was manna to Wallace.

After his flirtation with agnosticism in the 1840s and 1850s, Wallace's emerging theism in the 1860s and early 1870s dovetailed neatly with his spiritualist leanings. He endorsed Dale Owen's assertion of the concord between spiritualism and Christianity—once the latter was "shorn of parasitic creeds." Dale Owen also emphasized, as did Wallace, that spiritualism "teaches no speculative divinity" but only one that was demonstrated clearly by a vast body of phenomena. In *Footfalls on the Boundary of Another World*, Dale Owen (1861, 4) commented also on the difference between Swedenborg and Swedenborgianism. For Owen, Swedenborg—"the great spiritualist of the eighteenth century"—was a profound thinker whose writings "even at a superficial glance, must arrest the attention of the right-minded." Swedenborg's doctrine of the "constant influence exerted from the spiritual world on the material" and "his glowing appreciations of that principle of Love which is the fulfilling of the Law" were "of too deep and genuine import to be lightly passed by. To claim for them nothing more, they are at least marvelously suggestive, and therefore highly valuable." Owen had little use for the interpolations made by many of Swedenborg's disciples. He felt they had exaggerated Swedenborg's occasional mystical allusions into a full-blown and often incomprehensible system. Owen concluded that one must "appreciate Swedenborg outside of Swedenborgianism" (Owen 1861).[6] Wallace may have first learned of Swedenborg from Dale Owen's 1861 work. He would then have been prepared to share James's and Peirce's regard for Swedenborg when he met them during his North American tour in 1886–1887. Wallace followed Owen's injunction to separate Swedenborg from the Swedenborgians, just as he would later distinguish Marx from the plethora of Marxist disciples. Dale Owen's books provided Wallace with a compelling statement of spiritualist philosophy—integrated with Owenite socialist teachings—that appealed to his intellectual as well as ethical and political concerns. These concerns were crucial in shaping Wallace's developing worldview.

For Wallace and the many other Victorians who espoused spiritualist tenets, a question arises. Did spiritualism act as a surrogate faith, or religion, for them? Much has been written about the role of alternatives to fill the void for secularists when stark materialism proved disconcerting. Similar explanations have been advanced for Christians when their orthodoxy was tested by the doubts engendered by scientific theories and biblical criticism. In sociological terms, late-nineteenth-century spiritualism lacks the institutional attributes that are taken to characterize organized religion. There was no official spiritualist church. The creeds that emerged were sufficiently varied to defy simple categorization. Some spiritualist groups arranged services that bore a close resemblance to more traditional church services. For other groups, all formal paraphernalia of religious observance were rejected. To complicate the sociology of spiritualism further, its practice varied in different communities and among different social groups within the same community. Moreover, Victorian spiritualism was embraced by freethinkers, Nonconformists, church-going Anglicans, and, even on occasion, Roman Catholics. The manner "in which spiritualism became entangled with religion, organized and disorganized, exoteric and esoteric, is not easily summarized" (Oppenheim 1985, 62). Nor need it be. For the majority of adherents, it was the optimistic ethical and social message supported by scientific evidence and interpretation that was spiritualism's main drawing card. For Wallace, spiritualism alone did not function as a surrogate faith. He combined it with theistic and scientific convictions within a broader evolutionary teleology. Wallace's scientific theism was his faith.

The ethical implications of spiritualism were not the only elements that attracted Wallace. He, like increasing numbers of adherents—first in North America and then in Britain—found certain phenomena of spiritualism irresistible. Induced trance states and somnambulist manifestations by itinerant lecturers as spectacle had an immediate popular appeal. The enduring fascination with the marvelous and the mysterious, a constant in human history before, during, and after the scientific revolution(s) of the seventeenth and eighteenth centuries, had lost none of its power during the nineteenth century (Daston and Park 1998). As positivism and scientific naturalism gained increasing prominence, fascination with the marvelous and the occult increased as counterpoints, especially among the general public. (Parallels with the late twentieth and very early twenty-first centuries are not hard to draw.) Traditional religious authority, particularly among sects such as Unitarians and Quakers, grew more relaxed. Converts to such liberalized sects participated in a shift in perspective on the nature of spirit activity. Diabolic characterizations became less frequent as the emphasis on (usually benevolent) disembodied human intelligences grew. Some hoped that spiritualism might constitute a new, scientifically respectable form of revelation and thus modernize some forms of traditional religion creeds. Table tilting and other staples of séance phenomena, which ostensibly provided witnesses with the opportunity to have conversations with those who had "passed over," captivated many. No lesser mortals than Queen Victoria and the prince consort were caught up in spiritualism's popularity. Neither could resist the opportunity to converse with unseen spirits. After Albert's death in 1861, Victoria relied on the mediumistic powers of John Brown, her personal servant and confidant, to communicate with the departed prince. Such widespread immersion in spiritualism by the 1860s testifies to the excitement that characterized the movement's rapid growth. Sympathetic commentators referred to the "infectious" quality that drew people of all social classes to the various private and public arenas in which spiritualism flourished. The movement's equally vocal critics preferred the term "contagion." It is against this broader landscape that Wallace's infection must be situated. As both an articulate champion of spiritualism and a powerful scientific voice, Wallace was uniquely placed to assume a highly visible role in the inevitable controversies surrounding the claims and counterclaims of so striking and novel a feature on the mid-Victorian landscape.[7]

MESMERISM

Just as Wallace's approach to spiritualism was shared by many of his contemporaries, he was not alone in his interest in studying mesmerism. From the 1830s to the 1860s and beyond, mesmerism attracted a wide following. Since it took hold in Britain later than it did in the rest of Europe, mesmerism had the aura of being a new and fascinating science of life and mind to the early and mid-Victorians. Despite some initial public experiments that were easily discredited during the 1830s, mesmerism entered into the social, psychic, and institutional fabric of Victorian society by 1840. Far from being the marginal or pseudo-science portrayed by most historians until recently, mesmerism (and animal magnetism and hypnotism) permeated Victorian culture at all levels. Aristocrats, the industrial middle classes, factory workers, doctors and their patients were keenly interested in, if not always converted by, mesmeric claims and practices. In England, Ireland, and Scotland, as well as throughout the developing empire, mesmeric practitioners, from doctors to itinerant lecturers, forced both intellectual elites and the general public to consider seriously the status to be allocated their field. Recent scholarship has now made clear the prominent role mesmerism played in the debates as to what constituted appropriate scientific and medical authority. To dismiss mesmerism, along with spiritualism and phrenology, as fringe movements is to view the nineteenth century through the most opaque of Whig history lenses. The prominence of Victorian mesmerism reveals, once again, that the debates over authority in scientific, medical, and intellectual life generally were far more intense than the majority of twentieth-century historians have cared to admit. What counted as legitimate or "real" science in the second half of the nineteenth century was a very open question indeed. It is anachronistic to assume that there was a clearly definable and invincibly empowered orthodox or professional community of science in Victorian Britain. Mesmerism was central to the process by which traditional cultural—including scientific—assumptions and attitudes were tested, contested, and ultimately transformed. It is significant that mesmerism's period of greatest influence coincided with a period of major social and political reforms (culminating in the Second Reform Act of 1867) in Britain. Mesmerism provided an excellent forum for questions concerning the nature not only of the human mind and body but of the body politic as well. It brought to the surface issues of power and authority that were rarely acknowledged openly in mid-Victorian Britain. Mesmerism became one focus for tackling controversial claims about psychological and physiological influences on the functioning of society. It touched the emotionally charged matters of class, gender, and race (Winter 1998, 3–9, 306–309). All these questions had been occupying Wallace from his early youth. On his return to London in 1862, these questions assumed a new vivacity for the naturalist who had spent so many years in tropical jungles and villages.

Wallace and his brother Herbert, with whom he traveled for a time in South America, had been early converts to mesmerism. They were convinced that the chief phenomena produced on subjects in mesmeric trances were authentic and well documented. They also found that they possessed mesmeric powers themselves. Among the indigenous peoples of South America, Wallace and Herbert were able to induce the typical mesmeric responses of catalepsy, altered states of consciousness, loss of sensation, and partial paralysis. Thus, as early as the 1840s, Wallace had become convinced that such phenomena were objective reality "and by no means due to the imagination of the unusually stolid Indian" (Wallace [1905] 1969, 2:275–276). Wallace's activities must be viewed within the context of Victorian mesmerism in the expanding empire at midcentury. Colonial India was viewed as a vast social laboratory for studying a wide variety of issues in a setting where the contentious British debates might prove less intrusive. In addition to mesmerism, the introduction of scientific and technological practices, educational reforms, and the use of statistics in the study of social institutions found virgin territory (Baber 1996).

The Scottish surgeon James Esdaile was Britain's chief colonial advocate of mesmerism. Arriving in Calcutta in 1845, Esdaile set up the first of what came to be an expanding network of hospitals. In them, mesmerism had explicit medical objectives. Anesthetizing surgical patients was the most prominent. Esdaile succeeded in persuading Sir Herbert Maddock, deputy governor of Calcutta, to establish an official committee to evaluate his work. The committee members concluded that the mesmeric phenomena described and witnessed were genuine. But they argued against the wholesale introduction of mesmeric medical practices in India as premature. The committee members wanted more experiments, especially ones conducted using European subjects, before they could endorse institutionalized mesmerism in India. Apart from the central issue of verification, the debates about colonial mesmerism involved questions of superiority and inferiority and influence and submission between colonizer and colonized. Despite the committee's qualified endorsement, Esdaile and his supporters were successful in the campaign to have mesmerism accepted as a legitimate if minor procedure in hospital practice in India. Mesmerism became a focus of more intensive research and scrutiny, particularly before the introduction of chemical anesthesia (Winter 1998, 187–212). But Wallace was now (since 1862) back in England, not in some colonial setting. His attempt to investigate spiritualist as well as mesmeric phenomena was a far more complex task, socially, scientifically, and institutionally.

Mesmerism, like spiritualism, reflected and influenced the vigorous debates in mid-Victorian Britain concerning the definition and exercise of scientific authority. Although Wallace is more well known for his prominent role in the controversies surrounding spiritualism, he also entered the heated frays on mesmerism. His most significant personal contest was that with William Benjamin Carpenter. From the late 1830s until his death in 1885, Carpenter was one of Britain's most distinguished physiologists. He was a leader in the movement to further popular scientific education and assisted in the development of the University of London, where he served as registrar from 1856 until 1879. Carpenter was also one of the most relentless debunkers of mesmerism, spiritualism, and phrenology. From the early 1850s onward, in a stream of polemical articles, he hammered home his conviction that the phenomena of mesmerism and spiritualism were open to only two interpretations. Thought reading, automatic writing, and related activities, Carpenter insisted, could be either explained on the basis of demonstrable physiological hypotheses (such as involuntary but entirely mundane muscular movements) or dismissed as imposture or delusions (Oppenheim 1985, 241–244).[8] Clearly, a Wallace-Carpenter confrontation was inevitable.

WALLACE CONTRA CARPENTER

As in so many of his relationships with scientific contemporaries (and, often, adversaries), Wallace was able to detach controversy from personal animosity. Carpenter was one of Wallace's near neighbors in London. He frequently visited Carpenter in the evenings, when the latter was usually peering through his microscope. Carpenter enjoyed displaying the world of minute organisms and structures to Wallace, who appreciated Carpenter's enthusiasm if he did not always agree with some of Carpenter's specific explanations and conclusions. The two became "very friendly . . . and often walked across the Regent's Park into town together." When, some years later, Wallace and Carpenter entered into "a rather acute controversy upon mesmerism and clairvoyance," Wallace accused Carpenter of misstatements, evasions, or willful obscurantism. But he always maintained that his opposition to Carpenter's criticism of spiritualism and mesmerism was restricted to questions of fact and evidence rather than personal animus. Wallace's inference was clear: it was Carpenter who had made their debate more acrimonious that it need have been. At the BAAS meeting in Glasgow in 1876, Wallace was president of the biological section. The session, he noted, "was rendered rather lively by the announcement of a paper by Professor W. F. Barrett on experiments in thought-reading. The reading of this was opposed by Dr. W. B. Carpenter . . . but as it had been accepted by the section, it was read. Then followed a rather heated discussion; but there were several supporters of the paper, among whom was Lord Rayleigh, and the public evidently took the greatest interest in the subject, the hall being crowded." In his autobiography, Wallace added that the issues raised by Barrett's paper were pursued in the following decades by Barrett, Frederick Myers, Sidgwick, Edmund Gurney, "and a few other friends [who] founded the Society for Psychical Research [SPR], which has collected a very large amount of evidence and is still actively at work [in 1905]" (Wallace [1905] 1969, 2:42–43, 49).[9]

The relationship of mesmerism and spiritualism to the work of the SPR, and Wallace's role in that context, are complicated. Wallace became an honorary member of the SPR in the year it was founded (1882) and remained one until his death. He was asked on at least two occasions to assume its presidency but declined. A rift soon developed between Wallace and a number of members of the society as to the specific role to be accorded spiritualist researches within the SPR's broader mandate to investigate psychic phenomena generally. As the society developed, many of its members turned increasingly to investigating evidence drawn from areas such as thought transference, hypnotism, hallucinations, multiple personality, dreams, and abnormal psychology. In so doing, the initial spiritualist umbrella under which the society undertook many of its researches weakened or was discarded, and investigations became segregated into several independent divisions. Wallace's continued emphasis on mediumistic phenomena as a core of psychical research—particularly given his steadfast defense of many mediums accused of fraud—was looked on with growing disfavor by a number of members of the SPR. Wallace considered the demotion of spiritualist interpretations in the SPR's hierarchy of explanations for the phenomena it investigated misguided. By 1886, several prominent spiritualists, including William Stainton Moses and E. Dawson Rogers, left the society. Wallace stayed on but realized that the holistic philosophy of science, ethics, political reform, and theism he had articulated was not shared by all members of the SPR (Malinchak 1987, 145–178). The SPR, and the British scientific establishment generally, became more elitist and specialized in the last decades of the nineteenth century. Wallace's ambivalent relationship to the SPR, therefore, is testimony both to the strength of his egalitarian ethos and to the tensions inherent in defining professional science (Oppenheim 1985, 135–136; Winter 1998, 305, 346–347).

Wallace's complaint that Carpenter's antimesmerist and antispiritualist critiques were often tedious, irrelevant, and pugnacious was just. Carpenter's constant admonition that a rigorous scientific education would immunize individuals against what he termed the tricks of mediums clearly missed the mark, considering the number of eminent scientists who endorsed either spiritualism or mesmerism or both. Wallace's frustration with Carpenter's increasingly nasty diatribes was shared by many of Carpenter's spiritualist adversaries. Wallace was voicing a general sentiment when he pointedly remarked that, as far as Carpenter was concerned, "nobody's evidence on this particular subject is of the least value unless they have had a certain *special early training* of which . . . Dr. Carpenter is one of the few living representatives." Despite his personal vendettas, Carpenter was recognized by both allies and foes alike as Britain's most formidable denigrator of both spiritualism and mesmerism (Oppenheim 1985, 243–244 and 244n. 130). Wallace did confide to Barrett (9 December 1877) that he had been "advised by other friends not to waste more time on Dr. C." He left it to other spiritualist and psychical investigators to refute Carpenter's more outrageous attacks. Such refutation, Wallace indicated, would more than anything else serve to "lower Dr. C. in public estimation on this subject [because of Carpenter] being forced to acknowledge that what he has for more than thirty years declared to be purely subjective is after all an objective phenomenon" (Marchant [1916] 1975, 427).

THE WORLD OF SÉANCES

When Wallace attended his first séance in 1865, he was impressed with the "rapping and tapping sounds and slight movements of a table" (Wallace [1905] 1969, 2:276). Repeated séances, including several with the renowned English medium Mrs. Marshall later that year, exposed Wallace to a wide variety of physical and mental spirit manifestations. During the following years he continued to attend séances regularly and read voraciously in the spiritualist literature. Satisfied that the tests that he and others devised and executed excluded the possibility of collusion or deception, Wallace gradually became convinced both of the authenticity of these remarkable phenomena as well as of the spiritualist interpretation of them. At first, Wallace did not reject the possibility that the simpler séance phenomena, such as table rappings and table vibrations, might result from some involuntary force coming from mediums themselves. It was only when he observed "some very remarkable phenomena" at séances in 1866 and early 1867, presided over by the medium Miss Nichol (later famous as Mrs. Samuel Guppy, and still later as Mrs. Volckman), that Wallace became completely convinced that the phenomena he and others had witnessed "could not possibly have been produced by any of the persons present," including the medium. It was Wallace's sister, Fanny Sims, at whose home Miss Nichol was living temporarily, who first discovered her friend's unique mediumistic gifts in November 1866. Wallace and Fanny had numerous sittings with Miss Nichol subsequently, and her talents matured rapidly. Phenomena at her séances included "a very large leather arm-chair which stood at least four or five feet from the medium, [which then] suddenly wheeled [itself] up to her after a few slight preliminary movements." Henceforth, Wallace would remain unshaken in his conviction that séance phenomena could not be accounted for by any know physical forces and could only be attributed to spirit force (Wallace [1905] 1969, 2:279–280, 291–293). We can thus set Wallace's full-blown embrace of spiritualism at the end of 1866 or the very start of 1867.

Wallace had been struck by the mass of testimony accumulated since the advent of modern spiritualism, which he dated from 1848. In that year, the daughters of the Fox family of upstate (Hydesville) New York received intelligent communications via "mysterious knockings" (Wallace 1875b, 152–53). The "Rochester Rappings," as these auguries of spiritualism came to be called, quickly made celebrities of the Fox sisters Margaret and Kate. Spiritualist interest and séances spread rapidly across the eastern United States and then made their progress westward across the nation. In the heady mid-nineteenth-century atmosphere of mysticism, religious unorthodoxy, and social utopianism—as well as mesmerism and phrenology—Britain was fertile ground for the spread of spiritualism. In 1852 and 1853, two American mediums, Mrs. Hayden and Mrs. Roberts, visited Britain. Spiritualism's transatlantic crossing had commenced. It was the visits of Daniel Douglas Home to England, first in 1855 and again in 1859, that truly ignited the enthusiasm of British audiences. Home was perhaps the most remarkable if enigmatic of nineteenth-century mediums. Although eccentric enough to make him subject to caricature in *Punch* and other British periodicals, he was never proven to be a charlatan. Between 1855 and the early 1870s, Home's mediumistic powers attracted numerous followers. His audiences came from all strata of society. Home presided over séances for Napoleon III and Tsar Alexander II, British luminaries such as Sir Edward Bulwer-Lytton, Robert and Elizabeth Barrett Browning, and Mrs. Frances Trollope, as well as numerous men and women of modest or humble circumstances. Robert Owen, an iconic figure to Wallace, was one of Home's earliest visitors in 1855. The free trade crusader John Bright arranged a meeting with Home in 1864. Home's reported spirit manifestations produced both adoring disciples and vehement critics. He fully cooperated when scientists sought to investigate his powers. At a time when fraudulent mediums were none too rare, even so staunch a skeptic as the psychical researcher Frank Podmore grudgingly admitted that "Home was never publicly exposed as an impostor; . . . [and] there is no evidence of any weight that he was even privately detected in trickery." Home's successes, coupled with the growing tide of séances conducted by other mediums, made spiritualism a force whose momentum grew dramatically in mid-Victorian Britain (Oppenheim 1985, 10–16). It was in this atmosphere of intense interest in the alleged phenomena of spiritualism that Wallace turned his analytic scrutiny to a personal investigation of such phenomena.

Although spiritualism (and mesmerism) enjoyed the support of many individuals, the response from the emerging professional scientific community was, at best, mixed. During the 1860s, then, Wallace was faced with a professional conundrum. How best could he portray himself both as the brilliant scientist he clearly was *and* a supporter of spiritualism? He described his first acquaintance with spiritualist phenomena and the effect they produced on him in the "Notes of Personal Evidence," which later appeared in his *On Miracles and Modern Spiritualism* (Wallace 1875a).[10] Wallace approached the subject with an open yet critical mind. Since the phenomena he witnessed seemed devoid of manipulation or trickery, he felt it obligatory to investigate the claims made on behalf of spiritualism with the utmost seriousness (Wallace [1905] 1969, 2:276–277). This campaign was to prove an uphill battle. Wallace conducted his campaign relentlessly. Detailed personal observations and innovative inferences that he drew from them had characterized Wallace's earliest studies in natural history. That same combination motivated his investigation of spirit phenomena. It accounted, in part, for Wallace's persistence despite criticism from certain quarters. Even those closest to him commented occasionally on his frequent willingness to defend the mediumistic powers of individuals accused of fraud and deception (Raby 2001, 184–192).

Myers was perceptive when he said that Wallace was one of those whose "natures . . . stand so far removed from the meaner temptations of humanity that [they] thus gifted at birth can no more enter into the true mind of a cheat than I can enter into the true mind of a chimpanzee" (Myers 1895, 218). Wallace was, almost congenitally, incapable of seeing the worst in those in whom he believed. This incapacity did not preclude vigilant condemnation of those whom he considered guilty of the great social and ethical injustices he opposed so passionately throughout his life. Wallace was not unique in possessing characteristics that at times seem ambiguous. The Victorian period, no less than any other in history, abounded in individuals who were amalgams of a bewildering variety of traits. Wallace forged a compelling worldview from disparate elements within himself and the cultures in which he lived for nearly a century. Two beacons guided his life and career. One was the belief in natural law, under whose sway he included science and spiritualism. The other was the conviction that humans were capable, given a suitably reformed sociopolitical environment, of achieving that better state that his evolutionary teleology rendered plausible.

THE FORCE OF PERSONAL EVIDENCE

As previously indicated, Wallace stated that his early experiences with mesmerism predisposed him to study spiritualist phenomena. On numerous occasions in the 1840s, and then during his tropical journeys, Wallace witnessed mesmeric phenomena such as phreno-mesmerism, catalepsy, and "sympathetic sensation." More significant from the perspective of personal and psychological impact, Wallace found that he possessed mesmeric powers. When teaching at the Collegiate School at Leicester in 1844, he successfully mesmerized some students. Wallace was struck by "the sympathy of sensation" between his subjects and himself. Such sympathy was observed (and recorded) repeatedly and was to Wallace "the most mysterious phenomenon [he] had ever witnessed" (Wallace 1875a).

This sympathy manifested itself by tests such as placing a lump of sugar or salt in Wallace's own mouth, out of sight of the patient. The patient then "immediately went through the action of sucking, and soon showed by gestures and words of the most expressive nature what it was [Wallace] was tasting." In "Notes of Personal Evidence," Wallace explicitly rejected deception or trickery because of the precision and precautions with which he had prepared the experiments. He also rejected psycho-physiological explanations, such as power of suggestion on the part of mesmerist toward the patient. Wallace deemed such hypotheses to be "no explanation at all." Wallace's insistence on this point is extremely revealing about his state of mind regarding the phenomena he witnessed. From his earliest encounters with mesmerism and spiritualism, Wallace resented imputations of delusion or fraud. He felt that he, and those whose experiments he followed, were earnest seekers of accurate knowledge of phenomena that seemed beyond the pale of (then) accepted canons of science. The years spent among indigenous inhabitants of South America and the Malay Archipelago left their mark on Wallace's psyche and his conception and practice of science. Animist creeds and belief in the reality of spirits abounding in nature were fundamental precepts of those peoples. Wallace had shared their daily lives and experiences and absorbed their cultures. By the time he returned to London, Wallace was a naturalist for whom "mysterious" phenomena had become a significant component of a valid knowledge base. He was determined to see where such phenomena belonged in the explanatory framework of nineteenth-century science (Wallace 1875a).

Despite the similarities of the phenomena of mesmerism and spiritualism, Wallace maintained there were two fundamental differences. First, mesmerized subjects never had any doubts as to the reality of what they saw, heard, or felt. Moreover, they generally lost their memory of how they fell into mesmeric trances during which they were transported from "a lecture-room in London . . . on to an Atlantic steamer in a hurricane, or [into] the recesses of a tropical forest." Participants at séances, in contrast, were always fully aware of events at every moment and often critically examined apparatuses and took notes of the proceedings. Second, whereas the mesmerizer had the power of acting only on "certain sensitive individuals," there was no limitation to the number of persons who simultaneously witnessed the mediumistic phenomena at séances. Visitors to "Mr. Home or Mrs. Guppy," Wallace noted, "all see whatever occurs of a physical nature, as the record of hundreds of sittings demonstrate." He added that even many skeptics of spiritualist philosophy nonetheless testified to observing manifestations at séances. Wallace's insistence on these important differences between mesmerism and spiritualism did not imply that he viewed one set of phenomena as more objective or "real" than the other. He considered the conflation between mesmerism and spiritualism to be yet another tactic by which certain scientists sought to discredit both fields. Wallace's distinction between the two was intended to promote independent verification of spiritualism and mesmerism (Wallace 1872a).

Edward B. Tylor enters the picture at this point. Wallace had just reviewed Tylor's two-volume *Primitive Culture: Researches into the Development of Mythology, Philosophy, Religion, Art, and Custom* (1871). He was impressed with Tylor's massive documentation on anthropological issues so close to his own studies. But he had strong reservations about Tylor's explanations of primitive beliefs and practices, notably those relating to animism or the "doctrine of souls." Wallace rejected Tylor's underlying thesis that history's long record of human preoccupation with supernatural or miraculous entities or events was readily explained by characterizing them as "mere belief." Even more galling to Wallace, was Tylor's terming contemporary spiritualist claims as "survival of old beliefs." These claims, Wallace objected, "have been recently investigated by Mr. Crookes and other Fellows of the Royal Society, and are declared to be realities by members of the French Institute, by American judges and senators, and by many medical and scientific men" in Britain. Wallace judged Tylor guilty of the same prejudice that marked the work of scientific naturalists such as Tyndall and Carpenter. He was irked by their refusal to concede that "the so-called supernatural is not all delusion, and that many of the beliefs of all ages [including the nineteenth century] classed as superstitions, have at least a substratum of reality." Wallace did not suggest that claims made by proponents of spiritualism (and mesmerism) be accepted uncritically. The opposite, in fact, was his objective. His purpose in writing "The Scientific Aspect of the Supernatural," and other similar pieces from the mid-1860s through the 1870s, was to encourage scientists to broaden the scope of what were considered appropriate areas of inquiry. Even so powerful a work as Tylor's *Primitive Culture*, Wallace argued, failed to come to grips with the fundamental question, "How much of truth is at the bottom of the so-called superstitious beliefs of mankind?" Wallace, at this point in his career, was fighting for the freedom of scientists to search for the "underlying *facts*" that might well explain phenomena that he felt were too readily dismissed as superstition. Science, as Wallace was coming to conceptualize it, had first to divest itself of the self-imposed constraints that precluded an objective investigation of all fields, however odd or unsettling. As he wryly noted, it was pertinent to recall that the history of science was replete with examples of theories dismissed at first merely because they were opposed to existing orthodoxy. For Wallace, the search for truth demanded that widely attested phenomena must, at least initially, "be recognised as possible realities and studied with thoroughness and devotion and a complete freedom from forgone conclusions." Otherwise, he admonished, it would be "hopeless to expect a sound philosophy of [physical nature or] of religion or any true insight into the mysterious depths of our spiritual nature" (Wallace 1872d).[11]

THE SCIENTIFIC ASPECT OF THE SUPERNATURAL

It is within this context of open inquiry that Wallace composed a succinct account of the accumulating evidence regarding spiritualism. Drawing on his own experiences at séances as well as the large literature composed on both sides of the Atlantic, he published *The Scientific Aspect of the Supernatural* in 1866.[12] Wallace prefaced his account by arguing that many events deemed miraculous or supernatural because they appear to run counter to laws of nature are, actually, "natural" and can be shown to involve no violation of natural process, broadly defined (Wallace [1905] 1969, 2:280). To brand certain events incredible because they are inexplicable on then known natural laws was, Wallace insisted, tantamount to maintaining that man has "complete knowledge of those laws, and can determine beforehand what is or is not possible." The history of science, however, demonstrates the progressive and cumulative character of human knowledge. Wallace noted that "the disputed prodigy of one age becomes the accepted natural phenomenon of the next and that many apparent miracles have been due to laws of nature subsequently discovered." Five hundred years ago, he declared, the effects produced by the telescope and microscope would have been called miraculous by those ignorant of the laws of optics. Just a century ago, he continued, "a telegram from three thousand miles' distance, or a photograph taken in a fraction of a second, would not have been believed possible, and would not have been credited on any testimony." Closing in for the kill, Wallace invoked a more recent and closely related example. At the start of the nineteenth century the fact that surgical operations could be performed on patients in a mesmeric trance without their apparently "being conscious of pain was strenuously denied by most scientific and medical men in [England], and the patients, and sometimes the operators, denounced as impostors." By the middle decades of the century, Wallace asserted, these phenomena were more generally credited and recognized as a consequence of "some as yet unknown law" (Wallace [1886] 1875, 39–40).

For Wallace, the phenomena of spiritualism presented an analogous case. They could be shown to follow, not contravene, the course of nature. To render these manifestations "intelligible or possible from the point of view of modern science" required, Wallace suggested, "the supposition that intelligent beings may exist, capable of acting on matter, though they themselves are uncognisable directly by our senses" (Wallace [1866] 1875, 42–43). The activities of these disembodied intelligences were consonant with "the grandest generalisations of modern science, [according to which] light, heat, electricity, magnetism, and probably vitality and gravitation, are believed to be but 'modes of motion' of a space-filling ether." That spirits, intelligences of an "ethereal nature," could act on ponderable bodies and produce the varied physical effects witnessed at séances was, to Wallace, a legitimate and plausible deduction. Wallace was not the only one advancing the hypothesis that the [alleged] ether might be the medium linking the material to the spiritual world. William Crookes and Oliver Lodge, among other scientists, maintained similar suppositions (Trusted 1991, 158–161). Invoking a venerable Enlightenment argument that the faculty of vision and existence of light and color would be inconceivable to a race of blind men, Wallace maintained that it is "possible and even probable that there may be modes of sensation as superior to all ours as is sight to that of touch and hearing" (Wallace [1866] 1875, 44–45). In a like vein, Wallace four years later (1870) tackled David Hume and his own contemporary William Lecky, the highly respected

writer of the *History of Rationalism* and the *History of Morals,* on the subject
of miracles. Wallace then felt secure enough to proclaim that "Hume's arguments against miracles are full of unwarranted assumptions, fallacies, and
contradictions . . . [and that] the philosophical argument so well put by Mr.
Lecky and Mr. Tylor, rests on false or unproved assumption, and is therefore
valueless." That Wallace would challenge figures of such prestige, and in fields
other than his own areas of expertise, might easily be dismissed as audacity
or folly. But the stakes for Wallace were now sufficiently high to warrant an
assault on the very pillars of skeptical thought. Otherwise, he warned, "history
will again have to record the melancholy spectacle of men, who should have
known better, assuming to limit the discovery of new powers and agencies
in the universe, and deciding, without investigation, whether other men's
observations are true or false" (Wallace 1870a).

Wallace's claim was never that the alleged phenomena of spiritualism
be accepted uncritically. He urged that they be accepted as matters "to be
investigated and tested like any other question of science." The thrust of *The
Scientific Aspect of the Supernatural* lay in the evidence adduced by "persons
connected with science art, or literature, . . . whose intelligence and truthfulness in narrating their own observations are above suspicion." Wallace was
particularly sensitive to the charge that his advocacy of spiritualism was influenced "by clerical and religious prejudice" and detracted from his authority
as a student of natural history. Although theism would later come to permeate
his evolutionary teleology, Wallace always maintained an aversion to conventional religious dogmas. It is in this specific context that he stated that until
the time of his first personal acquaintance with the facts of spiritualism he had
been a "confirmed philosophical sceptic, rejoicing in the works of Voltaire,
[David Friedrich] Strauss [whose influential *Life of Jesus* (1835) denied the
supernatural character of Jesus' career and contributed to the "higher criticism" of the Bible] and [the German materialist philosopher and zoologist]
Carl Vogt." It was not by any preconceived opinions, Wallace asserted, but
only "by the continuous action of fact after fact, which could not be got rid
of in any other way," that he was "compelled" to accept spiritualism. He
placed great weight on the testimony of Augustus De Morgan (the English
mathematician), Nassau William Senior (the political economist), William
Makepeace Thackeray (the novelist), and other eminent figures. They all
had reported witnessing authentic spirit manifestations as diverse as table
moving, communications by raps, clairvoyance, and the production of flowers and other objects at séances. Wallace cited one witness to the playing of
the "Last Rose of Summer" on an (apparently) unassisted accordion, "but in
so wretched a style that the company begged that it might be discontinued"
(Wallace [1866] 1875, vi–vii, 49, 53, 82–87, 95–98).

WALLACE'S "FABRIC OF THOUGHT"

Wallace's main purpose in writing *The Scientific Aspect of the Supernatural* was to encourage objective assessment of spiritualism's evidentiary claims. But the theoretical and especially moral implications of that doctrine had already begun to permeate his "fabric of thought." He concluded with a description of the hypothesis according to which "that which, for want of a better name, we shall term 'spirit,' is the essential part of all sensitive beings, whose bodies form but the machinery and instruments by means of which they perceive and act upon other beings and on matter." At death, the spirit quits the body but still retains "its former modes of thought, its former tastes, feelings, and affections." Wallace claimed that under "certain conditions disembodied spirit is able to form for itself a visible [and, in some instances, tangible] body out of the emanation from living bodies in a proper magnetic relation to itself" and thereby communicate to persons either directly or through the agency of mediums. The significance of these communications, for Wallace, derived not in their imparting any "knowledge to man which his faculties enable him to acquire for himself," but in their moral use. Spirit manifestations were incontrovertible evidence of the "reality of another world . . . and of an ever-progressive future state." He emphasized the continuity between the character of the embodied and disembodied spirit. Spiritualist continuity was "in striking contrast with the doctrines of [traditional] theologians, which place a wide gulf between the mental and moral nature of man in his present and in his future state of existence." Wallace chided those critics who scoffed at the trivial nature of some of the events witnessed at séances. Such trivialities, he remarked were hardly "to be wondered at, when we consider the myriads of trivial and fantastic human beings who are daily becoming spirits, and who retain, for a time at least, their human natures in their new condition" (Wallace [1866] 1875, 107–110, 124). The study of spiritualist phenomena seemed, above all, capable of providing insights toward "the partial solution of the most difficult of all problems—the origin of consciousness and the nature of mind." Wallace maintained that rather than being incompatible with evolution, spiritualism completed his biological theory. It accounted for those human attributes that he considered inexplicable by natural selection. Spiritualism, he asserted, was a striking supplement to the doctrines of modern science. The organic world has been carried on to a high state of development and has been ever kept in harmony with the forces of external nature, by the grand law of "survival of the fittest" acting on ever-varying organizations. In the spiritual world, the law of the "progression of the fittest" takes its place and carries on in unbroken continuity that development of the human mind that has been commenced here (Wallace [1866] 1875, vii–viii, 114–116).

Wallace had a hundred copies of *The Scientific Aspect of the Supernatural* printed separately and sent them to those of his colleagues—including Huxley, Tyndall, and the positivist George Henry Lewes—whom he hoped to persuade to take up the subject seriously. Tyndall read the pamphlet "with deep disappointment." He wrote Wallace that, while he saw "the usual keen powers of your mind displayed in the treatment of this question," he deplored Wallace's willingness to accept data that were "unworthy of [his] attention" (Wallace [1905] 1969, 2:280–81). Huxley, to whom Wallace had described spiritualism as "a new branch of Anthropology," replied that although he was "neither shocked nor disposed to issue a Commission of Lunacy against you," he remained completely disinclined to investigate the alleged phenomena (Marchant [1916] 1975, 418). Huxley's dismissal of the compiled evidence as "disembodied gossip," which interested him as little as did the more mundane variety, particularly rankled Wallace. It typified the indifference or derision with which a number of his scientific associates, particularly the scientific naturalists, regarded his efforts. More gratifying was the attitude of Robert Chambers and others who had reservations about the scientific naturalists' attempt to define science. On receipt of Wallace's pamphlet, Chambers wrote (10 February 1867):

> I have received your letter and your little volume. It gratifies me
> much to receive a friendly communication from the Mr. Wallace
> of my friend Darwin's "Origin of Species," and my gratification is
> greatly heightened on finding that he is one of the few men of science
> who admit the verity of the phenomena of spiritualism. I have for
> many years *known* that these phenomena are real, as distinguished
> from impostures; and it is not of yesterday that I concluded they were
> calculated to explain much that has been doubtful in the past, and
> when fully accepted, revolutionize the whole frame of human opinion
> on many important matters. . . . How provoking it has often appeared
> to me that it seems so impossible, with such a man, for instance, as
> Huxley, to obtain a moment's patience for this subject—so infinitely
> transcending all those of physical science in the potential results! My
> idea is that the term "supernatural" is a gross mistake. We have only
> to enlarge our conceptions of the natural, and all will be right.
>   (Wallace [1905] 1969, 2:285–286)

Later that year (1867), while attending the BAAS meeting in Dundee, Wallace visited St. Andrews University. There he met Chambers and "had the great pleasure of an hour's conversation with him in his own house" (Wallace

[1905] 1969, 2:280, 285–286). Wallace held Chambers's *Vestiges* in high regard when he first encountered it some twenty years previously. And the two now shared a positive estimation of both the evidentiary status and the potential broader cultural and moral significance of spiritualism. Chambers's path toward spiritualism predated Wallace's. Chambers had begun attending séances early in the 1850s. Skeptical at first, Chambers—like Wallace—was increasingly convinced by personal observation that séance phenomena were untouched by trickery. They represented authentic spiritual agencies at work. In 1859, the year the *Origin* appeared, Chambers had complained that spiritual phenomena suffered from the same skepticism that condemned the developmental law of progress (Secord 2000, 495). Chambers's parallel between the obstacles placed by segments of the scientific community to accepting evolutionism and spiritualism mirrored Wallace's views.

Chambers was also deeply impressed with the mediumistic powers of Home. At a séance in 1860, Home placed Chambers in touch with a spirit who claimed to be his father. As a test, Chambers asked that the spirit play, on an accordion lying on the floor, his father's favorite English tune. The accordion immediately played "The Last Rose of Summer," which Chambers exclaimed was his father's favorite melody. He became so convinced of the reality of spiritualist phenomena and the authenticity of Home that he agreed to give a sworn deposition in 1867 attesting to Home's irreproachable character in a lawsuit brought against the medium. Chambers's testimony was crucial to Homes's successful defense. Like Wallace, Chambers enunciated, from 1853 until his death in 1871, an increasingly theistic evolutionary teleology. From the mid-1850s onward, Chambers moved overtly from the dominant scientific naturalism of his earlier career toward a conviction in human/spirit interaction and a belief in personal immortality (Home [1888] 1976, 146–147, 268; Millhauser 1959, 174–185). Finally, and not unexpectedly, both Chambers and Wallace had a complex relationship with a certain sectors of the scientific community.

WALLACE'S STATUS IN THE SCIENTIFIC COMMUNITY

This complex relationship is central in assessing the role of spiritualism, and later theism, in Wallace's life and career. Wallace was establishing a niche in the professional scientific community on his return to London. He was never a typical Victorian scientist—the term itself is an oxymoron. But Wallace wanted to have not only his scientific achievements but his emerging broader worldview accepted, or at least not derided, by his scientific peers. Later, especially from the 1880s onward, when Wallace's thought and activities were more deeply correlated with his social, political and theistic convictions, he could afford to be less sensitive to hostile criticisms. But such Olympian calm could not descend on Wallace until he fought the battles of the 1860s and 1870s over the scientific status of spiritualism and mesmerism.

Wallace was not fighting a solitary battle against some fictitious monolithic scientific community. A number of his scientific colleagues found séances and various spiritualist activities as congenial as did Wallace. He was closely connected, socially and intellectually, through these spiritualist activities, convictions, and writings with a highly supportive network of friends and associates. This network included scientists as well as many other individuals from all segments of Victorian society (Oppenheim 1985). But during the 1860s and 1870s, Wallace was stung by the dismissive tone of the scientific naturalist faction regarding attempts to secure legitimacy for spiritualism. Fear of losing professional standing, significantly, was a dominant motive in Darwin's friend and disciple George John Romanes's reluctance to admit publicly to his sympathies with spiritualism (letter from Charles Carleton Massey to William F. Barrett, 24 December 1881, in WFBP). Their different approaches to professional status were but one of several factors that provoked controversy between Wallace and Romanes on the subject of spiritualism (Kottler 1974, 180–182). Wallace's spiritualism, as well as his sociopolitical views, meshed with his earliest conception of the nature of science and the philosophical and cosmological framework of evolution. He refused to conceal his spiritualist affinities. Wallace thrust himself to the center of the debates surrounding spiritualism's status in the scientific community.

MILL VERSUS DE MORGAN

Claims and counterclaims for the alleged phenomena of spiritualism abounded during the 1860s. Wallace had been "urged strongly to make a personal investigation of the subject . . . [by] H. W. Bates, and Professor E. B. Tylor," among others. He was especially desirous of learning, "first-hand . . . of the frame of mind of eminent men upon this subject." As might be expected, eminent Victorians differed on this, as on so many other matters. Two representative "frames of mind" were those of John Stuart Mill and the brilliant mathematician Augustus De Morgan. They display the split among the intellectual elite on this question. Mill had been sent a tract in which it was stated that he, along with Ruskin, Tennyson, and Longfellow had become believers in spiritualism. The tract had been sent by a Mr. N. Kilburn of Auckland, New Zealand, who asked Mill if it were true. In a letter to Kilburn (18 March 1868), Mill tersely replied: "It is the first time I ever heard that I was a believer in spiritualism, and I am not sorry to be able to suppose that some of the other names I have seen mentioned as believers in it are no more so than myself. For my own part I not only have never seen any evidence that I think of the slightest weight in favour of spiritualism, but I should also find it very difficult to believe any of it on any evidence whatever, and I am in the habit of expressing my opinion to that effect very freely whenever the subject is mentioned in my presence." Wallace was not aware of this specific letter of Mill until Kilburn sent it to him in 1874, "or I might have mentioned the subject when I dined with [Mill] in 1870." But Mill's views were well-known, so it is likely that Wallace understood where he stood on the matter much earlier than 1874. In his autobiography, Wallace commented on Mill's position in the context of the climate of the late 1860s. "If," Wallace noted, "by 'any evidence whatever' Mr. Mill meant testimony of others, I myself, and most spiritualists, were in the same frame of mind when we began our inquiries; but as he used the word 'evidence,' he no doubt included personal evidence, and to decide beforehand that he would not believe it is very unphilosophical." Thus Wallace on one of mid-Victorian Britain's foremost philosophers! Wallace quickly added that Mill "only says *difficult*, not *impossible*, and here . . . I quite agree with him." During this period of intense debate, Wallace also "had letters from other men of various degrees of eminence of a much more satisfactory nature" (Wallace [1905] 1969, 2:283–284).

De Morgan, when he received Wallace's 1866 pamphlet, wrote him a letter that revealed a quite different frame of mind from Mill's:

> I am much obliged to you for your little work, which is well adapted to excite inquiry. But I doubt whether inquiry by *men of science* would lead to any result. There is much reason to think that the state of mind of the inquirer has something—be it internal or external—to do with the power of the phenomena to manifest themselves. This I take to be one of the phenomena—to be associated with the rest in inquiry into cause. It may be a consequence of action of incredulous feeling on the nervous system of the recipient; or it may be that the volition—say the spirit if you like—finds difficulty in communicating with a repellent organization; or, maybe, it is offended. Be it which it may, there is the fact. Now the man of science comes to the subject in utter incredulity of the phenomena, and a wish to justify it. I think it very possible that the phenomena may be withheld. In some cases this has happened, as I have heard from good sources.

Wallace felt that De Morgan's letter depicted the typical "scientific frame of mind, as manifested by Tyndall, Lewes, and W. B. Carpenter, with great perspicuity" (Wallace [1905] 1969, 2:284–285). Wallace faced a dilemma. He was concerned in this phase of his career with his scientific standing. He was also aware that some influential segments of the emerging professional community were constructing a definition of science that would preclude spiritualism and related areas from the sphere of "proper science."

De Morgan was not a confirmed spiritualist. He was, however, part of that group of prominent scientists, including Lord Rayleigh, Joseph John Thomson, and Balfour Stewart, who were completely willing to examine the claims of spiritualism critically and without preconceived judgments. In the preface to his wife Sophia's book *From Matter to Spirit* (1863)—Sophia was an avowed spiritualist—De Morgan argued that rigorous experiments coupled with an open mind were the most appropriate tools a scientist could, and should, deploy to investigate spiritualist claims. "Thinking it very likely that the universe may contain a few agencies—say half a million—about which no man knows anything," he wrote, "I cannot but suspect that a small proportion of these agencies—say five thousand—may be severally competent to the production of all the [spiritualist] phenomena, or may be quite up to the task among them" (De Morgan 1863, v–vi). De Morgan had been annoyed by Michael Faraday's 1854 lecture "Mental Training," which suggested that scientists should investigate novel occurrences with appropriate preconceived notions of what was or was not possible in nature. Faraday's dictum on the nature of evidence in scientific demonstrations struck De Morgan as overly restrictive. The novel and controversial phenomena of séances, and he had attended many, demanded in De Morgan's opinion an open-mindedness and impartiality, not a predisposition to ridicule.

Balfour Stewart shared De Morgan's views. Stewart was active in many fields—chemistry, meteorology, astronomy, mathematics, and physics as well as natural history (he was director of Kew Observatory before he assumed the physics professorship at Owens College). He declared preconceived ideas to be the greatest obstacle threatening the advance of nineteenth-century science. Stewart was a staunch antimaterialist, as well as council member, vice-president, and finally president of the SPR. He maintained an independent approach to the study of possibly unknown forces at work in the universe. In *The Unseen Universe or Physical Speculations on a Future State* (1875), Stewart carefully refrained from endorsing spiritualism. The book's coauthor, P. G. Tait, had earlier spoken contemptuously of those who displayed too facile a credulity with respect to spiritualist claims. Stewart's and Tait's intention was to present a modern cosmology acceptable to scientists and theologians. Both men believed in the existence of an invisible world that was linked to the visible world by bonds of energy. Though fuzzy on what these bonds of energy were, the coauthors' fundamental purpose was to place the argument for design and continuity in nature in an explicitly theistic context. More to the point, Stewart's and Tait's theism was cast in a framework that they hoped would be recognizable, and acceptable, to their scientific colleagues (Oppenheim 1985, 330–338). Wallace had a similar goal. His copy of a later edition of *The Unseen Universe* is marked by favorable annotations (in ARWL). Wallace also was deeply interested in the possible explanatory value of forces or energies that might be linked to spiritualist phenomena. In a reply to Huxley's dismissal of spiritualism as "disembodied gossip," Wallace had written (1 December 1866) that he had no wish to pressure Huxley into experimental inquiries for which he had "neither time nor inclination." But Wallace pointedly reminded Huxley that while he had as little regard "for the 'gossip' you speak of . . . I do feel an intense interest in . . . the exhibition of *force* where force has been declared *impossible,* and of *intelligence* from a source the very mention of which has been deemed an *absurdity. . . .* I believe that I can now show such a force, and I trust some of the physicists may be found to admit its importance and examine into it" (Marchant [1916] 1975, 418–419). Wallace's position on the evidentiary basis for spiritualist claims can be contextualized by looking at the spectrum of scientific opinion regarding séances in the late 1860s/early 1870s.

WALLACE AND WILLIAM CROOKES

The road taken by Wallace is illuminated by looking at the concurrent investigations of William Crookes, one of Britain's greatest chemists and physicists. Wallace and Crookes had attended several séances together, including one conducted by Home. At that séance, in early 1871, Wallace witnessed some "most wonderful phenomena." Despite his own and others' scrutiny, Home's unrivaled mediumship came through unscathed. Crookes, Fellow of the Royal Society (and later its president from 1913 to 1915), assisted in the measurement of a table's levitation, again produced by Home. Wallace reported these events to his close friend Arabella Buckley. He added that "Mr. Home courts examination if people come to him in a fair and candid spirit of inquiry." Wallace urged Buckley to read an article by Crookes that had appeared in the *Quarterly Journal of Science,* outlining the results of Crookes's investigations regarding spiritualist phenomena. Wallace declared that Crookes's "facts are most marvellous and convincing, and appear to me to answer every one of the objections that have usually been made to the evidence adduced" (Marchant [1916] 1975, 420–421). Given that Crookes's researches were, in the early 1870s, regarded as even more controversial and suspect than Wallace's own, Crookes may not have been the wisest choice for Wallace's praise. The similarities between them are striking. Both

Wallace and Crookes lacked formal university education, had brilliant scientific achievements, showed audacity in areas where others feared to tread, and undertook a quest to uncover a unified cosmos in which "Matter and Force seem to merge into one another . . . [to constitute] Ultimate Realities, subtle, far-reaching, wonderful" (William Crookes's address "Radiant Matter," in D'Albe 1923, 290).

Crookes was fascinated by the phenomena of spiritualism and psychic forces throughout his long and distinguished career. His discovery of the element thallium (1861) had secured his election to the Royal Society in 1863. His most active involvement with spiritualism occurred during the period 1870–1875. If anything, Crookes's spiritualist investigations made him, for a time, a more puzzling figure to the scientific community than Wallace. Crookes's experimental expertise was recognized as extraordinary by his scientific colleagues. His intense study of séances included some of the most elaborate procedures and apparatuses for assessing the validity of spirit phenomena. Yet that scientific scrupulousness was vitiated, in the eyes of many of his critics, by two major factors. First, there were, and continue to be, questions about the precise nature of the relationship between Crookes and the famous, and very lovely, medium Florence Cook. The two participated in a long series of séances in 1873 and 1874, and speculation was prevalent that Cook was Crookes's mistress (Hall 1963). Such speculation naturally led to charges of collusion at those séances. These allegations were potent, since Crookes claimed that spirit manifestations were one of the surest planks on which spiritualism was tested and experimentally verified. The repeated appearance of "Katie King," whom Crookes swore he clasped in his arms and photographed on several occasions, were the most celebrated of Cook's manifestations.

The second critique of Crookes, one often leveled at Wallace, was gullibility. There is little doubt of his personal fondness for Cook. Her career received a tremendous boost from Crookes's prestigious participation at her séances. His complete faith in her honesty and integrity may possibly have rendered him a malleable accomplice rather than the staunchly objective observer he believed himself to be. Though the verdict is still out, there is no doubt that Crookes, like Wallace, truly believed in the authenticity of séance events (Oppenheim 1985, 340–343). Wallace was a staunch defender of Crookes's experimental successes. He objected to the skeptics' dismissal of "Katie's" repeated manifestations and photographs. Wallace was incensed by the chorus of charges that the phenomena described by Crookes and numerous other investigators (including himself) were merely the result of sophisticated trickery by skillful and duplicitous mediums. In a letter to Romanes, with whom he conducted a long and argumentative correspondence on the authenticity of séance phenomena and related touchy subjects, Wallace vented his anger. He complained bitterly that to "me, and I believe to most inquirers, it will appear in the highest degree *unscientific* to reject phenomena that could not possibly be due to imposture, and to ignore the hundreds of corroborative tests by other equally competent observers, and then, after this, to call all such observers (by implication) fools or lunatics!" Not surprisingly, Wallace also wrote Romanes protesting that the latter's "assum[ing], without any attempt at proof, that [Wallace's] writings on vaccination and land nationalization showed incompetence and absurdity was appealing to ignorant prejudice, and was therefore both unscientific and in bad taste" (Marchant [1916] 1975, 321, 323).

As was often the case in highly publicized efforts for experimental valida-tion, the séances Crookes conducted with Home between 1871 and 1873 had their passionate supporters as well as vehement critics. Some contemporaries described them as laying the "foundation stone" of evidential support for the reality of psychical phenomena (Podmore 1897, 53). During this period, Crookes attended numerous séances presided over by different mediums. But it was Home's abilities that impressed him most. Home's séances prompted Crookes to assert "the existence of a new force, in some unknown manner connected with the human organisation, which for convenience may be called the Psychic Force" (Barrington, Goldney, and Medhurst 1972, 22). Wallace also lauded Home's significance in lending powerful support to the spiritualist camp in their bitter debates with skeptics (Wallace [1905] 1969, 2:286–290).

There is one striking dissimilarity between Wallace's and Crookes's en-dorsement of spiritualism. Wallace continued throughout his life to defend the authenticity of practically all mediums. Crookes, in contrast, was alert to the extent to which deceit characterized the practice of some mediums. His experiments with Mary Showers in 1874 and 1875 revealed the pat-tern of fraudulence behind the manifestations reported at her séances. This experience was sufficiently traumatic to contribute to Crookes's decision to desist from active investigations of séances after 1875. Wallace, in contrast, remained ready to come to the defense of mediums, both intellectually and legally. This Wallace/Crookes difference was paralleled by a similar contro-versy between Wallace and William James. In chapter 3, the notable analo-gies between Wallace's and James's epistemological outlooks were demon-strated. Despite their multiple affinities, however, Wallace and James found themselves on opposing sides of the fence regarding the authenticity of some prominent mediums.

WALLACE CONTRA JAMES ON MEDIUMS

That Wallace and James had differences in their approach to the study of spiritualism and psychic phenomena was inevitable. Both fields had numerous and disputatious factions on each side of the Atlantic. One of the most notorious disputes concerned Mrs. Hannah V. Ross. Hannah, assisted by her husband Charles, gave a series of widely publicized séances in Providence, Rhode Island, and Boston during the period 1883–1885. The Rosses and their supporters were adamant that Hannah was particularly gifted at effecting "full-form materializations" of both children and adult spirit manifestations. Her controversial séances were widely covered in publications ranging from the *Banner of Light*—a spiritualist periodical published in Boston and the main organ of New England mediums—to the *New York Times*. Though her séances were often raided and subjected to on-site inspections by various local authorities, Hannah Ross continued to enjoy the confidence of both James and Wallace. James had even been cited in one *Times* article (4 February 1887) as having pronounced Mrs. "Ross among the wonders of the nineteenth century" (James 1986, 402). Further Ross séances, however, were soon to break her spell on James, although not on Wallace.

The two went together to sittings with Hannah in Boston on 27 and 28 December 1886. Wallace told James that he "should much like to join [him] in a private séance with Mrs. Ross either at her house or yours if you can arrange a party of say 10, half ladies." James refused to accord blanket affirmations as to the authenticity of mediums. Wallace knew of James's reservations and wanted to show them groundless, at least with respect to Hannah. An ardent witness for the defense at many trials of controversial mediums, Wallace wrote James that he considered allegations "of fraud on *mere suspicion*" as "unreasonable & unscientific. You ask for *facts & proofs* on our side, but offer only *suspicions* on your side." Eager to vindicate Hannah's reputation, Wallace himself arranged for a private séance at her house. He advised James not to "have *violent* skeptics at the party or any who would behave otherwise than at a friend's house. Have half ladies if possible, and as many who have some medium power or know *something* of the subject as possible. . . . Pray do not suggest a personal search of Mrs. Ross. It is both valueless & utterly unnecessary" (Skrupskelis and Berkeley 1992—, 6:184–186).

James agreed to Wallace's requests and the two joined more than a dozen ladies and gentlemen, including several regular contributors to the *Banner*, for the two sittings. While no spirit approached James or Wallace, several other sitters were favored with materializations. The only two participants in these particular séances who enjoyed scientific prestige were Wallace and James. Subsequent Ross séances came under increased scrutiny. During January to

April 1887, when Wallace had already left Boston for Washington, several of Mrs. Ross's sittings were subject to raids by "concerned citizens." She and her husband became the subjects of a much-publicized scandal. James refused to be drawn into the scandal but did allow a letter from him to be printed in the *Banner of Light* (10 February 1887). He now believed the "Ross gang" was guilty of "roguery," if not outright criminal misdemeanor but said little more (James 1986, 29–32). Wallace, true to form, actively entered the dispute on the side of the Rosses. He demanded that their critics adduce definite proof of fraud rather than merely suggesting it. In any event, both Rosses were arrested, but only Hannah's husband was brought to trial in May. He was found not guilty by the jury in early June 1887. Despite her husband's judicial "vindication," Hannah was never to regain her standing in the spiritualist community. By the 1890s, materializations—whether genuine or fraudulent—were no longer fashionable. More prominent were trance and medical mediums (James 1986, 402–405). The Ross episode, although it pitted Wallace against James on this particular issue, did not deter either of them from continued mutual interest in the broader empirical investigation of psychical phenomena. But it does point to a certain depth of conviction—or, in the harsher terms of Wallace's critics, gullibility—on Wallace's part with regard to spiritualism. His role in the notorious trial of "Dr." Henry Slade was characteristic and highlights the difference between Wallace's public image as a hard-core spiritualist and the more cautious public stances of Crookes and James.

THE SLADE TRIAL

Slade was one of the most skillful but unscrupulous of the many Americans who came to Britain to promulgate the spiritualist cause. He arrived in London in the summer of 1876. His reputation and abilities as an expert in obtaining spirit messages written on slate tablets were sufficiently impressive to baffle, initially, some of the more dubious psychical researchers. E. Ray Lankester (professor of zoology at University College, London) and Dr. Horatio B. Donkin (physician at Westminster Hospital) detected what they regarded as blatant if sophisticated trickery at Slade's séances. They wrote letters to the *Times* and proceeded to press charges.[13] The case was held at Bow Street, site of one of Victorian London's major police offices and magistrates' court, in October and attracted a great deal of attention. Charles Carleton Massey, a noted barrister and advocate of spiritualism (he served on the first council of the SPR in 1882), was counsel for the defense. Massey, completely convinced of Slade's innocence, argued eloquently before the magistrate, as did Wallace as a key witness. Wallace considered Massey "one of the most intelligent and able of the Spiritualists," and one "whose accession to the cause is due, I am glad to say, to my article ["A Defence of Modern Spiritualism" (1874)] in the *Fortnightly.*" Wallace supported Massey's protest against the Treasury Department's having taken up the prosecution. Slade was charged under the Vagrancy Act of illegally employing "certain subtle craft and devices to deceive and impose on certain of her Majesty's subjects." Wallace objected to the Treasury's involvement on the grounds that "it is an uncalled-for interference with the private right of [scientific] investigation in these subjects" (Wallace to W. F. Barrett, letter dated 18 December 1876, in Marchant [1916] 1975, 426).

Neither Massey's eloquence nor Wallace's testimony carried the day in this instance. Darwin and E. Ray Lankester testified against Slade in the trial (Milner 1996). Slade was convicted and sentenced to a three-month imprisonment. He was spared that humiliation when his conviction was overturned because of linguistic ambiguities in the wording of the specific statute in the Vagrancy Act under which he had been brought to trial. Slade immediately left England to return to the more comforting spiritual atmosphere of America. But his defenders, including Wallace and Massey, continued to proclaim their belief in the authenticity of his mediumship. Massey outdid Wallace this time in their crusade to rehabilitate Slade's reputation. He translated Johann Zollner's *Transcendental Physics* into English in 1880 as a tribute to Zollner's own efforts to restore Slade's reputation. During 1877–1878, Zollner, professor of physical astronomy at Leipzig, investigated Slade. Unable to detect any deceptive techniques on Slade's part, although many other investigators had done so, Zollner published a laudatory account of Slade's mediumistic powers. The evidentiary claims of spiritualists would continue to embroil Wallace and a host of eminent persons on both sides of the bitter divide into the beginning decade of the twentieth century (Oppenheim 1985, 22–23, 31–33).

Despite his skepticism regarding the activities of certain mediums, Crookes remained fascinated by the phenomena associated with spiritualism. He became involved with the SPR, serving first on its council and then as its president in the late 1890s. He also joined the Theosophical Society (in 1882) and the Ghost Club. Wallace regarded theosophy as bordering on the irrational. He was upset by Crookes's espousal of it (Marchant [1916] 1975, 432–433). Crookes finally turned his attention away from physical manifestations to what he considered the less tainted and more promising studies of mental phenomena, such as telepathy. He also became overtly hostile, after 1875, to the spiritualist claim of communion with the dead—until, that is, the death of his wife in 1916. After sixty years of marriage,

Lady Crookes's passing left her bereft husband "prostrated with grief." Her death was sufficient to cause the elderly scientist to embrace a deep, totally uncritical spiritualist faith. Crookes resumed attending séances explicitly for the purpose of witnessing (the now highly comforting) physical manifestations. He was persuaded that his wife was often close by him at these sessions. Crookes also sought out the services of the "spirit photographer" William Hope, who duly captured Lady Crookes's spirit on film in December 1916. Crookes wrote an enthusiastic letter to Lodge reporting this treasured event, to which Lodge gently replied that Hope had a reputation for trickery and tampering with photographic plates. Crookes remained convinced that he had in his possession a spirit photograph of his late wife (Oppenheim 1985, 344–352).

The loss of a loved one functioned in a similar manner to reinforce Wallace's spiritualist convictions. The death of his sister Fanny, with whom he has always been extremely close, in 1893 was a severe blow to Wallace. Two years later, his brother John died in California (Raby 2001, 263). It seemed "unnatural and incredible that the living self with its special idiosyncrasies [one has] known so long . . . should (as so many now believe) have utterly ceased to exist and become nothingness!" For Wallace, death made one feel, "in a way nothing else can do, the mystery of the universe." With all his vast knowledge of and belief in spiritualism, Wallace still felt "occasional qualms of doubt, the remnants of [his] original deeply ingrained scepticism." Death of friends and family served as a powerful antidote to such qualms. Wallace took comfort in the support that his "reason" lent to the psychical and spiritualistic phenomena that demonstrated "that there *must* be a hereafter for us all" (Wallace to J. W. Marshall, letter dated 6 March 1894, in Marchant [1916] 1975, 436).

The comparison of Wallace and Crookes reveals much about the battles waged over spiritualism in the Victorian era. Both men approached the subject with an openness that—precisely because of their impeccable scientific credentials—infuriated those colleagues intent on securing elevated social and institutional status for the scientific community. Crookes and Wallace were perceived, in some quarters, as threats to the hard-won authority of science. Both, however, were sincere in their efforts to extend the domain of legitimate scientific inquiry. And both were subject to rebuke for this. The most common charge leveled against Crookes and Wallace was that they were woefully fragmented personalities. Carpenter attempted to paint both men with the same brush. He declared that Wallace and Crookes had a tragic duality that led them, all too often, to abandon their rational halves and give free rein to their "other" persona. It was the other half of their split personalities,

Carpenter insinuated, that led to their "lapses" into mysticism and credulity. Carpenter was particularly malicious in characterizing Crookes's spiritualist interests (Carpenter 1871, 1877b).

Crookes treated Carpenter's criticism with levity; Wallace felt such accusations merited serious reproach and rebuttal. Crookes teasingly urged Carpenter to use appropriate scientific terminology in describing the alleged dual personalities. He wanted his persona to be designated the "Ortho-Crookes and Pseudo-Crookes" (Crookes 1877a, 1877b). Wallace's response to Carpenter's accusation, and similar ones by Lewes, Tyndall, and Romanes, that "there were two *mental* natures in Crookes and Wallace—the one sane and the other lunatic!" was more serious. He completely rejected the notion that scientists who sympathetically investigated spiritualism were straying from the course of rigorous methodology and accurate empirical procedures. He proclaimed that neither he "nor any other well-instructed spiritualist" expected those reading their fully documented reports to become facile converts. What they did demand, Wallace insisted, was that skeptics entertain "doubt of their own infallibility on this question; we ask for inquiry and patient experiment before hastily concluding that we are, all of us, mere dupes and idiots as regards a subject to which we have devoted our best mental faculties and powers of observation for many years" (Wallace [1905] 1969, 2:318, 350).

THE "OTHER WALLACE"?

The other Wallace thesis plagued Wallace during his lifetime. It continues to be a standard characterization used by many historians. That characterization is, simply, misguided. There was no "other Wallace." He was an integrated personality whose worldview incorporated diverse fields and synthesized them into a comprehensive and compelling framework. Wallace was gullible on more than one occasion. But such gullibility did not detract from his ability to pursue an overarching goal: a theistic evolutionary teleology that melded science, politics, ethics, and social reform. The "two Crookeses" thesis has also proved tenacious. But that thesis has now been shown to be untenable. No more than Wallace was Crookes a schizoid personality. Crookes's life, however fragmented certain aspects might at first appear, was motivated by a coherent goal. Crookes's science and studies of the occult, his physical and psychic experiments, were "integral, intertwined parts of his lifelong fascination 'with other conditions of existence than the familiar.' They shared a place in his efforts to elucidate the still mysterious forces that shaped the universe" (Oppenheim 1985, 352–353).

Wallace had many allies in his serious interest in investigating spiritualism and publicizing the claims associated with it. Britain was awash in broad public and scientific support of that doctrine. Radicals, freethinkers, feminists, scientists, and other intellectuals embraced one or another of the versions of spiritualism circulating in Europe and North America in the 1860s and later (Oppenheim 1985; Barrow 1986; Owen 1990). Wallace's initial interest in spiritualism soon became part of his broader forays into sociopolitical activities. Like his mentor Robert Owen, and many other Victorian spiritualists, Wallace saw spiritualist teachings as having important ramifications for social and political reform (see chap. 5).

Spiritualism and a growing commitment to a theistic conception of nature posed little difficulty for Wallace emotionally or intellectually and, in fact, actually influenced his maturing articulation of the scope of evolutionary biology. Recent scholarly studies have shown that important groups existed within the ranks of professional scientists in the Victorian period whose members fully endorsed the notion that there was an integral religious dimension to science. The "North British" physicists, such as William Thomson, James Clerk Maxwell, and Peter Guthrie Tait (Smith 1998), the Christian Darwinists (Moore 1979; Livingstone 1987), and idealist natural philosophers such as T. H. Green, F. H. Bradley, and Edward Caird all appropriated science and evolutionary theory to construct theistic metaphysical systems (Otter 1996). They shared Wallace's interests in theistic and spiritualist matters and would not have considered it unscientific to do so. When the topography of Victorian science is viewed not merely from the perspective of Huxley and the scientific naturalists but also from the broader intellectual landscape, theistic science is recognized as a powerful paradigm in Wallace's era (Lightman 2001).

SPIRITUALIST NETWORKS

The number of people interested in spiritualism in Victorian Britain was sufficiently large to have fostered an extensive institutional network. There were newspapers, magazines, and organizations to advance spiritualist causes. From the 1850s onward, more than two hundred groups devoted to various aspects of spiritualism came into existence. These included London-based associations as well as numerous provincial societies. Such groups provided the institutional setting for meetings, debates, social contact, and séance rooms for spiritualists and psychical researchers. Membership figures for these organizations do not in themselves provide accurate figures for the total number of individuals actively involved in spiritualist pursuits. Working-class spiritualists, for example, could often not afford the membership fees (sometimes as high as a guinea) to join either London or provincial groups. Instead, they used places like the Mechanics' Institutes or their own homes to share their spiritualist interests. Fees were not the only barrier to joining established groups. A number of middle-class spiritualists, who could easily afford membership fees, preferred the privacy of their own homes so that they would not jeopardize their jobs or their standing in the community by open identification with explicit organizations. Spiritualism in Victorian Britain was pervasive, despite not always being deemed completely respectable. But the number of spiritualists, like Wallace, who "came out of the closet" by far exceeded those who remained covert. The precise number of spiritualists remains an elusive calculation for the historian. The methodology of counting suitable heads is compounded by other factors. The spiritualist population was a constantly shifting one. For some participants, spiritualism was a temporary phase, on their journey to other theological or metaphysical frameworks. The labeling of sociocultural categories is always a tricky proposition. Were Swedenborgians and theosophists, for example, spiritualists? Or were they breeds apart? (Oppenheim 1985, 50–51). The total number of active British spiritualists, therefore, can only be approximated. Estimates range from 10,000 to 100,000 (Gauld 1968, 77).

When numbers from the rest of Europe and North America are included, it becomes indisputable that spiritualism was a potent force in Victorian culture. Wallace's emphasis on the political ramifications of spiritualism was not uncommon. As early as the 1850s, the spiritualist society in the borough of Keighley "combined Owenite socialist ideas with preparations for the millennium" (Harrison 1969, 251). Wallace's conversion to socialism came later, with his reading of Edward Bellamy's *Looking Backward* ([1888] 1960). But his conviction that social reform was inseparable from the moral transformation of society permeated his worldview as it did that of most Owenites (Royle 1998). Wallace regarded Owen as his "first teacher in the philosophy of human nature and [his] first guide through the labyrinth of social science." He considered Owen the greatest of all nineteenth-century social reformers. Wallace wrote a perceptive critique of Owen's life, emphasizing the ambivalent results of the massive "experiment" at New Lanark. For twenty-six years, Owen managed the workers employed in the mills of New Lanark. He attempted to implement his theories on education, character development, and societal change among them. Significantly, Wallace attributed the ultimate failure of New Lanark not to any defect inherent in Owenite philosophy and practice. He blamed, instead, Owen's decision to give up the New Lanark property and spend his fortune and the remaining years of his life in attempting to establish too many similar communities in too many different countries. Given his own varied pursuits, Wallace was, in the end, a more focused individual than his revered Owen (Wallace [1905] 1969, 2:91–105).

Efforts to form a spiritualist organization on a national scale were mounted in Britain from the 1860s onward. Controversy quickly arose over the question of whether a national organization would be inimical to the spontaneity and intimacy that many local groups felt were essential to their practice of spiritualism. Bureaucracy, such as that required by a national organization, was deemed incompatible with the essence of their spiritualist faith by many (Nelson 1969, 103, 111–116, 119). But Victorian Britain was an organized and organizing society, and spiritualists gradually followed suit. The path to national organization was fraught with turf wars, fierce personality conflicts, and tensions between the London metropolis and the provincial periphery. British spiritualists were eager to attend social evenings as well as hear innumerable talks on their public duties. Wallace lectured and wrote on this latter issue. Spiritualist followers participated at séances, collected reports of phenomena, and helped stock spiritualist libraries. Gradually, national organizations, some more durable than others, developed. In August 1873, the British National Association of Spiritualists was formed at a meeting in Liverpool, sponsored by the local Psychological Society. In 1875 it moved to commodious London headquarters at 38 Great Russell Street. Its membership then surpassed 400, a sizable total for a spiritualist society at that time. But financial and other troubles plagued the BNAS. Members were negligent about paying their fees, and it was becoming increasingly expensive to hire professional mediums for séances. By 1882, the BNAS folded. Stainton Moses, the autocratic head of the BNAS, recognized that a greater concession to democratic structures of governance was mandatory if a new organization were to be made viable. The London Spiritualist Alliance (LSA) was the result. Launched in 1884 by a somewhat chastened Moses and a number of his close followers, the LSA enjoyed a wider network of support—Wallace was a member—and continues to exist today (as the College of Psychic Studies). Perhaps the most important organization to emerge from this period was the Society for Psychical Research (SPR), which held its first meetings in 1882. Although a number of prominent spiritualists were among the founding members, the SPR marked the emergence of a new kind of national organization. Its membership quickly came to include prominent professors, MPs, and fellows of the Royal Society—a feat unmatched by any previous spiritualist society. The SPR's *Journal and Proceedings* was markedly more professional and prestigious than any organ of the existing spiritualist press. Despite the patina of authority and respectability, the SPR was intimately involved with the world of London spiritualism. This intimacy was to influence the history of the SPR well into the early decades of the twentieth century (Oppenheim 1985, 51–57). Wallace's complex relationship with the SPR mirrored the intricate yet tense interaction between spiritualism and the emerging field of psychical research. It also was a stage in his personal journey toward a theistic evolutionary teleology.

WALLACE'S REVISIONIST SURPRISE: A HISTORIOGRAPHIC MYTH

Wallace's perceived public position as a leading proponent of natural selection as the dominant agent in human evolution was altered abruptly in April 1869. In a review of two new editions of geological treatises by Lyell, Wallace announced that man's intellectual capacities and moral qualities were not explicable by natural selection. As unique phenomena in the history of life, they (as well as certain physical attributes) required the intervention at appropriate stages of "an Overruling Intelligence." Wallace declared that this nonmaterial agency "guided the action of those laws [of organic development] in definite directions and for special ends" (Wallace 1869c). Wallace's response to Lyell's tenth edition of the *Principles of Geology* (1867–1868) was doubly significant. He applauded Lyell's long-awaited endorsement of evolutionism. The public statement of his own views on man, moreover, paralleled (though for different reasons) Lyell's extreme reservations concerning the role of natural selection in human evolution. Darwin was, not surprisingly, disappointed with both Wallace and Lyell (Bartholomew 1973, 300–303).

Ironically, it was Darwin's statement of the principle of utility that Wallace invoked to substantiate his claim of the limitations of the scope of natural selection. In the *Origin,* Darwin (fig. 8) argued that natural selection could produce neither a structure harmful to an organism nor a structure that was of greater perfection than was necessary for an organism at the particular stage of its evolutionary development (Darwin [1859] 1964, 201–202). Citing the culture of the "lowest savages" and, by implication, man at more remote periods in his history, Wallace maintained that the utility principle precluded natural selection as the agent responsible for four characteristic human features: the brain, the organs of speech, the hand, and the external form of the body. The brain of savages, Wallace noted, is of practically the same size and complexity as that of the average European. Savages could, under appropriate cultural conditions, be capable of the outstanding intellectual achievements of civilized man. Yet, the mental requirements of the lowest savages are "very little above those of many animals." His highly developed brain, Wallace concluded, was an organ of greater perfection than necessary for survival. According to the utility principle, natural selection would have provided the savage with an intellect only slightly superior to that of the apes. It cannot, Wallace emphasized, explain the complexity of the savage's brain. The hand of the savage is, similarly, an organ of greater refinement than

Figure 8. Charles Darwin a
year before his death (from
an 1881 photograph by
Elliott and Fry).

required and could not have been produced by natural selection alone. Man's
highest civilized accomplishments—art, science, and technology—were dependent on "this marvellous instrument." For Wallace, the savage's perfect
hand was evidence of provision by a higher intelligence of an organ that
would be fully utilized only at a later stage in human development. The erect
posture of the savage (and prehistoric man), "his delicate and yet expressive
features, the marvellous beauty and symmetry of his whole external form,"
are additional examples of modifications Wallace claimed were of no physical
use to their possessors. Early man's (comparative) nakedness, he suggested,
was disadvantageous. Wallace argued, again, for intelligent intervention and
provision in the evolutionary process. "The supreme beauty" of the human
form and countenance, though initially of no practical use, had (probably)
been the cause of man's aesthetic and emotional qualities. Wallace believed
these qualities could not have arisen if humans had retained the appearance of
an erect gorilla. He further suggested that human nakedness, "by developing

the feeling of personal modesty, may have profoundly affected our moral nature." Wallace applied analogous reasoning to the complex and delicate physical and mental apparatuses responsible for human speech, which appeared in advance of the needs of their possessors (Wallace 1869c).

The 1869 review concluded with the proposition that a "new standpoint [was possible] for those who cannot accept the theory of evolution as expressing the whole truth in regard to the origin of man." Wallace was careful to declare that the higher intelligence was consonant with the teachings of science. He now invoked domestic variation, the very analogy he had criticized Darwin for using so extensively in the *Origin*. Wallace stated that just as man had used the laws of variation and selection to produce fruits, vegetables, and livestock, so also "in the development of the human race, a Higher Intelligence has guided the same laws for nobler ends." In both cases, the "great laws of organic development" had been adhered to, not abrogated. Natural selection had been supplemented by conscious selection. In human evolution, Wallace concluded, "an Overruling Intelligence has watched over the action of those laws so directing variations and so determining their accumulation, as finally to produce an organization sufficiently perfect to admit of and even to aid in, the indefinite advancement of our mental and moral nature" (Wallace 1869c).

The majority of Wallace scholars have interpreted his 1869–1870 views as representing a volte-face with respect to his previous conceptualization of evolution. This presumed radical shift is usually attributed to Wallace's growing involvement with spiritualism in the period 1865–1870. If, however, his thoughts and writings from 1845 to 1870 are analyzed within the broader framework of Wallace's holistic approach to human evolution, a different picture emerges. Philosophical, ethical, and sociopolitical concerns had always informed his biological investigations. Wallace's modifications of certain causal explanations of human evolution were developments from, not repudiations of, his earlier, preliminary hypotheses. To use a geological metaphor, Wallace was not an intellectual catastrophist but an intellectual uniformitarian. If it is argued a priori that spiritualism and evolutionary science represent mutually exclusive conceptual schemes, then Wallace's acceptance of spiritualism would readily be seen as the cause of his rejection of natural selection in explaining man's higher faculties. Wallace's basic approach to the study of man and nature, however, was set in his mind well before he finally hit on natural selection. He maintained a consistent but evolving overall worldview in his writings over a span of seventy years. Neither natural selection nor spiritualism was a departure from this central vision. Momentous as the discovery of natural selection had been, Wallace was skeptical as to its competence to explain all of human evolution. He envisioned some additional explanatory model to resolve fully the question of human origins, their higher faculties, and their future evolution.[14]

WALLACE ON CAUSALITY

To understand why Wallace had from the outset of his career been receptive to the idea that other than strictly mechanistic forces were operative in the history of human evolution, it is helpful to discuss briefly his notion of causality. Wallace rejected, at least as early as the mid-1840s, the simple model of causality offered both by creationist theology and Lamarckian biology. He viewed the creationist argument that God had specially provided all the earth's creatures (including humans) with just what they needed to survive as equivalent to a first cause that explained nothing because it could explain everything. Wallace found equally wanting the goal-centered Lamarckian model that emphasized the immediate causes and specific effects of organic change. His reading of Chambers's *Vestiges of Creation* had deepened Wallace's belief that causation had to be conceived as more encompassing in its operation than merely acting to meet the immediate material needs and/or conscious desires of each individual organism. Put otherwise, Wallace from at least the mid-1840s began to think of evolution as progressive in the sense that such progress represented movement toward a system-level goal: the development of higher, "godly," beings. In the 1840s and 1850s, therefore, an intoxicating brew of ideas from Lyell, Owen, and Spencer as well as Chambers fermented in Wallace's brain. The result was a notion of causality that provided him with a teleology that encompassed biological, societal, and spiritual/psychic evolutionary change. The seeds were sown for what would become one of the cornerstones of Wallace's lifework: the enunciation of an evolutionary cosmology that incorporated an overriding "general design" of nature calling for "a model of its productions recognizing not merely the place of material things within it, but: (1) man's emotional and intellectual response to material things, and (2) the possibility of higher causes altogether" (Smith 1992, 20–30). As discussed in chapter 3, Wallace's evolutionary philosophy bore strong resemblance to ideas embraced in the pragmatism of James and Peirce. In particular, Wallace endorsed the view that formal languages such as logic and mathematics were intimately tied to human social systems of belief—subject, of course, to the proviso that such beliefs were consistent with the reality of the senses (Menand 2001, 199–200, 228–230, 356–358). Thus, in a causal hierarchy, "Will existed prior to force, which itself was prior to matter (and thus 'nature'). Causal continuity [for Wallace] was best addressed in terms of will, not matter. . . . The brain, for example, was not to be construed as the 'cause' of conscious awareness; rather, it was a structure that had evolved pursuant to consciousness." Wallace's exposure to the precepts and evidentiary claims of spiritualism in the 1860s, therefore, confirmed certain fundamental ideas about human biological and social evolution he had entertained since the mid-1840s (Smith 1992, 37–39).

Wallace's 1856 essay "On the Habits of the Orang-Utan of Borneo" sheds further light on his early views on evolutionary cosmology. In discussing the characteristics of this fascinating creature—so close in appearance to the human form—which he encountered in his Malay travels, Wallace challenged conventional wisdom about the huge canine teeth of the male orang. Naturalists generally assumed that the teeth were used for the purposes of defense against the tigers, bears, and other carnivorous animals of the tropical forests. Wallace's close scrutiny of the orang's habitat and activities convinced him that the canine teeth served no such purpose. Tigers "cannot climb trees, and [are] therefore quite unable to attack the orang, which never need descend to the ground, and very rarely does so." To emphasize the point, Wallace noted that in the rare event that a tiger did attack an orang, he would do so stealthily and from behind. "Let us imagine," Wallace continued, "a tiger springing upon the back of an orang who was walking on the ground; what could the animal possibly do, with those fearful claws deep in his back and shoulders, and those tremendous teeth firmly fastened in his neck? The vertebrae would probably be broken . . . [and] the tiger, knowing the strength of its prey, would be sure to strike at a mortal part, or obtain such a hold as could not be shaken off." Moreover, the native Dyaks were unanimous in telling Wallace that the orang, which lives solely and exclusively on fruits or other soft vegetable food, never either attacks or is attacked by tigers, bears, or other large predators (Wallace 1856b, 26–29). Wallace's point is starkly clear:

> Do you mean to assert, then, some of my readers will indignantly ask, that this animal, or any animal, is provided with organs which are of no use to it? Yes, we reply, we do mean to assert that many animals are provided with organs and appendages which serve no material or physical purpose. The extraordinary excrescences of many insects, the fantastic and many-coloured plumes which adorn certain birds . . . the colours and infinitely modified forms of many flower-petals, are all cases, for an explanation of which we must look to some general principle far more recondite than a simple relation to the necessities of the individual. We conceive it to be a most erroneous, a most contracted view of the organic world, to believe that every part of an animal or of a plant exists solely for some material and physical use to the individual, . . . to believe, in fact, that we

know the one sole end and purpose of every modification that exists
in organic beings, and to refuse to recognize the possibility of there
being any other. Naturalists are too apt to *imagine,* when they cannot
discover, a use for everything in nature: they are not even content to
let "beauty" be a sufficient use, but hunt after some purpose to which
even *that* can be applied by the animal itself, as if one of the noblest
and most refining parts of man's nature, the love of beauty for its
own sake would not be perceptible also in the works of a Supreme
Creator.

(Wallace 1856b, 29–32)

SIGNIFICANCE OF THE ORANGUTAN ESSAY

Wallace's 1856 essay clearly indicates that he was already thinking in terms
of a broader evolutionary teleology that had as one corollary the centrality of
"some general design which has determined the details, quite independently
of individual necessities." He declared: "The separate species of which the
organic world consists being parts of a whole, we must suppose some dependence of each upon all." He had already embarked on a program of following
the "indications of a general system of nature, by a careful study of which we
may learn much that is at present hidden from us." More pointedly, Wallace
announced that he believed "that the constant practice of imputing, right or
wrong, some use to the individual, of every part of its structure, and even of
inculcating the doctrine that every modification exists solely for some such
use, is an error fatal to our complete appreciation of all the variety, the beauty,
and the harmony of the organic world" (Wallace 1856b, 30–31). In 1856, of
course, this program was just beginning to take shape in Wallace's thought.
He would spend the remainder of his career in gradually developing the full-
blown teleological evolutionary cosmology that found its full articulation in
his later works, notably *Man's Place in the Universe* ([1903] 1907) and *The
World of Life* (1910a). (See chap. 6.) But many of his articles, reviews, and
books from the late 1860s onward provide abundant evidence that natural
selection was subsumed under a more comprehensive evolutionary design
(Smith 1992, 25 n. 83). A note appended to the 1856 orang essay amplifies
Wallace's reference to "the works of a Supreme Creator":

The talented author of the "Plurality of Worlds" [William Whewell]
has some admirable remarks on this subject. He says, "In the structure of animals, especially that large class best known to us, vertebrate animals, there is a general plan, which, so far as we can see,
goes beyond the circuit of the special adaptation of each animal to

its mode of living; and is a rule of creative action, in addition to the
rule that the parts shall be subservient to an intelligible purpose of
animal life. We have noticed several phenomena in the animal kingdom, where parts and features appear rudimentary and inert, dis-
charging no office in their economy, and speaking to us not of purpose, but of law." Again: "And do we not, in innumerable cases, see
beauties of colour and form, texture and lustre, which suggest to us
irresistibly the belief that beauty and regular form are rules of creative agency, even when they seem to us, looking at the creation for
uses only, idle and wanton expenditure of beauty and regularity? To
what purpose are the host of splendid circles which decorate the tail
of the peacock, more beautiful, each of them, than Saturn and his
rings? To what purpose the exquisite textures of microscopic objects,
more curiously regular than anything which the telescope discloses?
To what purpose the gorgeous colours of tropical birds and insects,
that live and die where human eye never approaches to admire them?
To what purpose the thousands of species of butterflies with the gay
and varied embroidery of their microscopic plumage, of which one
in millions, if seen at all, only draws the admiration of the wandering
schoolboy? To what purpose the delicate and brilliant markings of
shells which live generation after generation in the sightless depths of
ocean? Do not all these examples, to which we might add countless
others, prove that beauty and regularity are universal features of the
work of Creation in all its parts, great and small?"

                                                    (Wallace 1856b, 30–31)

The 1856 essay documents that Wallace was thinking in terms of "higher
causes," rather than only proximate causes, as part of the explanatory framework for his evolutionary cosmology. After his discovery of natural selection
in 1858, Wallace had a major and potent model of evolutionary change for
articulating many aspects of this developing cosmology. But the solution to
the thorny question of the origin of the higher human faculties could not, for
Wallace, be given entirely in terms of natural selection. When, after 1863, he
began to write more explicitly, and publicly, about human evolution, he now
saw pieces of the puzzle coming together. The stage was set for Wallace's integration of spiritualism with evolution. He regarded spiritualism and natural
selection as complementary components of a larger evolutionary teleology.
But he appreciated the fact that many of his scientific colleagues—notably
Huxley, Darwin, Tyndall, and Carpenter—would regard them as mutually exclusive. For tactical reasons, he chose to emphasize the utilitarian objections
against the total efficacy of natural selection in the 1869 review. He hoped they

would be read as a scientifically less contentious analysis of the limitations of natural selection than an overtly spiritualist critique would have been. In one sense, Wallace's tactic succeeded. A number of biologists, already dubious of the explanatory potential of natural selection as the sole mechanism of evolution, recognized the force of his utility critique (Bowler 1983, 28). For the next twenty years, Wallace publicly maintained that a utilitarian analysis was a major basis for his critique of natural selection in human evolution. However, he increasingly adduced theism, in addition to spiritualism, to explain fully man's unique features. The teleological and theistic imprint on his evolutionary views is substantiated by the fact that Wallace remained throughout the rest of the century a staunch advocate of natural selection as the main agent of animal and plant evolution. But theism assumed an important role in Wallace's reconceptualization of the scope and efficacy of natural selection (Fichman 2001b). In *Darwinism* ([1889] 1975), when he conceded that natural selection could account for many of the unique physical features of man, he still rigorously exempted human moral and intellectual qualities from its sway (Wallace [1889] 1975, 455; Kottler 1974, 162, 188–92). As we have seen, this exemption had its roots in Wallace's evolutionary speculations, and writings, from the mid-1840s to the mid-1860s (Smith 1992, 29–30, 48n. 163). His early thinking—most notably in the 1856 orangutan essay—on the need to explore, without prejudice, a wide range of causal agencies in human, as well as nonhuman, evolution matured into the evolutionary cosmology that Wallace expounded with increasing conviction.

National styles may also have played a role in Wallace's original emphasis on a utilitarian critique. Certain sectors of the British scientific community were uncomfortable with theistic evolution. In the United States, in contrast, many leading evolutionists, such as Gray and James Dwight Dana, maintained that the "biological solution does not exclude the theological." Wallace's theistic position in the 1870s and beyond was viewed as buttressing their contention that the hypothesis of evolutionary descent was fully compatible with the conviction "that humanity bore the image of God." Significant numbers of American scientists regarded Wallace's (and others') evolutionary theism as confirming the Christian belief "that the elements attesting to the special relationship between God and human beings resided in the fact that the human species possessed attributes—self-consciousness, reason, the moral sense, free will, and religiosity—that were different in kind from those of all other animals" (Roberts 1988, 176–177). One must be careful, however, in overemphasizing national differences. Overtly theistic statements on evolution in the United States became less common in the closing decades of the century. But this does not imply that theistic evolution was being rejected. As Ronald L. Numbers has recently suggested, "theistic evolution was undergoing privatization more than elimination." Although references to the divine became less visible in the scientific literature toward the close of the Victorian period, many American evolutionists retained their religious views (Numbers 1998, 40). Since Wallace had never adhered to any traditional institutionalized religion, he escaped the crisis of belief that afflicted so many of his contemporaries on both sides of the Atlantic. Theism, therefore, held no legacy of trauma for him. Consequently, Wallace could become more, not less, overtly theistic in the latter decades of his life. He grew more confident in expressing publicly views that he chose to deemphasize in the 1860s and early 1870s.

The 1869 review stands as a public watershed in Wallace's career. It was, as Darwin noted, an "inimitably good" exposition of natural selection, but one that concluded with those few remarks on man that made him "groan." Wallace expected Darwin's and others' reactions with "regard to my 'unscientific' opinions as to Man, because a few years back I should myself have looked at them as equally wild and uncalled for" (Marchant [1916] 1975, 199–206). Wallace's depiction of his views as "unscientific" is intentionally ironic. It is also historiographically important. Wallace had never deemed unaided natural selection to be a sufficient mechanism for all aspects of human evolution, nor that of certain other animals. His spiritualist writings in the several years prior to the Lyell review emphasized further the need to posit auxiliary agencies to explain human higher faculties. Whether Wallace's theistic rendition of evolutionary theory falls within the category of scientific or nonscientific concepts is related to current debates among historians of science and religion. In the rich and ambiguous context of Victorian philosophies of nature, a demarcation between the two categories was (and remains) elusive (Fichman 1997). Wallace was articulating the personal as well as metaphysical tensions that accompanied the rise of professionalized science. His evolutionary teleology became clearer still the following year, with the publication of *Contributions to the Theory of Natural Selection* ([1870] 1891).

CONTRIBUTIONS TO THE THEORY OF NATURAL SELECTION

Contributions is a collection of ten essays, the last two of which are germane here.[15] The penultimate essay is a basically unmodified reprint of Wallace's 1864 "The Origin of Human Races" (1864b), though Wallace did change the title to "The Development of Human Races under the Law of Natural Selection." Two textual alterations indicate the extent to which his views on human evolution reflected his maturing philosophy of nature. The final euphoric paragraph of the 1864 essay was replaced in the 1870 version by a far more qualified anticipation of the course of human development. Wallace asserted that the present period of world history was abnormal, the great advances of science often being perverted by "societies too low morally and intellectually to know how to make the best use of them." Natural selection alone could not secure any permanent moral or intellectual advance. Wallace declared it was "indisputably the mediocre, if not the low, both as regards morality and intelligence, who succeed best in life and multiply fastest." Yet Wallace, as so many Victorians, was committed to the belief that mankind was, however erratically, advancing to a more elevated moral and intellectual plateau. Since this advance could no longer be ascribed "in any way to 'survival of the fittest,'" Wallace was "forced to conclude that it is due to the inherent progressive power of those glorious qualities that raise us so immeasurably above our fellow animals, and at the same time afford us the surest proof that there are other and higher existences than ourselves, from whom these qualities may have been derived, and towards whom we may be ever tending." The other significant alteration in the 1870 version is the insertion of the phrase "from some unknown cause" in Wallace's explanation of the great advance in man's mental development at that period in his evolutionary history when his mind, rather than his body, became the major object of selection (Wallace [1891] 1969, 179, 185).

The final essay in *Contributions,* "The Limits of Natural Selection as Applied to Man," elaborated on the arguments sketched in the 1869 review. It made explicit Wallace's philosophical commitment to an evolutionary teleology. In rejecting a materialistic version of evolution, Wallace admitted that it will "probably excite some surprise among my readers to find that I do not consider that all nature can be explained on the principles of which I am so ardent an advocate; and that I am now myself going to state objections, and to place limits, to the power of natural selection." Focusing on two phenomena—the origin of consciousness and the development of man from the lower animals—the essay attempts to demonstrate, "strictly within the bounds of scientific investigation," that there exists a providential force responsible for the development of consciousness and those human characteristics that cannot be explained by natural selection. In a harsher portrait than he draws elsewhere, Wallace depicted

> the savage languages, which contain no words for abstract conceptions; the utter want of foresight of the savage man beyond his simplest necessities; his inability to combine, or to compare, or to reason on any general subject that does not immediately appeal to his senses. So, in his moral and aesthetic faculties, the savage has none of those wide sympathies with all nature, those conceptions of the infinite, of the good, of the sublime and beautiful, which are so largely

developed in civilised man. Any considerable development of these
would, in fact, be useless or even hurtful to him, since they would to
some extent interfere with the supremacy of those perceptive and animal faculties on which his very existence often depends, in the severe
struggle he has to carry on against nature and his fellowman.

The fact that all the higher intellectual and moral faculties do occasionally
manifest themselves in the primitive state indicates their latency in the large
brain of savage man. That this organ is much beyond his actual requirements
is substantiated by the fact that certain of the higher animals, with far smaller
brains, exhibit behavioral traits similar, if not identical, to those of the savage.
Wallace cited the ingenuity of the jaguar in catching fish, the hunting in packs
of wolves and jackals, and the placing of sentinels by antelopes and monkeys.
This evidence of continuity in psychological and behavioral processes from
the higher animals to early man had provided Darwin with some of the most
crucial support for his theory of human evolution by natural causes only.
Wallace now used that evidence for a radically different purpose. It served
as testimony that the large brain of savage man was "prepared in advance,
only to be fully utilised as he progresses in civilisation." The brain, Wallace
concluded, "could never have been solely developed by any of those laws
of evolution, whose essence is, that they lead to a degree of organisation
exactly proportionate to the wants of each species, never beyond those wants."
Wallace thus adduced the necessity of a "supreme intelligence" in explaining
not only the course of human evolution but also the origin of consciousness
itself. His main target here was Huxley's "celebrated article 'On the Physical
Basis of Life'" (Wallace [1891] 1969, 186–193, 206–207, 212).

WALLACE CONTRA HUXLEY, AGAIN

The origin of mental faculties such as "the capacity to form ideal conceptions
of space and time, of eternity and infinity—the capacity for intense artistic
feelings of pleasure, in form, colour, and composition, and for those abstract
notions of form and number which render geometry and arithmetic possible,"
presented equally formidable difficulties, according to Wallace. The capacity
to form abstract ideas, because they lie so "entirely outside of the world of
thought of the savage, and have no influence on his individual existence or
on that of his tribe," could not have been developed by the accumulation and
preservation of gradual mental variations, since such variations would have
been of no use in the struggle for existence. That such traits have occasionally
been found among certain savage races argues, again, for their future role,
not present utility. Wallace claimed this as further testimony to the action of

"some other power than the law of the survival of the fittest, in the development of man from the lower animals" (Wallace [1891] 1969, 199, 202–203).

Wallace deemed conscience, or moral sense, inexplicable by natural selection. The question of the moral sense was a complex one in Victorian culture. Its origin, its psychological force, and its relationship to diverse ethical norms were topics of great concern (Richards 1987). The theory of natural selection intensified the philosophical debates on morality. It focused attention on the relationship between instinctual and acquired (learned) behavior and between individual and group welfare and survival. Wallace was familiar with the competing schools of British moral philosophy. He rejected utilitarian explanations of the origin of morality, such as the Benthamite rational calculation of pleasures and pains. Wallace considered utilitarianism inadequate to account for the peculiar sanctity attached to actions that early man may have considered moral as contrasted with the very different feelings with which he regarded what was useful. The utilitarian sanction for truthfulness, he argued, is neither powerful nor universal. Its opposite, falsehood, has in "all ages and countries . . . been thought allowable in love, and laudable in war; while, at the present day, it is held to be venial by the majority of mankind in trade, commerce, and speculation." Wallace emphasized the difficulties with which truthfulness, practical and otherwise, has always been beset, and the many instances in which it has brought "ruin or death to its too ardent devotee." He concluded that considerations of utility could never have invested "it with the mysterious sanctity of the highest virtue,—could [never have induced] men to value truth for its own sake, and practice it regardless of consequences." Wallace advocated, instead, the intuitional theory, which postulates an innate moral sense, antecedent to and independent of experiences of utility. Depending on individual or racial constitution, and on education and habit—modified by custom, law, and religion—the acts to which its sanction are applied will vary (Wallace [1891] 1969, 196, 200–203).

Wallace closed his critique with an analysis of the origin of consciousness. He wanted to refute Huxley's assertion that "thoughts are the expression of molecular changes in that matter of life that is the source of our other vital phenomena." Wallace contrasted life—"the name we give to the result of a balance of internal and external forces in maintaining the permanence of the form and structure of the individual"—with consciousness. He granted that life may conceivably be regarded as the result of "chemical transformations and molecular motions occurring under certain conditions and in a certain order." He was adamant, however, that no combination of merely material elements, no matter how complex, could ever produce the "slightest tendency to originate consciousness in such molecules or groups of molecules." Wallace held matter and consciousness to be "radically unlike, exclusive, and incommensurable." The presence of consciousness in "material forms is a proof of the existence of conscious beings, outside of, and independent of, what we term matter" (Wallace [1891] 1969, 207–210). In an adroit stroke, Wallace argued that Huxley's materialist reductionism was inconsistent with "the most recent speculations and discoveries as to the ultimate nature and constitution of matter." Citing the theory that what is commonly called matter is actually an arrangement of centers of attractive and repulsive force, Wallace asserted that the special properties of matter (electrical chemical, magnetic) can be explained on the basis of the interaction between these force centers. Rejecting the standard materialist line that all matter is conscious, Wallace declared matter itself to be "essentially force, and nothing but force." Moreover, the various forces in nature—of which matter and consciousness are different manifestations—may be ultimately reducible to "will-force; and thus, . . . the whole universe is not merely dependent on, but actually is, the WILL of higher intelligences or of one Supreme Intelligence" (Wallace [1891] 1969, 207–212).

Wallace's position at this juncture in his career seems anomalous. He was an effective advocate of natural selection being a primary mechanism of evolution as well as a formidable opponent of a complete evolutionary naturalism. No aspect of evolutionary theory was more ideologically charged than that which dealt with humans, particularly their moral and intellectual attributes. The intense public interest and controversy engendered by the theory of natural selection could hardly have arisen if the question of man's descent from the lower animals was not perceived as an inextricable component of that theory (Ellegard 1958, 332). Wallace's views could scarcely be ignored and "Limits to Natural Selection" created a furor. He came under fire from both Darwinians and their opponents. Darwinians objected to his spiritualist interpolations, although they could not effectively repudiate all of his arguments on the insufficiency of natural selection. Opponents of evolutionary naturalism, while receptive to Wallace's position on human origins, felt that he still accorded too great a power to natural selection in the plant and animal kingdoms (Kottler 1974, 157–159).

Yet Wallace's apparently anomalous position is such only at a superficial level of analysis. By embedding natural selection within the framework of a theistic evolutionary teleology, Wallace had found the solution to the central question—the presence of human higher faculties—which had previously eluded him. Arguing that the "laws of organic development have been occasionally used for a special end, just as man uses them for his special ends," Wallace signaled that natural selection was, ultimately, subservient to other higher, directed powers (Wallace [1891] 1969, 213–214). Wallace's more fully articulated views of the 1860s—1870s would increasingly permeate his elaboration of an explicit evolutionary theism during the last three decades of his life. The American Protestant theologian James McCosh, a president of the College of New Jersey (later Princeton University) and prolific author of works on the harmony of science and Christian faith, was among those who regarded Wallace as having provided scientific evidence for divine intervention in evolution (McCosh 1890, chap. 6). In dissociating himself from a complete evolutionary naturalism, Wallace was joining Lyell, Gray, and the substantial group of scientists and laypersons who adhered to some type of theistic teleology (Hull 1973, 64–65).

Contributions to the Theory of Natural Selection is a crucial document in Wallace's intellectual and professional evolution. It confirmed his position as the eloquent champion of natural selection, save, of course, with respect to human moral and intellectual attributes. *Contributions* also signaled the explicit convergence of biological and metaphysical concerns in Wallace's evolutionary theory. Metaphysical concerns were to dominate the subsequent elaboration of his scientific as well as sociopolitical and ethical concepts. Wallace's integration of such disparate fields made (and makes) him a colorful and intriguing figure. It also had the effect of distancing him further from the professional scientific community. Nearly forty years later, Lubbock (then Lord Avebury), whose data regarding the cranial capacities of primitive man Wallace had used in *Contributions,* expressed how bothered many mutual colleagues still were by Wallace's controversial path of 1870. Writing to Wallace (1 May 1910) after reading his autobiography, Lubbock remarked that it "must be a source of very many pleasant memories to you to look back and feel how much you have accomplished. It surprises me, however, how much we [still] differ, and it is another illustration of the problems (?) of our (or rather I should say of my) intellect. In some cases, indeed, the difference is as to facts. . . . As to Spiritualism, [however,] my difficulty is that nothing comes of it. What has been gained by your séances, compared to your [scientific] studies?" (Marchant [1916] 1975, 438–439). Wallace would have unequivocally answered that much had been gained: a holistic evolutionary worldview. But he would also have admitted that his path exacted a toll. The making of this Victorian spiritualist was a journey replete with tensions.

FINANCIAL PROBLEMS

During the 1860s, the question of Wallace's financial security, and sagacity, became a matter of serious concern. Until his return from the Malay Archipelago, he had always managed to stay afloat economically. Wallace supported himself as a surveyor and teacher in his youth. Although his valuable Amazonian specimens were lost by fire at sea during transport back to

England, Wallace's excellent agent Samuel Stevens had "fortunately insured them for £150." This sum enabled Wallace "to live a year in London, and get a good outfit and a sufficient cash balance for my Malayan journey." The eight years in the Malay Archipelago, aside from their immense scientific significance, "were successful, financially, beyond my expectations." The rarity and brilliance of Wallace's exotic preserved birds and insects brought high prices from the deep pockets of London's exuberant collectors of natural history curiosities. Stevens, ever astute, had invested the proceeds in Indian guaranteed railway stock. Wallace, on his return from Malaysia in 1862, thus found himself in possession of about £300 a year in income. He sold some of his own extensive private collections after he had made what use of them was needed for augmenting his scientific hypotheses and evidence. Wallace could, therefore, have enjoyed a comfortable, if by no means affluent, yearly income if he had not succumbed to the lure of financial speculation. As he later expressed it in his autobiography, "owing to my never before having had more than enough to supply my immediate wants, I was wholly ignorant of the numerous snares and pitfalls that beset the ignorant investor, and I unfortunately came under the influence of two or three men who, quite unintentionally, led me into trouble." On the friendly advice of individuals who seemed more knowledgeable than he, Wallace invested in English and American railways, foreign securities, slate quarry properties, and, most heavily, in English lead mines. Though all his investment choices turned out poorly, none was more devastating than the lead mine speculation. How, Wallace pleaded later, could one then know that the enormous amount of silver mining in Nevada, where the ore contained lead and silver combined, would lead to the ruin of English lead mining? The massive exploitation of American silver mines produced lead as a waste product in the refining of silver for market. American entrepreneurs quickly realized that the lead waste was itself a potential source of great income. Large-scale exports of lead to Europe commenced and so lowered the prices of that commodity that British lead mines in particular became wholly unprofitable. The rapid and inexorable fall in lead prices began about 1870. "The result of all this," Wallace noted, "was that by 1880 a large part of the money I had earned at the risk of health and life was irrecoverably lost" (Wallace [1905] 1969, 2:360–363).

Victorian Britain was bristling with speculation and manias, economic as well as intellectual. It was Wallace's misfortune, literally, to be counted among the many who lost most of their capital in the booms and busts of the late-nineteenth-century global economy. Wallace also gambled intellectually in one notorious instance. He agreed to accept the challenge of John Hampden (a relative of Bishop Hampden), an upholder of the flat earth theory, to "prove the convexity of the surface of any inland water, offering to stake

£500 on the result." Sensing an easy victory, Wallace agreed to the wager. What Wallace had not counted on was Hampden's lunacy, vitriolic tenacity, and relentless public and legal invectives to dispute the experimental results (Garwood 2001). Hampden's campaign against Wallace's conclusive proof went on for fifteen years. Although Wallace won the original wager, he spent £700 to contest Hampden's ceaseless litigation, persecution, and public calumny. Wallace held himself partly to blame initially "for my fault in wishing to get money by any kind of wager." He should have more quickly realized the quackery with which he had permitted himself to become involved. The Hampden affair, he declared bluntly, constituted "the most regrettable incident in my life" (Wallace [1905] 1969, 2:364–376).

The steady stream of brilliant books, notably the highly popular *Malay Archipelago,* coupled with his lectures, reviews, and articles on an encyclopedic scope of subjects, yielded Wallace a more secure annual income. This enabled him to provide modestly for his family. An unexpected gift of £1,000 from a cousin of his mother's in 1878, at the time of Wallace's unsuccessful attempt to secure the appointment as superintendent of Epping Forest, eased Wallace's financial plight somewhat. This time, Wallace was more prudent with his capital. He invested it so as to bring a risk-free income of from £50 to £65 per annum. Still, Wallace's total income in the 1860s and 1870s was never more than "barely sufficient to support my family and educate my two children in the most economical way." It would not be until he was awarded a Civil Service pension of £200 in 1881 that Wallace would finally be free from the constant economic anxiety under which he labored for nearly two decades (Wallace [1905] 1969, 2:377–378; Marchant [1916] 1975, 248–251).

WHAT SUSTAINED WALLACE?

Wallace's career path placed him at times at the epicenter of the British scientific community and at other times at its margins. This shifting status had ramifications that intensified as he turned his formidable energies from spiritualism to political activism. What sustained Wallace throughout the years of controversy engendered by his commitment to spiritualism and other contentious domains? One reviewer of James Marchant's *Alfred Russel Wallace: Letters and Reminiscences,* published a few years after Wallace's death, highlighted those character traits that explain Wallace's persistence. According to the Reverend R. J. Campbell,

> Wallace had that rare gift, denied to many great men, of keeping
> up with his time in thought and feeling. He was never superan-
> nuated. . . . He lived on, and worked on, with an optimism and

abandon, a zest and enthusiasm, not easily found in men many years
his junior. He was an idealist in many fields. . . . A fearless champion
of unpopular causes, he never was afraid of being called a crank or
unpractical dreamer, nor did he shrink from the full consequences
of his principles. . . . Here we have a man who might have had the
most brilliant social and academic distinctions thrust upon him . . .
but who deliberately chose to live a life of retirement and comparative
poverty.

Campbell, a friend of Marchant's, noted one further element to explain Wallace's optimism and tranquility amid the storms surrounding him: "He was no self-blinded sentimentalist; he looked facts in the [face]; but he had a source of consolation lacked by his great fellow-worker, the German savant Ernst Haeckel, and that was his robust faith in God. After his early period of agnosticism he never could divorce his science from his religion, and often declared that both were equally based upon observed and reliable facts" (Campbell 1916). Although Wallace's God was not that of any orthodox traditional faith, his belief in an overruling Intelligence became more pronounced in the second half of his life. His encounters with spiritualism in the 1860s provided Wallace with an additional source for that optimism which sustained him from his earliest years.

———•◆•———

NOTES

1. Oppenheim (1985) is the classic statement of this new historiography. Several works prior to Oppenheim's had begun the task of rescuing these fields from the neglect they had endured, notably, Nelson (1969), Turner (1974), and Cerullo (1982).

2. Phrenology was the theory originated by the German anatomist and physiologist Franz Joseph Gall (1758–1828). By the early nineteenth century, it had come to be viewed by many as one avenue for the scientific study of the mind and mental faculties. Phrenology's main premise was that the mental powers of individuals consisted of separate faculties, each of which had its own organ and location in a definite region of the surface of the brain—the size and development of each organ indicating the degree of development of its particular faculty (e.g., benevolence, intellectual ability, self-esteem, veneration). By studying the external conformation of an individual's cranium, phrenologists believed they could determine the strength or deficiency of any particular faculty of the individual in question.

Mesmerism was the doctrine popularized by the Austrian physician Franz Anton Mesmer (1734–1815). Its main premise was that a hypnotic state, usually accompanied by muscular rigidity and insensitivity to pain, could be induced by an influence (originally called animal magnetism) exercised by the mesmerist operator over the will and nervous system of a patient or subject. Wallace believed, as did numerous Victorians, that spiritualism, mesmerism, and phrenology were complementary, not independent, modes of scientific understanding of the human mind. The particular mix of these three theories varied significantly from one practitioner to another (Cooter 1984; Winter 1998).

3. Alison Winter's analysis focuses on mesmerism, but her methodological framework is equally applicable to spiritualism.

4. Stafford 1984; Lynch and Woolgar 1990; Myers 1990; Camerini 1993; Porter 1995; Baigrie 1996; Rubino 1997. The capacity for maps to "look the same" but radically change in meaning over time became significant. So, too, did the notion that the meaning of a map resides not only in the map but also in relation to the written text of which it is a part and the larger historical context in which it appears.

5. "On the Varieties of Man" was presented at the Ethnological Society of London meeting of 26 January 1864 but was not published in full in the society's *Transactions* until 1865 (Wallace 1865b). "The Origin of Human Races" was read at the Anthropological Society of London (ASL) meeting of 1 March 1864 and published in the ASL *Journal* in the same year (Wallace 1864b).

6. Wallace's annotated copies of the two Dale Owen books are now in in ARWL.

7. Nelson 1969; Barrow 1986; Malinchak 1987, 48–95; Owen 1990; Noakes 1998. For older accounts that give an eyewitness flavor to the debates surrounding spiritualism, see Moses 1894; Podmore [1902] 1963; Doyle [1926] 1975.

8. Carpenter elsewhere depicts spiritualism and mesmerism as "Epidemic Delusions" (1877b).

9. For one of Wallace's several more pointed replies to Carpenter's criticisms, see Wallace, "The Curiosities of Credulity" (1878a).

10. "Notes of Personal Evidence" was added as a postscript to "The Scientific Aspect of the Supernatural" (1866) when that essay was published in Wallace's *On Miracles and Modern Spiritualism* in 1875; the version of "Notes" I have used comes from the 3d ed. of *Miracles* (1896a).

11. Some fifteen years later, during his tour of North America, Wallace formed a close friendship with the sociologist Lester F. Ward. The two shared common political views and botanical interests but had rather different attitudes concerning spiritualism (Wallace [1905] 1969, 2:117–118). Wallace read Ward's two-volume *Dynamic Sociology* (1883), which he described in his autobiography as a "masterpiece of elaborate systematic study of almost every phase of social science" (Wallace [1905] 1969, 2:117). Wallace took great exception, however, to Ward's citing Tylor's explanation of spiritualism as authoritative. On the inside back jacket of his "American Journal," Wallace complained (privately, in this instance) that Ward allowed "no word of the possibility, even, of *spiritual beings* being *realities*, who *do* manifest themselves occasionally to man! No reference to the vast mass of *evidence* in favour of such a belief, and to the thousands of educated and intelligent men who have been *forced* to the belief by *evidence* against all their prepossessions" (Wallace 1886–1887).

12. An expanded version was reprinted in *On Miracles and Modern Spiritualism* (Wallace 1875b, 29–137). All further references to *Scientific Aspect* are to the version in *Miracles and Modern Spiritualism* (Wallace [1866] 1875).

13. "The Prosecution of Dr. Slade," *Spiritualist Newspaper* 9 (6 October 1876): 110; Lankester's and Donkin's letters appeared in the *Times* (London), 16 September 1876, 7. Wallace had offered his own positive account of his experiences with Slade in "A Sitting with Dr. Slade," *The Spiritualist* (London), 9, no. 4 (25 August 1876): 42.

14. I am indebted to Charles Smith (1992) for the clear statement of the overall continuity, rather than abrupt disjunction, between Wallace's early and later thought; this essay is now readily available on "The Alfred Russel Wallace Page" (http://www.wku.edu/~smithch/).

All references to Smith's 1992 essay will be to the original. The version on the Web site was "lightly revised in October 1999."

15. The original title is *Contributions to the Theory of Natural Selection* (1870); I have used the reprinted version (with alterations): *Natural Selection and Tropical Nature: Essays on Descriptive and Theoretical Biology* (Wallace [1870] 1891). All references to this work will be from the 1969 reprint of the 1891 ed. and will be cited as Wallace (1891) 1969.

CHAPTER 5

Land Nationalization to Socialism:
The Ethics of Politics and the Politics of Ethics

The battles over spiritualism clarified the epistemological basis of Wallace's maturing evolutionary philosophy. He now felt ready, by the 1870s, to charge into the heated arena of public affairs. Social, political, and economic implications of evolutionary thought for the pressing questions of Victorian industrial society came to occupy a greater portion of Wallace's life and work. His impassioned stands on land nationalization, socialism, vaccination, and other incendiary topics reinforced his reputation as an innovative thinker in diverse fields. Wallace continued to be a prolific author of books and articles on biogeography, geology, and climate (especially in the new field of glacial phenomena). He was also an active and influential social reformer in a turbulent period in British history. Writings on societal issues flowed ceaselessly from his pen.

Characteristically, Wallace's provocative sociopolitical activism was played out against the backdrop of the tranquil environment of his home life. In 1881, Wallace and his family moved to "Nutwood Cottage" in Godalming, at that time known for its excellent school system and the charm of its scenery. During the next eight years as resident of that rural seat, where his children attended Charterhouse School, Wallace delighted in his garden and greenhouse. Quietly cultivating more than a thousand species of plants, Wallace drew on the calm (in his case) of family life to sustain him for the public battles he was to engage in with such passion and tenacity. Gardening was, for Wallace, a source of "pure enjoyment" throughout his life. He made very few experiments with the myriad of plant species he grew. Rather, it was the "exquisite beauty and almost infinite variety of the vegetable kingdom, which enabled me better to appreciate the marvel and mystery of plant life, whether in itself or in its complex relations to the higher attributes of man"

(Wallace [1905] 1969, 2:103–104, 203–204). Wallace's observations of plants were directly related to his increasingly explicit evolutionary teleology. The serenity he drew from natural history seemed to energize him for the political battles into which he plunged.

EVOLUTION AND IDEOLOGY

The relationship between evolutionary theory and social and political ideology is one of the most intensively studied aspects of the cultural context of Victorian science (Young 1985b; Bowler 1993; Lightman 1997). That Wallace should attempt to integrate his biological findings with his concerns for societal reform is hardly remarkable. What is notable is the degree to which Wallace made this integration a matter of public record. Unlike the majority of his scientific colleagues, who either confided their political opinions to their immediate circle of family and friends or chose the less visible forums of diaries and personal journals, Wallace's political and social views were a matter of public record. His outspokenness on key issues such as capitalism, poverty, and the "woman question" made him a fixture of Victorian cultural wars. By analyzing the fundamental concepts that underlay Wallace's political philosophy, it becomes clear why he simply could not remain silent on the burning controversies of his day. Wallace believed that it was the duty, not merely a peripheral activity, of the scientist to "go public" on questions of poverty, land nationalization, gender roles, and kindred explosive subjects. His passion to use whatever tools lay at hand—science, education, lecturing, spiritualism—to effect radical and lasting societal change is as much a part of his core vision as were his detailed scientific observations.[1]

Wallace's perspective on the public role of the intellectual has strong parallels with that of Spencer. Although the two disagreed on many points of theoretical and practical detail, they were united by a conviction that it was the duty of the scientist to engage directly in social and political controversies. For both Spencer and Wallace, ethics lay at the very base of any legitimate sociopolitical philosophy. Spencer had given Wallace a copy of *The Data of Ethics* (1879). In his preface, Spencer described the present age as one in which "I am the more anxious to indicate in outline, if I cannot complete, this final work, because the establishment of rules of right conduct on a scientific basis is now a pressing need. Now that moral injunctions are losing the authority given by their supposed sacred origin, the secularization of morals is becoming imperative." Spencer described the opposite poles of the Victorian ethical debates as between those who fear the decline of the "controlling agency" of the "divine commandments . . . [and] the guidance it yields" versus those who feel that there is no danger in a "vacancy left unfulfilled by any other *controlling* agency. . . . The one holds that the gap left by disappearance of a code of supernatural ethics, need not be filled by a code of natural ethics; and the other holds that it cannot be so filled" (Spencer 1879, iv). Spencer is clear that for all who belong to neither extreme camp there exists the hope and belief that a naturalistic code can fill the vacuum. Wallace's copy of *The Data of Ethics* is heavily annotated. In one note in the margin of the preface, Wallace queried: "IS IT CONTROLLING? THE CONTROLLING POWER IS THE KNOWLEDGE THAT EACH PERSON IS MAKING THEIR FUTURE HAPPINESS BY FOLLOWING THE DICTATES OF THIS PURE MORALITY." Wallace's annotations reflect his conviction that any new ethics must emanate from a combination of theism, spiritualism, and social reform, as well as from data drawn from evolutionary biology. Wallace was wholly sympathetic to Spencer's goal of establishing a new ethics for technological society. Unlike Spencer, however, Wallace insisted that any such new ethical system transcend the framework of narrow scientific naturalism.[2] It is here that the imprint of epistemological concerns is most manifest in Wallace's developing evolutionary teleology. Wallace had earlier established, to his satisfaction, that culture was not predicated, except in certain obvious aspects, on biologically determined behavior. He thus freed himself from the constraints of a strictly naturalistic explanation for the history of cultural developments from humanity's origins to the Victorian period. Wallace could now articulate an emergentist conception of both individual and societal evolution.

LAND NATIONALIZATION

Wallace's early introduction to radical social and political speculation, primarily through Owenite writings and teachings, instilled in him a critical attitude toward central maxims of British legislation and political economy. The years spent in land surveying, prior to the voyage to South America, provided him with a detailed knowledge of the laws and practices governing private and public property, although he did not then consider those laws egregiously unjust (Wallace [1905] 1969, 1:89, 158). Reading Spencer's *Social Statics* (1851), particularly the chapter on "The Right to the Use of the Earth," on his return from the Amazon turned Wallace's interests toward social and political reform (Wallace [1905] 1969, 2:235). His travels in the Malay Archipelago were undertaken with a heightened attention to anthropological and sociological data, in addition to the strictly biological. Wallace's prolonged residences in primitive communities in South America and the East compelled him to question whether Europe had attained that pinnacle of social and moral development that its undoubted scientific and material progress rendered axiomatic to many Victorians.

During these travels, Wallace had been repeatedly struck by the "remark-

able [fact] that among people in a very low stage of civilization we find some
approach to . . . a perfect social state [in which] . . . there are none of those
wide distinctions, of education and ignorance, wealth and poverty, master
and servant, which are the product of our civilization." He concluded *The
Malay Archipelago* ([1869] 1962), somewhat startlingly, with a denunciation
of the highly vaunted civilization of nineteenth-century Europe. Wallace's
1864 article on the origin of human races (as discussed in chap. 4) had echoed
sentiments of European racial superiority (1864b). But closer analysis of the
situation back home made Wallace far less certain of the accuracy of these
sentiments. Technical mastery over the forces of nature had brought about a
vast accumulation of wealth and an ever more prodigious international commerce. It had also, Wallace asserted, brought about those crowded towns and
cities that "support and continually renew a mass of human misery and crime
absolutely greater than has ever existed before." He now felt that Europeans,
rather than the so-called savages among whom he had lived, suffered under a "barbaric" social and moral organization, Wallace specified the abuses
engendered by private property as the primary culprit:

> We permit absolute possession of the soil of our country, with no
> legal rights of existence on the soil to the vast majority who do not
> possess it. A great landholder may legally convert his whole property
> into a forest or a hunting-ground, and expel every human being who
> has hitherto lived on it. In a thickly populated country like England,
> where every acre has its owner and its occupier, this is a power of
> legally destroying his fellow-creatures; and that such a power should
> exist, and be exercised by individuals, in however small a degree, in-
> dicates that, as regards true social science, we are still in a state of
> barbarism.
>
>   (Wallace [1869] 1962, 456–458)

These passages in *The Malay Archipelago*, the most popular of all his
books, marked Wallace's debut as an outspoken social critic. Land policy
was a critical focus for social controversy in England at the start of the 1870s.
Wallace objected to the clearances of the time, as well as past enclosures and
Irish landlordism. John Stuart Mill, who had turned increasingly to issues of
land reform by 1870, was impressed with Wallace's sentiments. Mill asked
him to become a member of the General Committee of his proposed Land
Tenure Reform Association. The association, the main object of which was to
claim for the state all future unearned increments of land values (the increase
in land value not deriving from any actual improvements by the owner) was
formed in 1871. Wallace attended its meetings until Mill's death in 1873

caused its dissolution (Wallace [1905] 1969, 2:235–238). After Mill's death, the question of land reform continued to occupy Wallace intermittently for the next several years. He refrained from offering any definite proposal for outright nationalization because of theoretical objections advanced by Mill, Spencer, and many of their followers. Although severe critics of the inequities of private property in the land, they opposed any reform that would entail what they regarded as a pernicious increase in state intervention.

Wallace's relationship with Spencer was extremely close. But there were major differences in their strategies for societal reform. "As you may suppose," Spencer wrote to Wallace on 25 April 1881, "I fully sympathize in the general aims of your proposed Land Nationalization Society; but for sundry reasons I hesitate to commit myself, at the present stage of the question, to a programme so definite as that which you send me. It seems to me that before formulating the idea in a specific shape, it is needful to generate a body of public opinion on the general issue, and that it must be some time before there can be produced such recognition of the general principle involved as is needful before definite plans can be set forth to any purpose." He told Wallace that persuasion should precede direct action. Once public opinion was roused, Spencer believed, "the land-owner [could] be distinctly placed in the position of a tenant of the State in something like the terms proposed in [Wallace's] schemes: namely, that while the land itself should be regarded as public property, such value as has been given to it should rest in the existing so-called owner." Spencer concluded his advice cum warning to Wallace by asserting that the "question is surrounded with such difficulties that I fear anything like a specific scheme for resumption by the State will tend, by the objections made, to prevent recognition of a general truth which might otherwise be admitted." Spencer made it clear that he felt strongly that the time was far from ripe for taking any concerted action to nationalize the land (ARWP MS. 46434, fol. 350).

WALLACE TAKES THE LEAD

The bitter controversy over Irish landlordism, which intensified during 1879–1880, provoked Wallace to assume a more dominant role in the agitation for land reform and cast aside such hesitations as those felt by Spencer (fig. 9). The ineffectualness of the proposals put forward to resolve the Irish situation convinced Wallace that state ownership of some kind was essential to remove the abuses of the existing land tenure system. Charles Stewart Parnell's too modest proposals for Irish problems finally forced Wallace to act decisively. In 1880, he severely criticized Parnell's program for Irish peasant proprietorship as not abolishing privilege but, instead, merely reshuffling some land titles

from a smaller to a larger minority. Wallace was now prepared to go public
with his plan for more radical and thoroughgoing and, hence, more lasting
systemic change (Gaffney 1997, 612–613).

All land, Wallace proposed, would revert to the state, while the improvements or increased value given to the land—such as buildings, drains,
plantations—would remain the salable property of the present owner (now
"state tenant"). The management of the land would devolve not to the state
but to the actual tenant proprietors. The publication of these views in an
article in the *Contemporary Review* immediately attracted the attention of
those who desired land reform but opposed increased state intervention in
land management (Wallace 1880). The Land Nationalization Society (LNS),
with a program based on Wallace's principles, was formed in 1881 with Wallace as its president (Wallace [1905] 1969, 2:239–40). At this stage of his
political career, Wallace was still a radical Liberal. He was not yet uncomfortable with Liberal domination of the land reform movement (Offer 1981).
Land Nationalisation: Its Necessity and Its Aims was published the following
year (Wallace [1882] 1906). Although he remained committed to the goals

Figure 9. Photograph of
Wallace at age fifty-five, two
years before the publication
of *Island Life* (1880) and
three years prior to his
election as president of the
newly formed Land
Nationalisation Society
in 1881.

of the LNS, Wallace's march toward socialism in the 1880s forced him to move beyond even radical liberal land reform strategies.

Wallace dedicated *Land Nationalisation* to "the working men of England." He intended it as a rigorous yet easily comprehensible demonstration that "the vast riches and the degrading poverty of [England,] which, in their terrible combination and contrast, are unparalleled in the civilised world," derive from its system of land tenure. Drawing on a mass of documentary evidence, including the reports of Parliamentary commissions, Wallace argued that private ownership in land necessarily produces evil results "of the most alarming magnitude." Moreover, the widespread pauperism, vice, and crime of large portions of the English laboring classes, "which strike foreigners with the greatest astonishment," are due not to any special ignorance or ill-conduct on the part of English landlords but are inherent in the system itself. Wallace declared that so long as the "highest teaching of political science" tells the great landlords "that their land is their property," they will necessarily act so as to increase the profits from their holdings. Every step taken to secure this end—the enclosure of common land, the eviction of tenants from their homes to convert farms into game preserves or smaller holdings into larger ones, or the outright appropriation of the added value given to the land by the labor of tenants—was, Wallace noted pointedly, "supported by the power and majesty of the law." The fact that many landholders were also magistrates further enhanced their power to coerce their tenants into conformity with their own political and religious opinions. Wallace compared this catalog of despotic powers over individuals to those "we are accustomed to look upon with horror when occurring in the Turkish or Russian Empires." He detested the right of English landlords, as absolute owners of the land, to destroy ancient monuments and to work, sell, export, and totally exhaust the (nonrenewable) mineral wealth of the country solely for individual profit without regard for the national interest or future generations (Wallace [1882] 1906, 100, 129–135, 176–179).

In contrast to the miserable condition of many of the agricultural and town laborers of England, Scotland, and Ireland, Wallace cited the widespread system of "occupying ownership" in Switzerland, Germany, Norway, Belgium, and France. In those countries, the occupier and cultivator of the land was also its owner, and the population was generally satisfied and thriving. Wallace concluded that "in order to effect a real and vital improvement in the condition of the great mass of the English nation, not only as regards physical well-being but also socially, intellectually, and morally," a radical change in the system of land tenure was required. Only if private ownership of the land as a source of income from its rent or for commercial speculation were abolished, and each cultivator of the land became its virtual but not absolute or unrestricted owner, would England possess the "healthy, moral and contented" population its great wealth would seem to permit.

Wallace emphasized that any reform that merely transferred absolute ownership of the land from existing landlords to existing tenants would be self-defeating. The new owners, being free to divide their holdings and sublet portions, would in time constitute a new privileged class and the worst abuses of landlordism would revive. Wallace's fundamental conviction that every citizen be given the opportunity to procure suitable land for his or her personal occupation, with permanent security of tenure, entailed that the state alone be the actual owner of the land. Subletting would be prohibited by law. His proposal for land nationalization necessitated that a "person must own land only so long as he occupies it personally; that is, he must be a perpetual holder of the land, not its absolute owner" (Wallace [1882] 1906, 18–19, 137, 182–183).

A Proposed Land Nationalization Act

To effect the transfer to state ownership, Wallace proposed a Land Nationalization Act. The act was based on the distinction he had earlier drawn between the inherent value of the land (depending on natural conditions such as geological formation, climate, aspect, surface, and subsoil) and the improvements added to the inherent value by the labor or outlay of the owners or occupiers. Wallace specified that on the date of the act's coming into operation, the state would assume ownership of the land. The state would be compensated for the use of nationalized land by payment of an annual " quitrent," determined according to the assessed inherent value of each plot. The improvements created by landholders (or their predecessors) would remain their absolute property and would henceforth constitute the "tenant right," to be retained by them or sold as they wished. Wallace opposed outright confiscation of landed property. He stipulated that each existing landowner, and "any heir or heirs of the landowner who may be living at the passing of the Act, or who may be born at any time before the decease of the said owner," be paid an annuity by the state equal to the same net income from the land derived prior to nationalization. He defended this temporary continued "existence [of] a class of pensioned idlers, living upon the labour of others, without the smallest exertion of body or mind on their own part," on the grounds that the property of living individuals (and their immediate heirs) be strictly respected by the state. Future descendants, Wallace declared, had no such proprietary rights to the land (exclusive of tenant right). He considered the presumed rights of inheritance one of the worst abuses of landlordism (Wallace [1882] 1906, 193, 198–199).

Existing tenants at the time the Nationalization Act took effect would be entitled to continue the occupation of their houses or farms on payment to the state of the annual quitrent. Each tenant also would have to acquire the tenant right to the property, by purchase from the existing landlord. As absolute owner of the tenant property, he would then be free, if he chose, to bequeath either all or part of it. For those unable to provide the sum necessary for purchase of the tenant right, Wallace suggested that loan societies or municipal authorities be empowered to advance the required sum, which would then be repaid by the tenant over some fixed length of time. Wallace insisted that such mortgaging be strictly limited to prevent anyone from endeavoring to farm more land than his capital and abilities warranted, that is, from farming under a perpetual mortgage. At the same time, there need be no upper limit to the extent of land any single state tenant could occupy. A wealthy individual might retain or purchase rights to a vast acreage. Since he could not sublet any portion of his tenant right, however, Wallace envisioned no reason for anyone retaining more land that he and his daily employees could feasibly operate. City dwellers who so chose could also exercise the universal right embodied in Wallace's program and select plots of available agricultural land or portions of commons or waste lands for their personal occupation in proximity to cities and towns. Such (presumably) salubrious dwelling places would, he maintained, "always produce health and contentment" and would, for those industrial workers who utilized their land only to produce food as a supplement to purchased provisions, provide some security in times of unemployment. Finally, Wallace suggested—but did not specify how—that urban residences be similarly nationalized, and the present occupiers of leasehold houses or rental premises be enabled to become their owners (Wallace [1882] 1906, 202–218).

The publication of *Land Nationalisation* catapulted Wallace to a prominent role in the public debate on land reform. This debate provided a major focus for the broader question of social and political reform in Great Britain during the late 1870s and 1880s (d'A Jones 1968, 55). Wallace's new role did not force him to "turn renegade to natural history" (Marchant [1916] 1975, 262). Rather, Wallace would continue to probe more critically the relationship of evolutionary biology to sociopolitical issues. As Robert M. Young has pointed out, it "is not in the least surprising that those who were interested in the relationship between man and nature should, with consistency, be concerned about workers and property, and conversely" (Young 1971, 223). Just as Wallace had earlier reassessed the scope of natural selection in human evolution, he now analyzed more thoroughly the use (or misuse) of evolutionary theory to buttress particular social and economic policies. His views on land nationalization were integral elements in Wallace's developing system

of social evolutionism, in which biological and sociopolitical convictions re-
acted on one another. Wallace's "invasion" of political economy is scarcely
the aberration that many of his scientific peers, such as Huxley and Darwin,
decried (somewhat disingenuously, given their own abiding concern with,
and benefits from, matters economic). He entered that domain by a route he
knew well: land economics. Like many other struggling young men, Wallace
had been a land surveyor. One of his greatest works, *Geographical Distribution
of Animals* ([1876] 1962), was permeated with the language and metaphor
of surveying and boundaries. Perhaps the most famous single boundary for
which Wallace is known, as discussed previously (chaps. 2 and 3), is the Wallace Line, through the Macassar Straits, in the Malay Archipelago (Camerini
1993). This line, which marks the faunal divide between the characteristic
animals of the Australian and Asian regions of the archipelago, first appeared
publicly in Wallace's 1863 essay "On the Physical Geography of the Malay
Archipelago" (Wallace 1863). But Wallace was also keenly interested in determining whether a similar boundary could be drawn to demarcate the different
human races he encountered in his travels in the archipelago.

A year later (1864b), Wallace described just such a racial boundary. This
second line marked the division of "the Malayan and all the Asiatic races,
from the Papuans and all that inhabit the Pacific." This ethnological boundary, Wallace indicated, ran a few hundred miles east of the faunal one (Moore
1997, 296–297). But this boundary was not merely an abstract creation: human racial distribution in the Malay Archipelago was directly linked to food
distribution patterns. Wallace's theorization and observations in the Malay
Archipelago were suffused with the specter of Malthus. Just as in the 1840s,
when Wallace surveyed Welsh farms and noticed that boundaries could be
drawn between the domains inhabited by the impoverished Welsh farmers
and those of the (slightly) more well-to-do English laborers, so too later in
the Malay Archipelago did he envision ethnological divides as reflective of a
struggle for food and other resources. Wallace's approach to ethnology was
based, in part, on cartography. But this cartography was rooted in economics
(Moore 1997, 300–307). From the outset of his career Wallace sought insights into the relationship not just between humans and nature but also
between humans and nature in relation to land (Gaffney 1997, 611).

ENTER HENRY GEORGE

While writing *Land Nationalisation,* Wallace had, in 1879, read *Progress and
Poverty* (1879) by the radical American economist Henry George. George's
thesis that material progress had engendered rather than alleviated human
poverty and misery paralleled Wallace's own claims. He regarded George's

work as "a most remarkable theoretical confirmation" of the inductive argument he had developed in examining the evidence of the actual condition of people under different systems of land tenure (Wallace [1882] 1906, 173). Wallace was struck by George's devastating critique of the pessimistic conclusions drawn by Malthusian political economists. He informed Darwin that George, who accepted the operation of Malthus's principle of population with respect to animals and plants, denied that "it ever has operated or can operate in the case of man, still less that it has any bearing whatever on the vast social and political questions which have been supported by a reference to it" (Marchant [1916] 1975, 260). As codiscoverer of natural selection, Wallace rejected George's complete disavowal of evolutionary biology with respect to human questions. He was fully sympathetic, however, to George's arguments against laissez-faire economic policies, which invoked Malthus as an unequivocal source. Wallace's views regarding Malthusian premises—and the conclusions frequently drawn from them—were complex. He urged Spencer and Darwin to read George's book.

Spencer informed Wallace that he had "already seen the work you name—*Progress and Poverty;* having had a copy, or rather two copies, sent me. I gathered from what little I glanced at, that I should fundamentally disagree with the writer, and have not read more." Spencer declared that he demurred "entirely to the supposition, which is implied in the book, that by any possible social arrangements whatever, the distress which humanity has had to suffer in the course of civilization could have been prevented. The whole process, with all its horrors and tyrannies, and slaveries and wars, and abominations of all kinds, has been an inevitable one accompanying the survival and spread of the strongest, and the consolidation of small tribes into large societies; and among other things the lapse of land into private ownership has been," Spencer emphasized, "like the lapse of individuals into slavery, at one period of the process altogether indispensable. I do not in the least believe that from the primitive system of communistic ownership to a high and finished system of State ownership such as we may look for in the future, there could be any transition without passing through such stages as we have seen and which exist now." Spencer's unwillingness to read George arose from his theoretical objections to the central thesis of *Progress and Poverty,* not from any belief that political action on the part of intellectuals was inappropriate. On the contrary, Spencer had already given his opponents "more handles against me than are needful." He was acutely sensitive to the fact that someone as politically influential as himself had to exercise prudence as to which actions to take and which not to take if he deemed the timing inauspicious (ARWP MS. 46434, fol. 350, Spencer to Wallace, 6 July 1881). One avenue Spencer did endorse was an "appropriate" degree of government intervention. In a subsequent letter, Spencer praised the "economic merits" of state control of certain key industries. He drew Wallace's "attention to the facts lately brought out by Sir Thomas Farrer, secretary of the Board of Trade." Farrer, Spencer felt, showed "that under the system of railway administration in England, which differs from that of France in that the companies are less under State control, and their lines are not eventually to lapse into the hands of the State, the amount of convenience to the traveller, both in economy, swiftness, and number of trains, is far greater than in France" (ARWP MS. 46434, fol. 350, Spencer to Wallace, 23 February 1884). Spencer was as much a "political animal" as Wallace.

Darwin, too, declined to read George's book but for very different reasons. In the last letter he sent Wallace [12 July 1881], Darwin said he would "certainly order 'Progress and Poverty,' for the subject is a most interesting one. But I read many years ago some books on political economy, and they produced a disastrous effect on my mind, viz., utterly to distrust my own judgment on the subject and to doubt much everyone else's judgment! So I feel pretty sure that Mr. George's book will only make my mind worse confounded than it is at present" (Marchant [1916] 1975, 261). Darwin's plaintive response underscores a fundamental difference between the two naturalists on the appropriate relationship between a scientist's professional studies and commitment to sociopolitical activity. Darwin kept his reflections on the political implications of evolutionary biology (relatively) private. He was gripped by obsessive fears for his and his family's "respectability." Darwin disliked those portions of Huxley's first public lecture on the *Origin,* in which Huxley advocated scientific expertise as the crucial ingredient in Britain's technological and imperial ambitions. Huxley's words were a self-serving plea on behalf of the new white-collar specialist. They were also prescient. But Darwin dismissed them as "flashy rhetoric [and] so much 'time wasted' " (Desmond and Moore 1991, 489, 627). Darwin's polite but famously ambiguous response to Marx's claim that he and Marx were intellectual soulmates is further evidence of Darwin's fear of participating openly in the public debates concerning the societal ramifications of evolutionary biology (Colp 1976, 1982). The contrast with Wallace could not be greater. Wallace—president of the LNS, outspoken exponent of spiritualism, and crusader for a host of sociopolitical causes—thrived on public controversy, particularly in those arenas at the juncture of science and politics.

Wallace found George's vision enunciated in *Progress and Poverty* potent and alluring. George, like his fellow journalist, Edward Bellamy—whose work was to influence Wallace's sociopolitical ideas profoundly—was alarmed at the materialism and ugliness they saw around them in the Gilded Age of 1870s America. Both came to maturity in the aftermath of the Civil War. They were appalled by the seemingly unstoppable triumph of an urban industrial order marked by increasing inequality of wealth and power. George and Bellamy were convinced that a dominant factor in these insidious developments lay in the twin vices of a growing monopoly of land by the oligarchs and an unchecked, and unhealthy, rapid urban growth. This powerful combination drove masses of farmers off their land and into the cities where they were quickly submitted to lives of poverty and degradation. The greatest evil, they both concluded, was the obliteration of religion and morality before the altar of material wealth and civic luxury and licentiousness. Part astute economic and political observation and part redemptive zealotry, the writings and speeches of George and Bellamy exerted an immense appeal on both sides of the North Atlantic. *Progress and Poverty* and Bellamy's *Looking Backward* (1888) were two of the runaway best-sellers of the late nineteenth century (Thomas 1983, 1–4).

Wallace's critique of competitive capitalism, implicit in *Land Nationalisation,* disposed him to imbibe the message of George and Bellamy. It was Charles Parnell who indirectly brought Wallace and George together. Parnell, an Irish nationalist, was a leading figure in the fight for Irish home rule. But, as noted previously, Wallace and George regarded his specific proposals for improving the laws concerning Irish peasant proprietorship as too modest and, ultimately, ineffective. Both were annoyed by what they considered Parnell's temporizing in Ireland. Wallace used the LNS to give George a platform when he toured Britain in the winter of 1883–1884 advocating his "single tax" program. George's first lecture was at a meeting of the LNS (Wallace [1905] 1969, 2:255–256). The reaction of George Bernard Shaw, then a young reporter of twenty-five, to George's appearance was typical. He and some of his avant-garde colleagues found George's appeals to such "dated Enlightenment notions as Truth, Justice, and Liberty" unfashionable. But Shaw was captivated by George's rhetorical skills. He declared that only an American could have seen in a single lifetime "the growth of the whole tragedy of civilization from the primitive forest clearing." Shaw was also impressed with the manner in which George illustrated his theory with references to urban land values and London's prohibitive rents. Then and there, Shaw enlisted as "a soldier in the Liberative War of Humanity."

Others in the audience, while not so certain of the efficacy of the complete package of George's specific remedies to socioeconomic problems, responded enthusiastically to what they termed his "Christian message." The future Fabian Sydney Olivier, who declared that while it was easy enough to snicker at George for his "deduction of the immortality of the soul from the sound theory of property in land," quickly added that was precisely to miss the point of the lecture. For all his "rhapsodical and unchastened style, strongly suggestive of the pulpit," George had at least brought the social question "into general notice of others than readers of Mill and Spencer, and for that I think he is to be thanked." George's message met with great success at many similar gatherings in the next few months in England, Scotland, and Wales (Thomas 1983, 194–198). As the socialist John Atkinson Hobson later recalled, however, the "Prophet of San Francisco" had to choose his audiences with some care. It was working-class people, Hobson noted, who were most prepared to give full assent to George's proposition that unqualified private ownership of land was "the most obviously unjust and burdensome feature in our present social economy." Leading intellectuals such as John Ruskin and Frederick Harrison deemed George's lectures and message admirable. But his appeal, according to Hobson, lay preeminently with those groups of "largely self-educated, keen citizens, mostly nonconformists in religion, who carried forward a radical, freethinking tradition whose roots lay in an eighteenth-century moral economy" (Hobson1897; Barker 1955, 415–416).

WALLACE AND HOBSON'S *PROBLEMS OF POVERTY*

Hobson (1858–1940) is best known for his critiques of the economic bases of imperialism. He was a humanistic critic of current economics, particularly those theories of capitalism that posited exclusively materialistic definitions of value. In 1887, Hobson moved to London. He met the journalist William Clarke, who invited him to join the Fabian Society. An active member, Hobson wrote two books for the organization, *Problems of Poverty* (1891) and *Problem of the Unemployed* (1896). Other works of his during this period were *Evolution of Modern Capitalism* (1894) and *John Ruskin: Social Reformer* (1898). C. P. Scott, editor of the *Manchester Guardian,* recruited Hobson as correspondent in South Africa. While reporting on that country, he developed the idea that imperialism was the direct result of the expanding forces of modern capitalism. Returning to England in 1900, Hobson went on a national lecture tour. A strong opponent of the Boer War, he condemned it as a "conflict orchestrated by and fought for the preservation of finance capitalism at the expense of the working class." Over the next few years, Hobson published works exploring the links between imperialism and international conflict. In *Imperialism* (1902), Hobson argued that imperial expansion was driven by a search for new markets and opportunities for investment overseas. His vivid writings helped Hobson obtain an international reputation. He influenced leaders as diverse as Lenin, Trotsky, and David Lloyd George, whose "People's Budget" of 1909 included certain of Hobson's views (Brailsford 1948). It was inevitable that Wallace would admire so kindred a thinker.

Wallace read Hobson's *Problems of Poverty* closely; his copy of the book

is heavily annotated. Hobson's goal, as announced in the preface, "to establish on a scientific basis the study of the 'condition of the people,' " was
identical to that which permeated Wallace's sociopolitical analyses. Hobson
decided not to include the "larger proposals of Land Nationalization and
State Socialism . . . because it was impossible to deal, however briefly, even
with the main issues involved in these questions" within the confines of the
book's focus on exposing the causes of poverty. But his strategy to focus on
a specific audience—that "of the citizen-student who brings to his task not
merely the intellectual interest of the collector of knowledge, but the moral
interest which belongs to one who is a part of all he sees, and a sharer in the social responsibility for the present and future of industrial society"—delighted
Wallace. Among the many passages Wallace underlined and marked with
exclamation points, two express that sense of moral as well as intellectual
outrage which characterized Wallace's increasingly public cultural critiques.
The first is Hobson's blunt challenge to the proponents of late Victorian
industrial capitalism:

> We are not at present concerned with the requirements of the indus-
> trial machine, but with the quantity of hopeless, helpless misery these
> requirements indicate. The fact that under existing conditions the
> unemployed seem inevitable should afford the strongest motive for
> a change in these conditions [Wallace's underlining]. Modern life
> has no more tragical figure than the gaunt, hungry labourer wander-
> ing around the crowded centres of industry and wealth, begging in
> vain for permission to share in that industry, and to contribute to that
> wealth; asking in return not the comforts and luxuries of civilized life,
> but the rough food and shelter for himself and family, which would
> be practically secured to him in the rudest form of savage society.

The second passage in Hobson's book that commanded Wallace's complete
accord asserted that the "conscious socialist is he who, recognizing in theory the nature of this social property inherent in all forms of capital, aims
consciously at getting possession or control of it for society, in order to solve
the problem of poverty by making the wage-earner not only a joint-owner of
the social property in land but also in capital" (Hobson 1891, v–vi, 17, 198;
Wallace's annotated copy is in ARWL).

Hobson's view of George as a key figure in "spreading the word" was also
an apt description of Wallace. Wallace's and George's careers were linked
by two notable traits. The first is their moral fervor and the second is the
hostility that their socioeconomic theories engendered among academic and
professional economists. Like Wallace, George was neither a churchgoer nor

a conventional Christian. But he was a deeply religious man with a powerful faith in a benevolent deity and the immortality of the soul. In language reminiscent of Wallace's theism, George declared his deity "not a God who is confined to the far-off beginning or the vague future, who is over and above and beyond men, but a God who in His inexorable law is here and now; a God of the living as well as the dead; a God of the market place as well as of the temple; a God whose judgments wait not another world for execution, but whose immutable decrees will, in this life, give happiness to the people that heed them and bring misery upon the people who forget them" (George 1906–1911, 1:252). Wallace possessed a copy of Godfrey Blount's *The Blood of the Poor: An Introduction to Christian Social Ethics.* That Wallace had a copy of Blount's small tract reinforces the parallels between Wallace's and George's thought. Blount's preface announced that although he was "no Political Economist in the technical sense," he still felt obliged to write as a concerned citizen on economic matters. His objective was clear: to demonstrate "the absolute dependence of all forms of energy on food, and [the] consequent indebtedness [of] honest business . . . to the agricultural life." His second major goal was to show why "religion must not be disassociated from the facts of life." A main target of Blount, as it was for Wallace, was the degradation of the agricultural worker, whose life was increasingly embedded in urban and industrial networks of "mischievous" monetary standards of value. Blount's tract lacked rigor and can best be deemed another utopian prescription for social renewal. But its twinning of religion and economics reflected the views of many Victorian social critics, including Wallace (Blount 1911, 9–10, 14–15; Wallace's annotated copy is in ARWL).

Wallace's critique of capitalist theory drew on his rejection—or redefinition—of two of the central pillars of that theory: the definition of "capital" and of "wealth." In his presidential address to the LNS in 1891, Wallace told his audience that "money is not capital. A man may have a houseful of money, but it is not capital till it is converted into tools or raw materials, and even then it will not be productive capital unless it can command both labor and intelligence to use it." In Wallace's view, shared by a number of his contemporaries, "capital, like all wealth, is created and grows solely by means of labor and intelligence applied to land or to the products of the land. It is one of the misleading errors of political economists," he declared, "that the three factors of wealth are land, labor and capital. The true factors, as long ago pointed out by our vice-president, Mr. Volckman, are land, labor and intelligence." Wallace stated his reformist manifesto:

> It follows that, when the workers have free access to the *land,* and as
> the *labor* is certainly all their own, they only need the *intelligence* to

produce in a very short time all the capital that is needed. It would be an insult to the working men of England to suppose that they have not the necessary intelligence;—it would moreover be contrary to all experience and all history, for whence has usually come the intelligence that has created the wealth of England, if not mainly from the ranks of the workers? Hargreaves and Arkwright, Watt and Stephenson, were not landlords or capitalists, though these last have derived much of their wealth from their inventions. Under present conditions the value of both capital and land to their respective owners, depends entirely on the amount of labor they can command. Both land and machinery are worthless, if left unused, while the former sometimes, and the latter always rapidly deteriorates in value. The landlord and the capitalist are therefore absolutely dependent upon the laborer, and if the laborer can be put in such a position as to be independent of them, he really becomes their master, instead of being, as now, their slave, and they will have to come to him and beg him to enable them to make something, however little, out of their property.

(Wallace 1891c, 19–20)

Wallace's 1891 address—with its echoes of George and Hobson—forcefully expressed his conviction that scientific and industrial advance, if unaccompanied by economic and political reforms of a radical nature, would only serve to perpetuate historical inequities. Wallace had learned by direct experience in his early teens the potential power, despite their current degradation, of those who worked the land. His surveying activities exposed him to both the dignity and the plight of the agricultural populations of England and Wales. The intimate contact he established during twelve years among the indigenous, mainly agricultural, peoples of the Amazon and Malay Archipelago intensified Wallace's identification with land and labor. The zeal for societal reform, sparked by his early exposure to Owenite teachings, was a striking trait in the thought and character of the youthful Wallace. That zeal, when merged with his later scientific studies, produced the mature critic. First through land nationalization and then through conversion to socialism, Wallace used his formidable intellectual powers to preach reform. He attacked the pernicious confusion between "real and fictitious wealth." Real wealth, the product of labor, Wallace believed was all too easily transformed by "our fiscal and legislative arrangements" into fictitious wealth. Private investments yielded unearned income that served to perpetuate the morally bereft and politically explosive exploitation of the "labouring poor [by] an ever-increasing class of idle rich" (Wallace 1900c, 2:256–257). The term "preach" is not merely rhetorical. For Wallace, the necessity to effect economic and political change had a religious significance. Wallace's vision was a moral and cultural regeneration of Victorian society based on his conception of the philosophical implications of a theistic and teleological evolutionary biology. Wallace's biological socialism emerged from his critical observation of the world of nature and of humans, what he would ultimately term "the world of life."

WALLACE AND THE PROFESSIONAL ECONOMISTS: JEVONS AND MARSHALL

By the late 1880s, Wallace had become a well-known figure in sociopolitical and economic controversies. He was highly respected among the land nationalists and segments of the general population, especially the laboring classes. But what was his reputation among the professional political economists? Like George, Wallace's political and economic theories were not always favorably received among academics. This is not surprising, given that neither possessed professional credentials in those disciplines. More pointedly, for Wallace, it has been suggested that Victorian political economy "only became a full-blooded social science . . . ironically [when it began] moving away from, not toward, natural history" (Schabas 1997, 87–88). Would Wallace's stature as an evolutionist help or hinder his reputation among the professional economists? A look at Wallace's relations with two of Britain's preeminent economic theorists, William Stanley Jevons and Alfred Marshall, is instructive. Unlike George, Wallace was accorded a measure of respect for his economic views among the professional economists.

The leader in Britain for the drive to erect a rigorous basis for a science of economics was Jevons. His *Theory of Political Economy* (1871) is generally credited as the major signal of the Marginal [Utility] Revolution of the 1870s. Jevons "called for a radical transformation of the conceptual foundations and methodological principles of the classical theory of Ricardo and Mill." According to Jevons and his followers, "value was determined by utility, not labor. The distribution of goods and services was the result of individual deliberations at the margin, not the incessant struggle between laborers, landlords, and owners of stock. Jevons also campaigned for the adoption of mathematics, particularly the calculus." He set in "motion the program for a unified mathematical theory" that would come to underlay the new Victorian science of economics as taught by Alfred Marshall, among others (Schabas 1997, 73).

Wallace agreed with some of what he had read by Jevons. But he rejected the attempt to create a science of economics that took mathematics, not biology, as it primary methodological tool. Wallace believed the mathematical focus would divert attention from what he regarded as the essential elements in societal reform. Mathematical abstraction, he objected, would diminish the primacy to be accorded to the actual realities of labor and land. It would also shift analysis away from efforts to terminate the incessant struggle between laborers and capitalist landlords and the beneficiaries of unearned wealth from stocks and other socially divisive financial instruments of exchange and income. Wallace possessed a copy of the second edition (1878) of Jevons's small introductory treatise *Political Economy*. Wallace did not take exception to Jevons's objective in putting "the truths of Political Economy into a form suitable for elementary instruction." Nor did he disagree with Jevons's preface, which stated that there "can be no doubt that it is most desirable to disseminate knowledge of the truths of political economy through all classes of the population by any means which many be available. From ignorance of these truths," Jevons warned, "arise many of the worst social evils—disastrous strikes and lockouts, opposition to improvements, improvidence, destitution, misguided charity, and discouraging failure in many well-intended measures." The fact that Jevons's tract was published in the series of Science Primers, under the general editorship of "Professors Huxley, Roscoe, and Balfour Stewart," was likely one reason why Wallace purchased the book. Wallace also endorsed Jevons's claim that the current system of land tenure in England, Scotland, and Ireland was, "in fact [a] feudal system." Jevons asserted that the "laws should be made not for the benefit of any one class, but for the benefit of the whole country. The laws concerning landlord and tenant have, however, been made by landlords, and are more fitted to promote their enjoyment than to improve agriculture." It was not Jevons's description of Britain's economic inequities to which Wallace objected but his proposed solution. Unlike Wallace, Jevons was willing to grant that there was no inherent evil in "few landlords with great rent-rolls," rather than many "small landlords receiving small rents"—as long as the larger landlords were constrained by laws that guaranteed an equitable distribution of the wealth thus produced. Wallace simply could not share Jevons's faith in the ability of a system of large landholdings to effect social justice and economic equity (Jevons 1878, 5, 92–95; Wallace's annotated copy is in ARWL).

It was not the concept of rent itself but the prevailing practice of "competition rent" that Wallace opposed. Setting the rent of a particular holding at the highest price the market would bear was, as Wallace stated in his 1893 presidential address to the LNS, "undoubtedly the [best] case . . . from a landlord's or speculator's point of view—considering the money income to be got from the land to be everything, the well-being of the tenants nothing. . . . But," Wallace told his audience, "from our point of view—looking at the cultivation of the land as leading primarily to the well-being of the tenants . . . it seems to me to be the very worst mode possible." Instead of free market competition, Wallace proposed an alternative method of setting rents: "the valuation of lots by an expert. . . . These lots, with the rents thus determined, would then be open to selection, either on the system of 'first come first served,' or if thought fairer, of a ballot for the order of choice on certain fixed days." Wallace concluded that either method, "supposing the valuation to be fairly made," would reduce, and ultimately eradicate, inequality of opportunity, since no tenant could complain that "either by chance or through any other cause, some of the tenants were paying higher rents than others" (Wallace 1893a).

This 1893 LNS address came three years after Wallace's public declaration as a socialist. He emphasized that a primary goal of societal reform was that all citizens should "be able to form and keep a home; to be . . . as secure in that home, so long as they pay the moderate ground-rent for the land, as if they were the actual owners of the freehold, subject only to the payment of taxes." Wallace wanted "the new tenants under land nationalisation to be really free holders in the old sense—free men holding land from the community never to be interfered with so long as they continued to pay the moderate dues and to be law-abiding citizens." Wallace's goal, after his conversion to socialism, remained that which had been a key motivation for his life's work from his early youth: "to secure the equal well-being of the whole of the industrial community, and . . . initiate that progressive improvement, with the diminution and ultimate abolition both of enforced idleness and of undeserved poverty" (Wallace, 1893a).

In the context of the rapidly evolving systems of political and economic philosophy in late Victorian Britain, any facile dichotomies between thinkers who adhered completely either to socialist or to capitalist ideologies are unhelpful. Wallace's relationship to Jevons's work, limited as it may be, underscores the necessity to recognize that of all elements of Victorian culture that were in flux, economic and political theory and action were among the most complex. There were many links, both personal and theoretical, between individuals who found themselves on different sides of a malleable ideological divide. It is instructive, if somewhat ironic, to note that the Mechanics' Institute in Neath, which Wallace had helped design and construct in 1846, was funded in part by an uncle of Jevons (Wallace [1905] 1969, 1:245–246). British intellectuals such as Wallace, who pronounced themselves "socialist," propounded versions of that doctrine that incorporated some elements of capitalist political economy that still permeated most of late Victorian thought. Wallace's conversion to socialism is crucial to an understanding of his mature evolutionary teleology (Stack 2000, 692–694). His socialism—rooted in land nationalization—is one more element in his overarching vision of progressive, and purposeful, societal evolution. Here, Wallace remained true to his Owenite training.

Avid proponents of industrial capitalism in the early nineteenth century also supported land nationalization. In their view, the abolition of private landlordism would lower costs of production to the capitalist as well as attack aristocratic power. Wallace had quite a different motive for espousing land nationalization: he wanted not to cheapen the costs of capitalist enterprises but eradicate them. As Greta Jones has recently argued, "Wallace followed Owen in seeing the establishment of egalitarian communities as the nucleus of a new and more just society which would eventually replace competitive capitalism altogether. Even in the changed political and social circumstances of the late nineteenth century, Wallace still maintained his original utopian vision of the self-governing community based on Owenite principles." Indeed, when land nationalization became associated with George's socialism in the 1880s, Spencer quickly moved away from the first idea because of his antipathy to socialism. Wallace was disgusted by Spencer's defection from the land nationalization cause (Jones 2002, 75).

CLOSE ENCOUNTERS: MARSHALL ON WALLACE AND GEORGE

Even allies as close ideologically as Wallace, Hobson, and George differed on certain theoretical and practical issues. But these differences were insignificant compared to their common goals. Wallace and George were further united by their passionate and highly public missionary-style reformism. Indeed, George's visits to Cambridge and Oxford were near disasters. His university audiences were offended by his negative references to upper-class privilege. At Oxford, a clique of Tory undergraduate rowdies jeered as George attempted one of his messianic moral exhortations. Those audiences were also outraged by what they regarded as the heresies of a "confused meddler" in the precincts of academe. The formidable neoclassical English economist Alfred Marshall rose at Oxford to declare George's arguments for land nationalization to be unsound and demeaning to the rising scholarly discipline of economics. Marshall was only one of George's academic critics, on both sides of the Atlantic, who expressed amusement and annoyance at what they regarded as George's confusion of ethics and economics. For his part, George decried the old and new university establishments' dependence on big money and their tendency to teach dominant business values in the guise of disinterested learning. On his return to the United States, George would abandon the intellectual elite as a privileged minority, who distanced themselves from the rest of society and, in his view, shirked their responsibilities to the common people. George appealed, rather, to the tradition of individuals like Ralph Waldo Emerson and Walt Whitman whom he regarded as the true social critics and ablest humanitarian reformers. The reformer's allies, George asserted, were not to be found in the ascendant professional communities of late-nineteenth-century America. His allies were those "divine average" men and women who agreed with him in condemning the seductions of wealthy industrial society and sought a new moral order (Thomas 1983, 198–201).

Wallace had his own run-in with Marshall in 1883, the year following the publication of Wallace's *Land Nationalisation*. While still professor at University College in Bristol, Marshall delivered three lectures on the evenings of February 19, February 26, and March 5, 1883. This course of lectures was an explicit attack on George's *Progress and Poverty.* They constitute a cogent, if not altogether convincing, attempt to counter George's powerful role as a propagandist for land reform. George's influence, both in England as well as his native United States, was at its peak in the early 1880s.

Progress and Poverty appeared in an English edition in 1881, one year after its initial publication in America. The English edition sold close to 100,000 copies in three years—a highly impressive number. More offensive to Marshall was the fact that George entered the British lecture circuit during 1882–1884. He gave extensive and popular lecture series, and his ideas, and the controversies they engendered, received widespread press coverage. Marshall's Bristol lectures had as their main objective the demolition of George's theories and reputation. The first lecture argued that there had been—in direct contradiction to George's message—a vast rise in the standard of living of English workingmen during the nineteenth century. Marshall's second lecture presented, in as popular a form as possible given the inherent complexity of the subject, his general marginal productivity theory of distribution. The third and final lecture was a sustained attack on nationalization of the land.

Marshall's lectures elicited two letters from Wallace contesting Marshall's evidence and approach. Marshall responded. Their exchange demonstrates Wallace's growing reputation as a political activist (Stigler 1969).[3] While Marshall despised George both on intellectual and on personal grounds, he treated Wallace's views with respect. Marshall engaged Wallace not as a professional economist (which he was not) but as an eminent biologist whose views and quantitative arguments demanded serious attention. Marshall's lectures indicated that he was willing to entertain certain fairly radical land reforms. These included the limitation of all sales of state lands to hundred-year leases and the purchase by the state of survivor interests in land. These two particular planks in land reform agendas do not differ materially from those put forward by Wallace. Marshall's second letter to Wallace states that "Mr. Wallace cannot desire [certain land reform measures] more heartily than I do" (Stigler 1969, 186, 216). The irony here, of course, is that it was Wallace's eminence as an evolutionary scientist that prompted Marshall's markedly more temperate response to Wallace's critiques. After all, was not Wallace a scientist, rather than (like George) a poet? In the discussion that followed Marshall's second Bristol lecture, one questioner asked why Marshall had termed George a poet. The questioner then "called attention to the fact that Mr. George advocated nationalisation of the land as a remedy for poverty, and asked how it was that Mr. A. Wallace, an able man, came to the same conclusion. Professor Marshall said Mr. Wallace's proposal was much more reasonable than that of Mr. George. He did not call Mr. George a poet because he said erroneous things. He was a poet because he was poetic, and he was not a man of science because he said erroneous things" (Stigler 1969, 199). Marshall missed the main point that Wallace refused, as fervently as did George, to demarcate science from ethical and moral considerations.

Wallace encountered George again, during his tour of North America. Arriving in New York on 23 October 1886, Wallace stayed with the journalist A. G. Browne, who had called on Wallace during the summer at Godalming. In the cab to Browne's house, Wallace found himself accompanied by George, whom Browne had specifically invited to meet his houseguest. Wallace spent much of his remaining time in New York with George, who was then embroiled in a contentious race for mayor of New York City. Wallace addressed a gathering of George's followers on 25 October, informing them of reform movements in England. The New York mayoral campaign was closely monitored nationally. Although George was not successful in this bid, he came in with a surprisingly strong showing with the second largest vote in the 2 November election, edging out third-place Theodore Roosevelt. The healthy George vote, which pleased Wallace greatly, prompted a writer for the *Boston Evening Transcript* to declare, on 3 November 1886, that George now "holds the balance of political power in [New York] city" and "can defeat either Democrat or Republican who in the past or future has or shall give offence to the labor element. He is admitted to be a greater man today than if he had been elected mayor, and in the future will be feared and courted by both parties."

Given Wallace's own merging of spiritualist and socialist convictions, it is of particular interest that on Wallace's last full day in New York City (27 October), George spoke before a group of spiritualists. Arguing against his opponents' claims that a mayoral victory for him would bring anarchy and chaos to society, George drew a parallel between his own political platform of equitable distribution of wealth, including land, with the spiritualists' belief that "we go from this life to another life. You believe also, as I believe, that wherever we go we will find something to maintain our existence—that in the other life there are things for our support somewhat analogous to" those available in this life—when suitable reforms are achieved. George concluded his brief talk with the hope that the mayoral campaign would be the means of bringing the issue of equitable distribution of wealth into the open. A *New York Herald* story on 28 October 1886 reported that George emphasized that "the land belongs, as Thomas Jefferson says, in usufruct to the people: that everyone that comes into this world has the right to a foothold and to all the materials necessary to maintain life and an equal right to all that nature has provided for all such things." Allowing for a large dose of rhetoric in George's talk at the near climax of the mayoral contest, it is pertinent that George publicly drew a link between spiritualism and political reform. Just as Wallace had drawn enthusiastically on George's socioeconomic theses, George—in this week together in New York—had imbibed a modicum of Wallace's fervent spiritualism.

The harmony between George's and Wallace's public statements aroused admiration as well as hostility. George's theory in regard to land and its ownership, as codified in *Progress and Poverty,* came under scathing attack in an article in the New York *Independent.* The author, Hugh P. McElrone (editor of *The Catholic Mirror*), attacked George's proposals for reform as corrosive of the "principles upon which the American Republic is founded, nay, the principles upon which rests the whole fabric of Caucasian civilization." George was guilty not only of Godlessness but also of proposing a remedy for social and political ills that would set "loose the dreadful contagion of Communism" (McElrone 1886). Such intemperate attacks as this would become less common as George's reform ideas gained ever greater currency both in North America and in Great Britain. Wallace and George became closely linked in the public arena as men of similar hopes for societal reform. George's belief that the nineteenth "century was closing in darkness, that the principle of democracy, which triumphed in 1800 with the ascendancy of Thomas Jefferson to the presidency of the United States, might be conquered by the Hamiltonian principle of aristocracy and plutocracy in 1900" was one of the great spurs to his continued political activism. A similar note of ambiguity, if not despair, as to the prospects of an unreformed social order was the leitmotif of Wallace's *The Wonderful Century: Its Successes and Its Failures* ([1898] 1970). Both men were convinced of the absolute necessity of fundamental change in the emerging industrial world of the dawning twentieth century.

NORTH AMERICAN TOUR: 1886–1887

Wallace's reconnecting with George was but one aspect of his North American tour of 1886–1887. His sojourn in Boston, immediately following the several days spent in New York, was crucial in exposing Wallace to new philosophical currents (see chap. 3). New York and Boston were only the first two stops in what proved to be an eventful year in Wallace's life. Visits to Washington, D.C., and San Francisco, among other cities and towns, enabled Wallace to gather more general observations and experiences relating to the socioeconomic and political climate of North America. The details of these visits provide valuable insights into Wallace's frame of mind at this significant stage in his own evolution. During his tour, Wallace met many of the United States' most distinguished scientists as well as numerous leading political, social, and intellectual figures. In addition to speaking on scientific subjects, Wallace lectured on his sociopolitical views. He also made the acquaintance of America's leading spiritualists and temporarily resumed frequent attendance at séances. The single most lucrative lecture he gave was not on evolutionary biology but on spiritualism. Wallace delivered "If a Man Die, Shall He Live Again?" to an enthusiastic San Francisco audience of more than a thousand persons (Wallace [1905] 1969, 2:129, 160).

Since the death of Darwin in 1882, Wallace was England's elder statesman of evolutionary biology. Nonetheless, his financial status was, as ever, not commensurate with his fame. When he embarked from London, it was aboard "a rather slow steamer in order to have a cabin to myself at a moderate price." Wallace had originally hoped that his tour would extend to New Zealand, Australia, and South Africa. The fact that his London to New York passage made him "sick and unwell almost the whole time" moderated his expectations. Adding to the physical discomforts of travel—Wallace was no longer the intrepid explorer who had once surmounted the obstacles and discomforts of the Amazon basin and the Malay Archipelago—was his unfortunate choice of an American agent. The agent was lax, and there were costly gaps between Wallace's lecture engagements (Wallace [1905] 1969, 2:105–107). A major impetus for Wallace's tour was "the prospect of clearing 1,000 [pounds] by a lecturing campaign [which], though it would require a great effort, [offered] the chance of earning a lot of money which would enable" him to pay off his mortgage and leave at least some modest estate for his family (Marchant [1916] 1975, 392). The tour of the United States and Canada, while not notably successful financially, did afford Wallace the firsthand opportunity to explore the curiosities of North American flora and fauna, geology, and geography. Equally important, for his increasingly significant political and social activism, was Wallace's direct contact with the varied cultures of the human inhabitants of Britain's former American colonies. These contacts with scientists, statesmen, politicians, educators, and journalists as well as segments of the general public in North America would shape certain of Wallace's cultural writings and opinions on his return to England.

Before leaving Boston permanently, Wallace made a number of regional side trips. In late November, he visited the eminent Yale professor of paleontology Othniel Marsh in New Haven, Conn. Wallace described his host as "very wealthy, [who] had built himself an eccentric kind of house, the main feature of which was a large octagonal hall, full of trophies collected during his numerous explorations in the far West." Marsh's house was situated near Yale College's Peabody Museum, which contained an extremely large collection of fossil skeletons, chiefly of mammals and reptiles of America. Wallace was deeply impressed on seeing "these wonderful remains of an extinct world." Especially striking were the huge bones of the atlantosaurus—nearly 130 feet long and thirty feet high and, at the time, generally regarded as the largest land animal that had ever existed—and the remarkable horned dinosauria. Most significant for the statesman of evolution was the Peabody's nearly complete series of links connecting the modern horse with the ancient eohippus and hyracotherium. Wallace also met James Dwight Dana, the celebrated Yale geologist. He was taken on a walking tour of New Haven, which Wallace described as one of the most picturesque and pleasing cities he visited during his American travels. Wallace gave several lectures in Baltimore, including one titled "Island Life" at the recently founded Johns Hopkins University. He called on Hopkins's President Gillman, who introduced him to a number of professors and students of different disciplines. In the next few days Wallace spoke, formally and informally, on evolutionary theory. When asked by a group of psychologists as to his own interest in the unsettled problems in that field, he candidly replied that he paid little attention to them. Wallace "was only interested in the question of how far the intellectual and moral nature of man could have been developed from those of the lower animals through the agency of natural selection, or whether they indicated some distinct origin and some higher law" (Wallace [1905] 1969, 2:112–114).

Leaving Boston at the end of December, Wallace arrived in Washington, D.C., on 31 December 1886. Except for several brief excursions, he spent the next three months in the nation's capital. Once again, his visit was a rich mixture of scientific, political, and spiritualist undertakings. He was especially pleased to be able to meet with the many scientists in government departments and at the Smithsonian Institution. Wallace did not give his Lowell lectures but did read two nonspecialized papers. One, before the Woman's Anthropological Society, dealt with the "Great Problems of Anthropology." The other was "Social Economy *versus* Political Economy." In this second paper, Wallace presented his developing reformist views and increasingly strident critique of competitive capitalism. It was, in his words, "altogether too revolutionary for many of my hearers." A reporter from the *Washington Post* wrote that it was "astounding that a man who really possesses the power of induction and ratiocination, and who, in physical synthesis has been a leader of his generation, should express notions of political economy, which belong only or mainly to savage tribes." The *Post* article typified the reaction of some of Wallace's North American—as well as British—colleagues to what they regarded as an unfortunate, and misguided, splitting of his interests into "sound" scientific and "unsound" (or unpalatable) sociopolitical domains. Wallace attributed the poor reception of his political talk in Washington to the fact that "there was hardly a professed socialist in America." In his autobiography, written in 1905, Wallace would comment that "in the eighteen years that have elapsed since this paper was read an enormous advance in opinion has occurred, and to-day, not only to a large proportion of the workers, but to thousands of the professional classes, the views therein expressed would be accepted as in accordance with justice and sound policy" (Wallace [1905] 1969, 2:128–129).

WALLACE AND LESTER WARD

One of the high points of Wallace's stay in Washington was the close personal and intellectual bond he formed with the sociologist Lester Frank Ward. Though united by their socialist leanings, Ward proved recalcitrant to the force of Wallace's spiritualist convictions. As Wallace diplomatically put it, Ward was "an absolute agnostic or monist, and around this question our discussions most frequently turned. But as I had a basis of spiritualistic experiences of which he was totally ignorant, we looked at the subject from different points of view" (Wallace [1905] 1969, 2:118). Ward was sympathetic to Wallace's emerging evolutionary teleology. Ward's sociology was dualistic, not monistic, in one major respect. He stressed the distinction between physical, or animal, purposeless evolution and mental, human evolution decidedly modified by purposive action. Ward's purposive agents were human beings and their cultures, not Wallace's spiritual intelligences. But as did Wallace, he severed social principles from simplistic, direct biological analogies. Ward was among the first and most influential of those American thinkers who attacked the laissez-faire individualism of conservative social Darwinism and the conservative use of science generally (Commager 1967, xxxi–xxxiii, 74–77).

Wallace particularly admired Ward's 1893 *Psychic Factors of Civilization* (Wallace [1905] 1969, 2:117). Ward, on his part, agreed with Wallace that natural selection was incompetent to explain the higher faculties of humans. In *Pure Sociology* (1903), Ward cited Wallace's *Contributions to the Theory of*

Natural Selection ([1870] 1969) and *Darwinism* ([1889] 1975) as providing the definitive arguments for curtailing the theoretical role of natural selection in human evolution. But the "psychic factors" Ward invoked as the agents responsible for the origin and development of the higher human faculties were themselves products of human evolutionary advances. These psychic, or "telic" (Ward's preferred terminology), factors served in Ward's sociological theory to demonstrate that the "law of mind" is distinct from the "law of nature." Telic forces, themselves the product of human cultural evolution, prodded further evolutionary advances. Both Ward and Wallace shared a conviction in an ultimately purposive, directed evolutionary process. But whereas Wallace embraced a theistic evolutionism, with interventionist providential guidance, Ward's system relied on psychic forces that arose solely from the evolutionary process itself (Ward 1903, 490–492, 497–499).

Ward's notion of progressive evolution, as was the case with LeConte, also contained traces of Lamarckism. Wallace remained adamant in his condemnation of neo-Lamarckian intrusions into evolutionary biology in the closing decades of the nineteenth century (Wallace 1908a). But he shared Ward's and LeConte's social progressionism. In the case of LeConte there was an additional link to Wallace. LeConte argued that the theological implications of evolution were consonant with human progress in his *Evolution and Its Relation to Religious Thought* (1888). The paleontologist Edward Drinker Cope was another evolutionary theist, whose *Theology of Evolution* (1887) made explicit his conviction regarding the divine origin of the consciousness underlying evolution (Bowler 1983, 126–127).[4] Despite the theistic affinities between Wallace and LeConte and Cope, it was Ward with whom he formed the closest personal bond. Their shared faith in socialism as a precondition for progressive human evolution, and their mutual scientific interests (Ward was also a paleobotanist), united Wallace and Ward as intellectual comrades.

Several letters from Wallace to Ward afford additional evidence of their close political affinity. Ward was less activist than Wallace and less comfortable with publicly styling himself a socialist. He preferred to promote the cause of equality of opportunity for the masses from the academic pulpit. He coined the term "sociocracy" to characterize his belief that social and economic inequities should be abolished but within a framework of meritocracy (Stern 1935). Ward's proposed path to a socialist future thus differed slightly from Wallace's. These letters, because they were written to an American confidant, show Wallace in a less guarded moment. Two are particularly revealing.

The first was sent to thank Ward for sending a copy of *The Psychic Factors of Civilization*. It merits quotation because of the critical comments Wallace made regarding Spencer:

I have read the *third part* through, carefully, & think your exposition of the scientific character of *Socialism* as opposed to Herbert Spencer's *Individualism* exceedingly forcible, and calculated to do much good. I have also looked through & read a good deal of the *first & second* parts, which however being so purely psychological does not interest me so much. Chapter XVII on *Social Friction* is however an exception . . . & would come better in the 3rd part. If these were embodied together, with a good deal more of concrete illustration, it would form an excellent work on the Scientific Basis of Socialism, which would have great value as a weapon against the individualist school, and would enlighten many who are now blinded by the prestige of Spencer & the Political Economists. . . .

No doubt we *are* advancing on the very lines you point out as the true ones, but only empirically, and so much in the very teeth of the popular political economy that politicians only give way to it as a concession to the demands of the populace. I think I shall try to make known your doctrine in the form of a popular review article, though it will be a difficult job.

How dreadfully Herbert Spencer has fallen off in his *Justice* [1891]. Parts of it are so weak and illogical as to be absolutely childish. You have no doubt seen H. George's severe criticism of it.

(Wallace to Ward, 21 November 1893, in Stern 1935, 378–379)

Wallace had his differences with Spencer, and these differences augmented, particularly after Wallace's public declaration as a socialist. But he never characterized Spencer so harshly in his published writings. In his autobiography, Wallace referred to Spencer's *Justice* as a work that showed that equality of opportunity has "a broad foundation in the laws of nature." His sole, and mild, criticism is that "Spencer himself did not follow out his principles to their logical conclusion" (Wallace [1905] 1969, 2:272). One would be hard-pressed to recognize that Wallace is referring to the same book.

The second letter, acknowledging receipt of Ward's *Outlines of Sociology*, displayed a side of Wallace not often revealed. After telling Ward it was one of his best works on sociology—"Never was sound teaching on the subject more wanted, and wise legislation, if we are not to be soon plunged into a revolution"—Wallace noted that he had been avidly reading "Mr. Wyckoff's papers on the Workers in Scribners' Magazine." These prompted Wallace to ask himself how much longer laborers would continue to work at difficult and dangerous trades for pitiful wages and with no prospect of rest and comfort in their old age. Viler still was the fact that even getting work on such scandalous terms was becoming difficult if not impossible for increasing segments of the labor force. Wallace's dismay and disgust induced him to tell Ward, in one of the few writings extant in which Wallace alludes at any length to his family, that

> the whole miserable system—or want of system—has also been brought more vividly before me by my son's experience in America where he has now been a year and a half. He has had the best education I could give him in Electrical Engineering—3 years in College and 3 years in the workshops & at various jobs. So far, in America, he has been able to get nothing but labourer's or lineman's work at moderate wages, but the bosses always keep them at high pressure for nine hours a day, after which of course they are not fit for much but eating and sleeping. . . . He is a thorough Socialist, and makes friends with most of the men he works with, but after a job, they often have weeks or months of idleness before they get another. What a terrible thing it is that under the present social system, the vast majority of workers, however steady and well educated, have, and can have, no prospect but a life of toil and an old age of poverty or worse—and this when the work actually done, if properly organized, would provide not only necessaries but comforts for all, with ample leisure and a restful old age. Surely the coming century must see the end of the existing system of cut-throat competition, and wealth-production based on the misery & starvation of the millions!
>
> (letter from Wallace to Ward, 12 October 1898, in Stern 1935, 379)

Wallace was no armchair naturalist, nor was he an armchair socialist. He had explored nature in the field. He had experienced economic and political inequities firsthand, through his own experiences and those of his brothers and now his son. This letter, written close to the time of *The Wonderful Century* and a decade after the North American tour, tells us more about the personal sources of Wallace's cultural critiques than many of his more well-known books and articles. The combination of theoretical and practical factors shaped Wallace's sociopolitical views as profoundly as they did his scientific ones.

SENATOR STANFORD

Of all the high profile individuals Wallace met in Washington, it was Senator Leland Stanford who best exemplified what Wallace regarded as the paradoxes of America. Wallace's ambivalent reaction to meeting and being befriended by Stanford is testimony to Wallace's own tensions at this critical stage in his career. He was introduced to the extremely wealthy Stanford and his wife by Mrs. Beecher Hooker. The sister of Henry Ward Beecher and Harriet Beecher Stowe, she, like the Stanfords, were ardent spiritualists. The Stanfords informed Wallace that they had both "had long-continued intercourse" with their only son, who had died at age sixteen three years before in Florence. Their communications with their son occurred at séances with "several different mediums, and under circumstances that rendered doubt impossible." Wallace concluded that the senator, "a man of exceptional ability and intellectual vigour, . . . would hardly be imposed upon in such a matter." Wallace again encountered the Stanfords at the home of Charles Nordhoff, author of the influential 1875 book *The Communistic Societies of the United States* (Wallace [1905] 1969, 2:119, 165–166). Nordhoff gave Wallace a copy of his work, inscribed "To Dr. A. R. Wallace, with the author's friendly regards, Washington, Feb. 1887." Wallace read Nordhoff's account of his visits to various communities including those at Oneida, Bethel, Aurora, and several Shakerite communes with great interest. He marked several passages with triple underlining. Nordhoff's accounts described actual communities predicated on philosophies that had become highly congenial to Wallace. Nordhoff's rosy depiction of the egalitarianism, security, frugality, and healthfulness of rural communes resonated deeply with Wallace' personal and social values (Nordhoff 1875, 393–395, 402–403; Wallace's annotated copy is in ARWL).

Wallace had been introduced to Nordhoff by Major John Wesley Powell, director of the U.S. Geological Survey. Powell was a noted anthropologist as well as a geologist. He and Wallace quickly formed a friendship based on their mutual interests. Wallace was also given privileges at the prestigious Cosmos Club. In its comfortable confines he encountered some of Washington's, and America's, most eminent "scientific men and women." Many people had told Wallace that his *Malay Archipelago* had "first led them to take an interest in natural history and its more general problems." Powell took Wallace to a meeting of the Literary Society, held at Nordhoff's home. Wallace met "hosts of people who were really too polite and enthusiastic— 'proud to meet me;' 'honour and pleasure never expected;' 'read my books all their life!' etc.—leaving me speechless with amazement!" The paper read that particular evening was "by Mr. Kennan, describing his recent visit, on his return from Siberia, to Count Tolstoi, the great Russian novelist, philanthropist, and non-resisting nihilist. It was a very clever, sympathetic, and suggestive picture of a man described as a 'true social hero—one of the Christ type.' " Wallace was deeply interested in Tolstoy's life and political philosophy. He owned more than thirty books by the Russian sage [now housed at Edinburgh University Library]), many of which he annotated extensively.

Wallace often dined, thereafter, at Nordhoff's home (Wallace [1905] 1969, 2:118–119).

Wallace became part of Washington's social circuit. He particularly enjoyed Senator Stanford's group of (mainly wealthy) spiritualist friends. He was invited to stay with the Stanfords for a few days at their luxurious country house at Menlo Park when he later visited California. That state, especially the vibrant city of San Francisco, was home to a sizable population of spiritualists. Wallace's visit to Menlo Park was intriguing. He regarded Stanford as a striking exemplar of the advocates of spiritualism. Wallace's host impressed him further with a drive to see some of the other millionaires' residences and with a visit to the site of the grand university he and his wife were building to the memory of their son. To the nascent socialist, however, all was not as it appeared. The Stanfords' country house and their spectacular mansion in San Francisco were tended entirely by hordes of Chinese servants. The senator assured Wallace "that the Chinese had been the making of California, doing all kinds of domestic work, gardening, and shop-keeping when every European was rushing after gold." Stanford had initially endured much public hostility because of his opposition to the anti-immigration laws of California and through his large-scale employment of Chinese servants and workers. He believed that he had now lived that down by demonstrating how productive immigrant, as well as American, labor was in making his adopted state "the richest part of the Union." As a gracious and dazzled guest, Wallace did not voice at that time his antipathy to his host's practices. He did assert later, in his autobiography, that Stanford's view as to the general well-being of all the inhabitants of California was fallacious:

> He looked at the world, just as our legislators do, from the point of view of the employer and the capitalist, not seeing that *their* prosperity to a large extent depended on the presence of a mass of workers struggling for a bare subsistence. At the very time of our interview [in that summer of 1887] the actual fruit-grower could hardly earn the scantiest subsistence, because he was dependent on the middlemen and railway companies to get his crop to market, and because the very abundance of the crop often so lowered prices as to make it not pay to gather and pack. Since then, year by year, the unemployed and the tramp have been increasing in California as in the Eastern States, while San Francisco reproduces all the phenomena of destitution, vice, and crime characteristic of our modern great cities. But neither capitalists nor workers yet see clearly that production for *profit* instead of for *use* necessarily leads to those results. The latter class, however, thanks to the socialists, are rapidly

learning the fundamental principle of social economy. When they
have learnt it, the beneficent and peaceful revolution will commence
which will steadily but surely abolish those most damning results of
modern (so-called) civilization—insanitary labour, degrading over-
work, involuntary unemployment, misery, and starvation—among
those whose labour produces that ever-increasing wealth which their
employers are proud of, and which their rulers so criminally misuse.

Although Wallace delighted in the spiritualist fervor of his California ac-
quaintances, the specter of social inequity that he witnessed firsthand at the
Stanfords was never far from his thoughts (Wallace [1905] 1969, 2:165–169).

Wallace's remaining weeks in North America did little to lessen the impact of the vivid picture of social and economic inequity that both he and
George detested. Wallace spent his final days in California experiencing both
the natural beauty and the somewhat eccentric human behavior on display
in that state. He went with his brother to the region around Santa Cruz, then
known as a health resort on the Pacific coast south of San Francisco. He
saw not only the forest tracts that housed some of the finest specimens of
redwood still left in southern California but also the luxuriance of California
gardens. He was particularly struck by the common scarlet geranium, which
"grew into large bushes, forming clumps six or eight feet high [in] a mass of
dazzling colour." Returning to Stockton for a final week, Wallace witnessed
a Fourth of July celebration. At the town hall, a schoolboy read the Declaration of Independence and then the civic "oration" was delivered. Both
were "pretty good in substance, but declaimed with outrageous vehemence
and gesture. Then a patriotic poem was recited by a lady, but two crying
infants and exploding crackers outside much interfered with the effect." The
rest of the day was "a kind of small and rough carnival," with fireworks,
constantly exploding crackers, and processions of animals, clowns, crowds
of people, and members of all the trades and professions, including firemen
and members of the army corps (Wallace [1905] 1969, 2:169–170). After
this exuberant display of Americana, Wallace left Stockton that evening to
begin his journey back eastward across the continent.

EVOLUTION RUN AMUCK?

Wallace had only one more lecture engagement, at the Michigan Agricultural
College on 29 July 1887. He had ample time to explore the effects of rapid
industrial expansion in many regions in the foothills of the Sierra Nevada and
the Rocky Mountains that were once pristine. The effects of hydraulic mining
were already reducing the once fine and fertile valleys to a waste of sand,

gravel, and rock heaps. Wallace was struck by the size of the operations of the great Nevada silver mines, those same mines that were rendering English mining no longer competitive or profitable. Wallace was also struck by the majesty of the Rocky Mountains, with their exquisite wild flowers, many of which were species "quite new to me, and of very great interest." He also noted the "smooth, rounded forms of the rocks here [which] are plainly due to the effects of glaciation." Wallace's journey eastward took him to Lake Tahoe, Reno, and Salt Lake City, in whose environs he saw flowery valleys that "equalled the finest of the European Alps." He also recorded that, "as compared with Switzerland, the Rocky Mountains are very poor in snow-clad peaks and high alpine scenery, but are quite equal, and perhaps even superior, in the number, extent, and grandeur of its can[y]ons or deep valley gorges" (Wallace [1905] 1969, 2:169–178). By 18 July, Wallace had reached Denver and, by 27 July, Omaha. Arriving in Chicago, he was struck by the numerous houses being constructed in bare open country—"indications of a land 'boom,' such as are continually got up by speculators." Chicago itself seemed "enveloped in a smoky mist worthy of London itself." It appeared to Wallace as a London run riot, disfigured by the railroad companies and intense urban manufacturing concentrations. Like all new American cities he had seen, Wallace described Chicago as an architectural as well as sociological hybrid.

Slums coexisted side by side with handsome shops and palatial town-houses. In the intense heat of a Midwestern summer, screeching engines of various types and function poured "out dense volumes of the blackest smoke, and at this time of year the grass is dried up, and the trees all blacker than in London." Wallace noted, with obvious relief, that he did not have any "busi-ness to keep me in Chicago." Memories of the tropics, and even memories of London, conjured up far more pleasing images to this North American traveler. Wallace endorsed the opinion of a "writer in 'The Century' for June 1887, [who] well says—'A whole huge continent has been so touched by human hands that over a large part of its surface it has been reduced to a state of unkempt, sordid ugliness'" (Wallace 1886–1887, journal entry for 29 July 1887). His stay at the Michigan Agricultural College was decidedly more pleasant. After giving his final lectures on "Darwinism" and "Colours of Animals," Wallace enjoyed "the botanical garden attached to the college, the library, and the insect collections, which latter were very fine."

The last phase of his North American journey took Wallace first to Kingston, Canada. There, he was a guest of the parents of his friend Grant Allen, at their delightful old country house on the shore of Lake Ontario. Wal-lace beheld trilliums in flower for the first time, boated around the celebrated Thousand Islands and spent one night at the elegant Thousand Islands Hotel.

Viewing the St. Lawrence River from its broad verandah, he was struck by the "varied and beautiful combinations of rock, wood, and water, hardly to be surpassed." He paid particular attention to the rock masses or ridges, all of which displayed "striae, furrows, and deep scorching" which followed the general direction of the St. Lawrence valley. "This," Wallace asserted, "is the most conclusive indication of ice-action as opposed to other causes." The delights of the Canadian landscape ended on his entry into Montreal. Its appearance was "much spoilt by factory chimneys and the usual but quite unnecessary pall of smoke." Once again, Wallace drew what for him was now a dogmatic conclusion: "For all this unsightliness in almost every city in the world, land monopoly and competition are responsible." Wallace embarked from the port of Montreal aboard the steamer *Vancouver*, bound for Liverpool with one brief stop at Quebec City (Wallace [1905] 1969, 2:180–191).

Wallace had found that "over the larger part of America everything is raw and bare and ugly." He attributed this to "a want of harmony between man and nature." The softening effects of "human labour and human occupation carried on for generation and generation in the same simple way" were nonexistent in the rapidly industrializing United States. Wallace was appalled that the "slow and gradual utilization of natural forces allowing the renovating agency of vegetable and animal life to . . . clothe the whole landscape in a garment of perennial beauty" had been sacrificed to the mania for industrial growth. America was not unique. Wallace commented that his beloved England was in the process of creating ugliness and destroying both natural and human beauty. "But in America," he felt, "it is done on a larger scale and with a more hideous monotony." The blame, Wallace declared, lay not with "our American cousins [but from] the evil traditions inherited from us." The greatest of the evils bequeathed by Europe to the (recent) emigrants to North America was "the ingrained belief that *land* [could] be . . . monopolized by capitalists or by companies." To Wallace, this was a recipe for disaster. It left "the great bulk of the people as absolutely dependent on these monopolists for permission to work and to live as ever were the negro slaves of the south before emancipation." More so than in England or Europe, Wallace condemned North American land speculation as the vast organized business curse of purported progress.

In North America, Wallace believed he had witnessed evolution run amuck. "A nation formed by emigrants from several of the most energetic and intellectual nations of the old world, [included] from the very first all ranks and conditions of life—farmers and mechanics, traders and manufacturers, students and teachers, rich and poor." Wallace emphasized that "the very circumstances which drove them to emigrate led to a natural selection of the *most* energetic, the *most* independent, in many respects the *best* of their several nations." Such a population would necessarily, he averred, develop the virtues as well as the prejudices and vices of the parent stock in an exceptionally high degree. In words that would echo later in *The Wonderful Century,* Wallace declared that America showed the potential catastrophe of unbridled scientific and economic growth. To him, America seemed an experiment of civilization in the making. The experiment was one in which the cult of capitalist progress had obliterated any constraints, ethical as well as socioeconomic. The only hope, Wallace would learn shortly, lay in the fact that it was an American, Edward Bellamy, whose "books first opened the eyes of great numbers of educated readers to the practicability, the simplicity, and the beauty of Socialism." Wallace did not declare himself publicly as a card-carrying socialist until 1890. But socialist sympathies, dating from his youthful Owenite experiences, are evident from the 1840s onward. The 1886–1887 tour clearly served to enhance his socialist proclivities. Wallace returned from his year in North America with an invigorated sense of his mission as a crusader for social and political reform (Wallace [1905] 1969, 2:191–199).

WALLACE AND GEORGE IN BRITAIN

The writings and personality of George, one of America's most effective reformist firebrands, had made a deep impression on Wallace. George's zeal was contagious, and Wallace was not immune. Wallace and George locked arms in their struggle against privilege and their hopes for a soon to be realized utopian future. That class privilege was a provocative issue in late Victorian Britain is obvious. In the discussion period following Marshall's third Bristol lecture in 1883 (mentioned earlier in this chapter), one questioner had declared that "the Professor [Marshall] gave them political economy for the rich and middle classes and not for the poor" (Stigler 1969, 211). Class bias, from the other end of the spectrum, was at the core of the scorn George's lectures provoked in elite British quarters. When George had lectured in Oxford on 15 March 1884, before a rather different audience from the one to which Marshall spoke in Bristol, the meeting ended in chaos. George possessed the ability to inspire loathing as well as inspiration among various audiences. His Oxford listeners, who included Marshall, accorded George and his theories scant support. The account of that meeting betrays an antipathy to George, which he reciprocated. One Oxford undergraduate haughtily reminded George that there "were ladies present." George had just remarked that if only one class of society owned the land, "that class ruled the people who ruled the men of England. Who were the men to whom they applied the same title that they applied to the Deity, your Lord? . . . There were on this island to-day men who trembled in the presence of their landlords almost as slaves." Those comments, among numerous similar ones, elicited vehement cries of "shame" from the audience. It is abundantly clear from the transcript of the meeting that the Oxford audience found George, not the elite of England, shameful (Stigler 1969, 217–226, esp. 223).

Wallace was never so overtly confrontational as George. But they pursued a common cause. Wallace was always proud of his modest roots. In a postscript to his presidential address to the LNS at the annual meeting in 1885, Wallace stated that it was his early and close relationship to the land that instilled in him a lifelong opposition to private ownership of that precious component of nature. Wallace had been forced to leave school for financial reasons at the age of fourteen. He worked as an apprentice land surveyor with his elder brother in both England and Wales. The experience of living on the land during some of the most impressionable years of his life gave Wallace an intimate "knowledge of peasant life and an interest in agriculture." It also gave him a desire to reform existing British class structures (Wallace 1885c).

So close was Wallace and George's public relationship that in Britain, for many years, George's single tax and Wallace's land nationalization were linked as parts of the same political program. This linkage had significant impact on British politics. The Fabian Bernard Shaw attributed his conversion to socialism to his purchase of *Progress and Poverty* (at Wallace's land nationalization meeting). He declared that when he was "swept into the great Socialist revival of 1883, I found that five-sixths of those who were swept in with me had been converted by Henry George" (Henderson 1911, 152–153). That Shaw and so many others were converted to socialism by the nonsocialist George is explicable when George's objections to socialism are clarified. He opposed not the goals of socialism but what he, and many of his generation, saw as the socialists' means to those ends—theories of materialism, doctrines of class conflict, and predictions of revolution. George singled out what he perceived to be the loss of individual freedom inherent in any socialist scheme. Late-nineteenth-century socialism was not a single movement, and George had become fixated with one particular variant of socialism. Shaw, and Wallace, would find socialism as interpreted by George equally abhorrent. Wallace's socialist vision explicitly precluded all those doctrines that so repelled George.

What George saw as the best answer to the problem of poverty in the industrial age, the single tax, or appropriation of all ground rent through taxation, was itself a utopian remedy. The single tax was designed to provide a complete solution to the problem of land in a rapidly industrializing era. It avoided a choice between individualism and collectivism by retaining a nominal right of private property. This left improvements on the land in the hands (and pockets) of the owners, while retaining the land itself (and its confiscated rents) for the benefit of the entire community (Thomas 1983, 117–122). George's message was simple and alluring. He did "not propose either to purchase or confiscate private property in land. . . . Let the individuals who now hold it still retain, if they want, possession of what they are pleased to call *their* land. Let them buy and sell, and bequeath and devise it. We may safely leave them the shell, if we take the kernel" (George [1879] 1938, 405). To Liberal Prime Minister Herbert Asquith, a friend of Wallace, George and Wallace were two arms of a pincer movement. Asquith's view, which saw the remedies of the two social reformers as complementary utopian visions, was accurate. When (after 1890) Wallace became an outspoken socialist, he continued to support George's single tax movement. Upon George's death in 1897, the single tax movement dominated land reform efforts in Britain from 1895 to 1914, and beyond. However, when land reform finally came in the Town and Country Planning Act (1947), aspects of the act recalled more of Wallace's ideas than those of George (Gaffney 1997, 614–615).

So great was George's charisma that he was asked to run for mayor of New York again in 1897. Though only fifty-eight years old, George's health had long been precarious. It was against both his doctors' orders and his family's wishes that he agreed to this second run at the mayoralty. His basic optimism, despite his frail condition, surfaced at the prospect of another electoral campaign. He declared to his sons that the "great, the very great advancement of our ideas may not show now, but it will. And it will show more clearly after my death than during my life. Men who now hold back will then acknowledge that I have been speaking the truth." George did not live to complete the campaign. He died of a stroke on 28 October, just five days before the election. His body lay in state in central Manhattan, with more than a hundred thousand mourners passing his bier. The obituary in the *New York Times* eloquently summarized the thoughts of many:

> Profoundly tragic as is the death of Henry George at this moment, it can truly be said that his life closed in the noblest services to his ideals, fitly rounding a career that from the start has been singularly worthy. . . . Whatever we may think of the theory he worked out, no one can dispute its benevolent spirit. . . . He was the most unselfish of men. He coveted neither wealth nor the leisure so dear to the thinker. Ambition in the ordinary sense did not move him, and though he dearly loved the sympathy of his fellow-men, the usual rewards of popularity left him indifferent. His courage, moral and intellectual, was unwavering, unquestioning, prompt, and steadfast
>
> (Schwartzman 1997)

George's obituary could well have been describing Wallace's own personal characteristics.

THE WONDERFUL CENTURY?

Wallace's North American tour and his close ties to George encapsulated the conundrum of his later years. He would continue to find a coterie of scientists, associates, and friends who endorsed Wallace's increasingly explicit evolutionary theism. Its strong doses of socialism and spiritualism elicited enthusiastic reactions from the adherents of those widespread movements. But certain of his colleagues, scientists and others would publicly bemoan Wallace's embrace of what they perceived as crusades that tarnished his scientific reputation. A review of Wallace's *The Wonderful Century: Its Successes and Failures* ([1898] 1970), which appeared in the *American Historical Review* more than a decade after his North American tour, captured the sense of unease that Wallace's broadening public stance was engendering in many of his admirers. The reviewer noted that the "book, though suggestive and interesting as the product of a mind distinguished for its accomplishments in the field of physical science, is yet disappointing to one who looks to it for a well-balanced discussion of its main theme. The first part presents a series of discussions of the inventions and discoveries of the age, but the second portion is an extraordinary exhibition of hobby-riding, in which phrenology, spiritualism, opposition to vaccination, and universal panaceas for poverty play a part so exaggerated that, in spite of the author's eminence in his own field, it is impossible to take the whole book seriously as an estimate of nineteenth-century civilization" (*American Historical Review* 1899). The ambivalent response to Wallace's articulation of the sociopolitical and moral implications of evolutionary theory turned on two main factors.

As Wallace more openly identified himself with radical movements and ideologies, he emerged as a particularly troublesome voice to some of his contemporaries. His scientific prestige made Wallace a force to reckon with, even in sociopolitical matters. His views in this domain were anathema to those who believed that science and technology fueled a chariot of unalloyed Victorian cultural "progress." The second half of *The Wonderful Century* is a sophisticated deconstruction of many of the icons of late-nineteenth-century materialism. Propagandists for the unmitigated virtues of competitive capitalism were not amused. Furthermore, Wallace was paving the way, if not intentionally then at least by the tenor of his emerging evolutionary worldview, for a partial demotion of his professional scientific stature. As the writer in the *American Historical Review* (1899) put it, with perhaps unintended irony, the "best passage in the book is the history of the writer's own codiscovery with Darwin of the principles chiefly associated with the latter's name. His candor and generosity in recognizing Darwin as the principal discoverer are admirable." The canonization of Darwin, safely dead and (presumably) politically correct, had begun. The marginalization of Wallace, equally, had begun. That Wallace entitled the book that emerged from his Lowell lectures *Darwinism* ensured that this process would continue into the closing decades of the twentieth century. Spencer made an astute comment when he received a presentation copy of the book from Wallace. He wrote Wallace that he regretted "that you have used the title 'Darwinism,' for notwithstanding your qualification of its meaning you will, by using it, tend greatly to confirm the erroneous conception almost universally current" (ARWP MS. 46434, fol. 350, Spencer to Wallace, 18 May 1889). Spencer was alluding to the canonization of Darwin. He was also referring to the pejorative connotation of "Darwinism" as implying unbridled capitalist competition. Wallace, like Spencer, rejected such ruthless readings of the term. It thus remains even more curious, psychologically as well as semantically, as to why Wallace chose that title for his exposition of evolutionary science.

THE SOCIALIST COMES OUT OF THE CLOSET

Returning to England in August 1887, Wallace set to work on *Darwinism*. He intended it as a popular treatise, and an answer to three decades of criticism of natural selection. The book is replete with irony and paradox. Wallace had supplemented natural selection in pivotal stages of human evolution. He was vulnerable in his self-appointed role as defender of that mechanism as the key to species change. The book provided an eloquent defense of natural selection but within the framework of a theistic evolutionary teleology. The paradox resolves itself when we recall that Wallace never regarded natural selection alone as wholly adequate to explain all aspects of the evolutionary process—particularly of humans. It required nearly four decades for him to articulate fully his complex conceptualization of evolution—an articulation that was permeated by ideological, philosophical, theistic, and sociopolitical convictions. The clearest expression of Wallace's Weltanschauung is not in *Darwinism* but, rather, in a series of articles on sexual selection as an agent of evolutionary dynamics. Those articles made explicit the convergence of epistemological and activist concerns that transformed Wallace into one of the Victorian era's most incisive cultural critics.

In June 1889, Wallace moved to Parkstone and busied himself once again with gardening and rural life. But sociopolitical issues were his paramount concern. The following year, Wallace wrote an article on "Human Selection" for the *Fortnightly Review* (Wallace [1890] 1900c). He called it "the most important contribution [he had] made to the science of sociology and the cause of human progress" (Wallace [1905] 1969, 2:209). "Human Selection" contained his first public declaration as a socialist. Wallace's socialism was voluntaristic rather than doctrinaire. He declared that "compulsory socialism is to me a contradiction in terms—as much as would be compulsory friendship" (Wallace [1905] 1969, 2:268).

In the years immediately following his initial public advocacy of land nationalization, Wallace thought it the best solution to the abuses of unregulated private ownership of property. His youthful flirtation with Owenite socialism had been tempered "by the individualistic teachings of Mill and Spencer, and the loudly proclaimed dogma, that without the constant spur of individual competition men would inevitably become idle and fall back into universal poverty." In 1889, this philosophical and political tension between the conflicting claims of Owenites and Spencerians was resolved for Wallace by his reading of Edward Bellamy's Socialist utopia *Looking Backward* ([1888] 1960; Wallace 1891b). Wallace regarded Bellamy's book as a definitive repudiation of every "sneer, every objection, every argument [he] had ever read against socialism." He realized that Bellamy's utopia was just that: a vision of an ideal future state, not a manual for achieving socialism. Wallace also recognized that different nations would require different routes to a socialist future. Wallace's nondoctrinaire view of socialism is underscored by a reprint of the article on his conversion that he pasted into his copy of the eighteenth edition of *Looking Backward*. This article, titled "A Distinguished Convert"—and which appeared in the *New Nation* (1891b)—included extracts from a letter Wallace had written to Richard T. Ely of Johns Hopkins University, author of *An Introduction to Political Economy* (1889). It also included an addendum that Wallace felt necessary to make his own views clear to both American and British audiences (Wallace 1891b). Ely, while sympathetic to socialism, had pointed out what he perceived as certain weaknesses of the doctrine. In his letter, Wallace noted that Ely's criticisms would have appeared cogent to him only a year earlier (1888). But, he now asserted, such criticisms were all answered in *Looking Backward:*

> From boyhood—when I was an ardent admirer of Robert Owen—I have been interested in socialism, but reluctantly came to the conclusion that it was impracticable, and also to some extent repugnant to my ideas of individual liberty and home privacy. But Mr. Bellamy has completely altered my views on this matter. He seems to me to have shown that real, not merely delusive liberty, together with full scope for individualism and complete home privacy, is compatible

with the most thorough industrial socialism—and henceforth I am heart and soul with him. It is, however, a long way to such a goal, and your [Ely's] book will, I think, help men to a knowledge of the evils that have immediately to be remedied. I cannot myself see how the greatest of the evils of our present system—its involuntary idleness and consequent pauperism—can ever be got rid of under the system of unrestricted competition and capitalism, with labor as a marketable commodity.

The *New Nation* article is crucial to understanding Wallace's road to socialism (Wallace 1891d).

First, it contains a statement of Wallace's relationship to socialism and to Bellamy that Wallace explicitly wanted brought to the attention of a wide audience. Second, it publicized a fundamental component of Wallace's conception of socialism: the sanctity with which he clothed the concepts of individualism and personal "home privacy." Chapter 17 of *Looking Backward* deals with organizational aspects of Bellamy's utopia. In a key passage, Dr. Leete, one of the novel's protagonists, is asked whether the socialist government would deprive "small minorities of the people to have articles produced for which there is no wide demand . . . merely because the majority does not share it." Leete replied that as long as there was "a popular petition guaranteeing a certain basis of consumption," then "our officials [who] are in fact, and not merely in name, the agents and servants of the people" would effect appropriate measures to ensure a limited production of the desired commodity. Any other course would, Leete emphasized, "be tyranny indeed . . . and [one] may be very sure that it does not happen with us, to whom liberty is as dear as equality or fraternity." The importance of this passage to Wallace is manifest. He marked it by triple vertical lines in the margin. Wallace specifically underlined the phrase "liberty is as dear as equality or fraternity" (Wallace's annotated copy of *Looking Backward*, 137–138, ARWL). The entire tenor of Bellamy's utopia resonated with Wallace's deepest feelings on the sanctity of his home and personal privacy. Wallace detested dogmatic socialist agendas. He had nothing but scorn for Enrico Ferri's *Socialism and Positive Science (Darwin-Spencer-Marx)*. Wallace possessed a 1905 English translation of an 1896 French edition of Ferri. On the inside front cover of his copy, Wallace dismissed Ferri's book as "vague & badly constructed, . . . not worth translating." Ferri's assertion that there was a single, scientifically valid, socialist doctrine—that enunciated by Marx—angered him. Wallace was adamant that socialism was in accord with evolutionary science. He was equally adamant that there was no single path to socialism (Wallace, annotated copy of Enrico Ferri, *Socialism and Positive Science*, in ARWL).

WALLACE ON MARX

In *The Revolt of Democracy* (1913a), Wallace indicted technologically potent Victorian society with failing to deal effectively with what he saw as its greatest blight:

> As President of the Land Nationalisation Society for thirty years, I have given much attention to the various inquiries by Royal Commissions, by Parliamentary Committees, or by private philanthropists, into Irish evictions and Highland clearances, sweating, unemployment, low wages, unhealthy trades, bad and overcrowded dwellings, and the depopulation of the rural districts. These inquiries have succeeded each other in a melancholy procession during the last sixty years; they have made known the almost incredible conditions of life of great numbers of our workers; and they have suggested more or less ineffective remedies, but their proposals have been followed by even less effective legislation when any palliative has been attempted.

Wallace applauded the intention of Sir Henry Campbell-Bannerman, who became liberal prime minister in 1905, to change "this attitude of negation of all his predecessors . . . and so to legislate as to make our native soil ever more and more 'a treasure-house for the poor rather than a mere pleasure-house for the rich.' " But the three great national strikes of 1911–1912—of the railway, mining, and coal workers—made it clear to Wallace that even with the good intentions of the liberal governments there would be no solution to labor unrest until the "reasonable claims of the workers" are satisfied by radical reform of the governing institutions of the country (Wallace 1913a, 1–2, 7–8, 12–13). It is in this general sense that Wallace respected Marx as a major social critic and reformer.

He endorsed Marx's conviction that laborers should receive "the whole produce" of their labor. But Wallace could not abide what he deemed the pernicious aspect of Marxian doctrine (especially as developed by the "German socialists")—"that indefinite increase of . . . governmental interference with labour and industry." In an article of 1884, Wallace had offered his definition of the legitimate role of government intervention. It was "to secure peace from external foes, and safety from internal violence; . . . [to] give free and speedy justice between man and man; . . . [to] secure to all alike free access to the land and all natural powers; [and to] abolish every monopoly of individuals and classes." Wallace believed that when labor was freed from the shackles imposed on it by capitalist overlords, then all members of a liberated social order would work collectively, but equally, "to realise the best social state which, *in its present phase of development,* humanity is capable of." Wallace was optimistic that the "distant future will take care of itself; let us try," he advised, "to improve the future that is immediately before us" (Wallace [1884] 1969, 2:248–249).

Wallace saw a voluntaristic/collectivist society, as sketched in the 1884 article, as his political testament. It was "the true system of *laissez-faire,* now so much abused as if it had failed, when really it has never been tried." When Wallace reprinted the article in his autobiography twenty years later, after he had become an avowed socialist, he still considered his take on "true" laissez-faire "to be logically unassailable." Further, he considered the principles guiding the LNS to be compatible with major aspects of Marx's own thinking. He declared one LNS proposition to have anticipated "the main thesis of Marx." That proposition maintained that it "is out of the pauper and floating masses who have been separated from the land, and have consequently no option between starvation and selling their labour unconditionally, that capital is originally formed, and is, thereafter, enabled absolutely to dictate to the very labour that creates it, and to defraud that labour of those surplusses [*sic*] which ought to remain wholly with the latter." Wallace borrowed this LNS proposition verbatim from an 1856 work by a medical doctor, Robert Dick, titled "On the Evils, Impolicy, and Anomaly of Individuals Being Landlords and Nations Tenants" (Wallace [1905] 1969, 2:241–242, 244, 248). What are we to make of Wallace's juggling of Marx, Spencer, and Dr. Dick?

First, Wallace always trod across disciplinary lines. His quest to uncover a unifying thread among the vast array of thoughts and actions of the late Victorian era impelled him to do so. He strode from biology to philosophy to politics to religion and back again with a boldness that at once impressed and bewildered (or repelled) his contemporaries. Initially, Wallace's opinions were widely sought on sociopolitical matters because of his renown as codiscoverer of natural selection. Later, the mature Wallace came to be regarded as an innovative cultural critic in his own right. His "invasion" into political economy has been deemed recently by one economic historian as injecting a fresh voice into certain perennial debates about the production and distribution of wealth in society. Though not ranking with other more theoretically influential "interlopers" in that formidable domain—such as Adam Smith (a philosopher), François Quesnay (a physician), David Ricardo (a broker and sometime MP), Thomas Malthus (a clergyman), Marx (a sometime journalist) and John Stuart Mill (a customs official and sometime MP)—Wallace shared with them a conviction in the necessity and worth of interdisciplinarity (Gaffney 1997, 610).

A second essential point in assessing Wallace's reputation as a cultural critic is that he used certain terminology in idiosyncratic fashion. In his lecture "Social Economy *versus* Political Economy," given in Washington, D.C., Wallace sought to distance himself as far as possible from "the old 'political economy.'" He considered it "effete and useless, in view of modern civilization and modern accumulations of individual wealth. Its one end, aim, and the measure of its success, was the accumulation of wealth, without considering who got the wealth, or how many of the producers of the wealth starved. What we required now," Wallace argued, "was a science of 'social economy,' whose success should be measured by the good of all." This speech got a mixed reception in 1887. Eighteen years later, however, when Wallace referred to it in his autobiography, he noted that public opinion had "advanced" so dramatically that many more members of the working and professional classes supported his views (Wallace [1905] 1969, 2:129). In elaborating his mature socioeconomic proposals, Wallace sought to repudiate most of the central tenets of classical political economy.

WALLACE'S CRITIQUE OF POLITICAL ECONOMY

The list of Wallace's divergences from classical political economy is large. (1) He opposed free trade and favored imposing tariffs in accordance with the tariffs that foreign countries imposed on the home country. He also maintained that free trade imposed external costs, in the form of environmental degradation. (2) Wallace supported introducing minimum wages. He recommended "a very high minimum wage for really necessary or useful work." (3) Wallace condemned interest and profit income. He believed lending at interest should be illegal, except for personal loans of fixed duration. Not even such personal loans would be enforceable by law; the loans would be made at the risk of the lender. (4) Wallace championed nationalization of the land. He called private property in land "barbarism" masquerading as "civilization." He scorned the free trade in land movement, which sought to abolish entails and to make conveyancing cheap and expeditious. Wallace believed the elimination of such market imperfections would only lead to a further concentration of land ownership. (5) Wallace advocated prohibition of the export of coal and iron. He believed unhindered export would raise the price of these products for British workers. (6) Wallace endorsed the management by state authority of all industries "essential to public welfare." (7) He believed "capital" was the "tyrant and enemy of labour." And, finally, (8) Wallace urged the provision of bread free to "anyone who was in want of it." The overriding goal of Wallace's sociopolitical philosophy was to find a remedy for what he (and many but not all of his contemporaries) regarded as the deplorable plight of the working class (Coleman 1999, 6–8, 10).

The similarity between certain of Wallace's and Marx's critiques of classical political economy explains Wallace's esteem for Marx. On the issue of free
trade, Wallace's language is Marxist. He regarded free trade as one example
of that ideological legerdemain by which more powerful nations sought to
sanctify their exploitation of less powerful nations. Wallace later criticized
the Dutch system of colonial management—from which he had benefited so
greatly in the Malay Archipelago—as one that "farmed" the natives and their
societies "for the benefit of the mother country." Not merely the Dutch but
all European colonial powers were, in Wallace's opinion, opening the way
to greater global commercial activity in the guise of introducing "advanced
civilisation." By the late 1870s, Wallace had come to view such globalization
as a recipe for disaster. It would only "lead to the gradual alienation of the land
to capitalists, give an unnatural stimulus to the population, and inevitably introduce the evils of feverish competition, pauperism, and crime, from which
the [subject] country has hitherto been comparatively free" (Wallace [1879]
1883, 300–325).[5] Wallace's developing critique of British, as well as European and North American, society led him explicitly to condemn Victorian
colonialism as pernicious. The seeds for this critique had been sown during
Wallace's early tropical journeys.

Wallace's constant emphasis on the role of the individual in a socialist
society is consistent with his position as political activist and evolutionary
biologist. Wallace was one of the contributors to a collection of essays on
socialism, *Forecasts of the Coming Century by a Decade of Writers*, edited by
Edward Carpenter (1897). Wallace's essay was "Reoccupation of the Land."
William Morris, Bernard Shaw, and Carpenter were among the other contributors. The essay Wallace found most provocative was by his fellow naturalist Grant Allen: "Natural Inequality." Wallace and Allen had a history
of controversy, especially with respect to their opposed theories of animal
coloration. Allen's advocacy of free love was also anathema to Wallace, as
were Allen's eugenics proclivities. Allen linked free love to eugenics in his
assessment of marriage. To Allen, marriage was a legal encumbrance that
militated against the freedom to choose temporary partners for the explicit
purpose of procreating the "fittest" offspring. Wallace called Allen's proposals
"detestable" (in his account of them in "Human Selection") and a certain
recipe for the destruction of family life and long-term parental affection, two
of the most sacred tenets of Wallace's own personal life. However innovative
in other domains, Wallace was highly conservative in his views on marital and
sexual matters. Wallace did find that he and Allen were united on one crucial
point: the compatibility of socialism with individualism.

He marked the following passage in Grant's essay with triple underlining: "I have never been one of those who hold that Socialism is opposed

to individualism. On the contrary, I believe that Socialism will encourage
and develop individuality. I am a Socialist just *because* I am an individualist.
Who are even," Allen asked, "now the best Socialists among us? The least
individual? Not a bit of it; the most markedly individualistic and idiosyncratic
temperaments in Britain." Allen could have been describing Wallace. Allen's
citing of Spencer's dictum that "the character of the units determines the
character of the aggregate" in this passage which so impressed Wallace, points
again to the powerful impact Spencer's views exerted on Wallace throughout
his life. Wallace fully appreciated that, as biologists and socialists, he and
Allen were united by their expert knowledge of evolutionary science in their
espousals of socialism. Wallace underlined Allen's statement "that one of the
great points of Socialism will be this, that while it will seek to redress such
natural inequalities as feebleness, ill-health, loss of limb or organ, deficient intellect, or deficient moral sense, it will seek to develop to the utmost all better
inequalities, such as conspicuous strength, health, manual dexterity, mental
ability, virile or feminine faculty, . . . [and] artistic, scientific, or literary ability, . . . and to utilise them to the utmost for the good of the community. In
short, while discouraging all false betternesses, it will encourage and make
the most of all true ones" (Wallace's copy of Edward Carpenter 1897, 137–
138, in ARWL). Wallace urged Allen to write a socialistic novel. He felt Allen
was well-equipped for the task as a "well-known and talented writer" and an
articulate advocate for socialism (Wallace [1905] 1969, 2:272–274).

BIOLOGICAL SOCIALISM

As a field naturalist, Wallace could hardly deny that individual differences
were a fact of evolutionary life. His biological socialism is realistic as well
as utopian. Economic and social inequities should and would be abolished
by the transition to a socialist state. But other inherent human differences,
such as those delineated by Allen, were biological raw materials that could be
put to communal advantage within a voluntaristic framework. The emphasis
Wallace placed on the role of voluntary female choice in his mature concept of
human sexual selection was crucial. It enabled him to argue for a noncoercive
approach to effecting permanent biological improvements within a socialist
society. Carpenter's essay in *Forecasts of the Coming Century*, "Transition to
Freedom," outlined a form of voluntaristic socialism that, "broad enough
and large enough to include an immense diversity of institutions and habits,"
accorded with Wallace's vision. One of Carpenter's footnotes caught Wallace's attention. Carpenter warned that the growth of the modern millionaire
was a "serious evil. Now that any man endowed with a little low cunning and
tempted by self-conceit and a love of power has a good chance of making

himself enormously rich, Society is in danger of being ruled by as mean a set of scoundrels as ever before in History. And nothing less than a complete transformation of our monetary system will enable us to cope with this danger." Wallace doubly underlined this footnote. He wrote in the margin of his copy that "the abolition of inheritance will do it best, quickest, & most equitably" (Carpenter 1897, 190). Wallace would later present an expanded account of his view that the abolition of inheritance of wealth would gradually lead to the entire disappearance of millionaires in his 1905 article in the *Clarion*, "If There Were a Socialist Government—How Should It Begin?" (Wallace 1905a).

The addendum that Wallace requested in the *New Nation*'s (28 March 1891) account of his conversion to socialism provides further insight into his pragmatic conception of socialist politics. He wanted to spell out his view that the course of reform could not be the same in Britain as it was in the United States or in any other nation. Since Britain's landed aristocracy still retained "much of the power they possessed in feudal times," Wallace deemed its continued existence the prime "source of the profound class-distinctions that still prevail among us." He asserted that land monopoly "must be got rid of before we can make any real progress" toward socialist reforms. The primary remedy for Britain's sociopolitical malaise—though not necessarily for other nations—was, Wallace reiterated, "the nationalization of the land." After his declaration as a socialist, land nationalization still remained "the first step . . . by which the power and prestige of the great landlords will be destroyed. The workers," Wallace declared, "having free access to land, will then be able to accumulate capital and to establish co-operative industries among themselves. They will thenceforth be independent of both landlord and capitalist, and the step will not be a difficult one from partial and local to complete and national cooperation." The *New Nation* article concluded with the statement that "Mr. Wallace is one of the best known followers of Darwin, and in fact himself made some of the discoveries which are popularly attributed to Darwin" (Wallace 1891b). Wallace would have found the journalist's remark delicious. Wallace had recently published *Darwinism*, in which he desisted from making the same claim himself.

If Wallace was coy in *Darwinism*, he was not in "Human Selection" ([1890] 1900c). That 1890 article was more than Wallace's public proclamation of his socialism. It was, he asserted, also his "first scientific application of [that] conviction." In preparing "Human Selection," Wallace had immersed himself in a wide spectrum of socialist literature. He read William Morris's *News from Nowhere*. Wallace thought it "a charming poetical dream, but as a picture of society almost absurd, since nobody seems to work except at odd times when they feel the inclination, and no indication is given of any organization of labour." Laurence Gronlund's *Our Destiny* (1890), in contrast, struck Wallace as a "beautiful and well-reasoned essay on the influence of socialism on morals and religion." It was also a favorite of Shaw's. Gronlund's earlier *Co-operative Commonwealth* (1884) was "an exposition of constructive socialism, which has given us in its title the shortest and most accurate definition of what socialism really is" (Wallace [1905] 1969, 2:266–268).

But it was Bellamy's novel that exerted the greatest impact on Wallace, as it did on numerous thinkers on both sides of the Atlantic. The publication of *Looking Backward* in 1887 propelled Bellamy overnight from the literary wings into the very center of American reform activism. His book sold 60,000 copies in the first year. Sales climbed swiftly to more than 100,000 in the next year, as editions appeared in England, France, and Germany. *Looking Backward* was, however, not intended as textbook socialism. It was a work of utopian fiction. Bellamy's inspirational eloquence was as powerful a draw to his readers as was the harmonious socialist society he depicted. In a letter Wallace published in *Land and Labor* on his support of land nationalization and socialism, he characterized Bellamy's "remarkable book" as setting forth "for the first time—so far as I know—[a] practicable and altogether unobjectionable scheme of socialist life" (Wallace 1889). Wallace was so "captivated" by *Looking Backward* that after his initial reading of the English edition in early 1889, he gave it a "second almost immediate perusal." This was an honor he had previously bestowed on only one other book, Spencer's *Social Statics*. He read Bellamy's tract for an unprecedented third time (in 1890), in order to refresh his memory "on certain suggestions which seemed to me especially admirable" (Wallace [1905] 1969, 2:266–267). The novel attracted followers as disparate as litterateurs, political mavericks, retired army officers, Theosophists, Christian socialists, feminists, and a curious assortment of visionaries and reactionaries. Wallace shared with Bellamy, as both did with George, a growing aversion to what they all perceived as the shallowness and sham that drove the forces of a predatory capitalism in the late nineteenth century. Wallace, Bellamy, and George offered complementary critiques of the secularizing force set loose by Victorian materialism and called on new theistic resources to counter and contain it (Thomas 1983, 36, 42, 262, 266–275).

"Human Selection" is Wallace's most explicit statement of what he believed to be the appropriate relationship between evolutionary biology and political ideology. He opened "Human Selection" by noting that in "one of my latest conversations with Darwin he expressed himself very gloomily on the future of humanity, on the ground that in our modern civilization natural selection had no play, and the fittest did not survive. Those who succeed in the race for wealth are by no means the best or the most intelligent and it is notorious that our population is more largely renewed in each generation from the lower than from the middle and upper classes" (Wallace [1890] 1900c, 1:509). Wallace's politics were antithetical to Darwin's, but he agreed that there was an undoubted check to progress in social evolution. Wallace dismissed as possible solutions to this evolutionary dilemma any proposals based solely on beneficial environmental influences, such as education or hygiene. Though these could produce improvements in any given generation, Wallace held that they could not of themselves lead to a sustained improvement of humanity. Implicit in such proposals was the Lamarckian concept "that whatever improvement was effected in individuals was transmitted to their progeny, and that it would be thus possible to effect a continuous advance in physical, moral, and intellectual qualities without any selection of the better or elimination of the inferior types" (Wallace [1890] 1900c, 1:510). The inheritance of acquired characteristics was still accepted by many evolutionists. Under the rubric of neo-Lamarckism, it underlay certain biologically oriented reformist speculations in the 1880s and 1890s (Stocking 1962). Wallace rejected all versions of Lamarckism, old and new. He maintained that the researches of Galton and August Weismann had demolished the theory of the inheritance of acquired traits. According to Weismann, the hereditary material (germ cells in the ovaries and testes that produce egg and sperm) cannot be modified by changes undergone by the remaining body cells (comprising the somatoplasm [Robinson 1976, 234–237]). Wallace accepted Weismann's influential but controversial hypothesis and concluded that there remained "some form of selection as the only possible means of improving the race" (Wallace [1890] 1900c, 1:510).

WALLACE CONTRA EUGENICS

Wallace detested what he termed "artificial selection," under which he included such schemes as Galton's eugenics. Among Galton's proposals was "a system of marks for family merit." Those individuals who rated well in health, intellect, and morals would be encouraged, by state subsidies, to marry early and raise large families. While such "positive eugenics" might increase slightly the number of excellent human specimens, Wallace argued that it would be socially ineffective and evolutionarily insignificant. Positive eugenics would leave the bulk of the population unaffected and fail to "diminish the rate at which the lower types tend to supplant . . . the higher" (Wallace [1890] 1900c, 1:513). Given the limited knowledge of human inheritance, Wallace declared that artificial selection was not only scientifically dubious but also culturally pernicious. Eugenics, by perpetuating class distinctions, would postpone social reform and afford quasi scientific excuses for keeping people "in the positions Nature intended them to occupy." Negative eugenics, or the prevention or discouragement of procreation by those deemed unfit, seemed to Wallace "a mere excuse for establishing a medical tyranny. And we have enough of this kind of tyranny already. The world does not want the eugenist to set it straight. . . . Eugenics is simply the meddlesome interference of an arrogant scientific priestcraft" (Marchant [1916] 1975, 466–67).

Wallace had initially welcomed Galton's *Hereditary Genius* (1869) as providing important data in support of evolutionary biology. His later antipathy to Galton's eugenics stemmed from what he regarded as its cultural authoritarianism. Wallace also distanced himself from Galton because of the latter's general lack of empathy, if not utter rudeness, to those whom he regarded as not of his social standing. Galton's views on gender were a further irritant to Wallace. Galton met, and was polite to, important females such as the novelist George Eliot, the social reformer Beatrice Potter Webb, and the celebrated Florence Nightingale. But he never regarded them as intellectual equals. Galton was also insulting to ambitious but socially less privileged men. He treated his highly talented exploring assistant in Africa, Charles Andersson, and the quite famous Henry Morton Stanley with disdain. Galton publicly minimized the significance of Stanley's now-legendary meeting with David Livingstone at Lake Tanganyika in 1871. He dismissed Stanley's recounting of his spectacular exploring feats in the Congo (which won him a knighthood) as "essentially a journalist aiming at producing sensational articles." Finally, Galton's descriptions of his own African explorations are replete with crude and arrogant comments about indigenous peoples whom he likened to baboons, pigs, and dogs. While Galton was hardly alone among Victorian explorers in expressing such Eurocentric attitudes, he was among the most extreme (Fancher 1998, 108–109, 112). Such character traits were anathema to Wallace. His habitual empathy, as manifested in his views on women and his deep appreciation of the cultures of indigenous peoples, separated him from Galton both on scientific and personal levels.

To Wallace, neo-Lamarckism, eugenics, and capitalist apologias extracted from evolutionary theory were not merely biologically dubious. They also proceeded from fundamentally objectionable social premises. Wallace maintained all were founded on class distinctions and economic inequities. Their advocates ignored or failed to confront the central fact that Victorian culture frustrated, rather than facilitated, genuine evolutionary advance. For Wallace, the key to permanent human betterment lay in the operation of a benevolent biological selection. Socialism, he believed, would provide the sufficient—and necessary (Wallace [1905] 1969, 2:266)—condition for that selective force:

> When we have cleansed the Augean stable of our existing social orga-
> nization, and have made such arrangements that all shall contribute
> their share of either physical or mental labour, and that all workers
> shall reap the full and equal reward of their work, the future of the
> race will be ensured by those laws of human development that have
> led to the slow but continuous advance in the higher qualities of
> human nature. When men and women are alike free to follow their
> best impulses; when idleness and vicious or useless luxury on the one
> hand, oppressive labour and starvation on the other, are alike un-
> known; when all receive the best and most thorough education that
> the state of civilization and knowledge at the time will admit; when
> the standard of public opinion is set by the wisest and the best, and
> that standard is systematically inculcated on the young; then we shall
> find a system of selection will come spontaneously into action which
> will steadily tend to eliminate the lower and more degraded types of
> man, and thus continuously raise the average standard of the race.
> (Wallace [1890] 1900c, 1:517)

Socialism, by removing disparities of wealth and rank, would eliminate the
economic and political prejudices that, Wallace claimed, dominated the selection of reproductive partners in Victorian society. In their place, mate choice
would focus on those higher moral and intellectual traits often neglected
(or rendered subservient) in competitive capitalist society (Wallace [1890]
1900c, 1:526).

SEXUAL SELECTION: WALLACE'S EARLY VIEWS

That the selective process Wallace envisioned as the key to further human
evolution is a form of sexual selection is, at first sight, surprising. One of the
major theoretical disagreements between Wallace and Darwin had stemmed
precisely from Wallace's refusal to accord scientific status to female choice
as an agent of evolution. In the *Origin*, Darwin briefly introduced the theory of sexual selection to account for certain animal characteristics whose
occurrence did not seem explicable by natural selection. "When the males
and females of any animal have the same general habits of life, but differ in
structure, colour, or ornament," Darwin maintained that such sexual dimorphism arose not from "a struggle for existence, but [from] a struggle between
the males for possession of the females." He further indicated that sexual selection includes two distinct processes. First, in certain species, particularly
polygamous ones, there is an actual (or threatened) combat between males for
the privilege of mating. Those males possessing variations that better equip

them for combat will succeed in competition with their rivals and leave the most progeny (who inherit those variations). Thus, Darwin suggested, arose the antlers of male deer, the spurs on the legs of certain male birds and the huge mandibles of male stag beetles. Second, there are species in which the males possess musical organs, bright coloration, or ornamental appendages (such as the elaborate tails of the male birds of paradise). Darwin claimed that such traits had developed because the females were more attracted to males of striking appearance. He saw "no good reason to doubt that female birds [for example], by selecting, during thousands of generations, the most melodious or beautiful males, according to their standard of beauty, might produce a marked effect" (Darwin [1859] 1964, 88–89).

Initially, Wallace conceded some minor evolutionary role to sexual selection. He granted that male rivalry accounted for "the development of the exceptional strength, size, and activity of the male, together with the possession of special offensive and defensive weapons." But Wallace designated it "a form of natural selection which increases the vigour and fighting power of the male animal, since, in every case, the weaker are either killed, wounded, or driven away" (Wallace [1889] 1975, 282–283). The second part of Darwin's hypothesis, female choice, struck Wallace as extremely dubious (Wallace [1905] 1969, 2:18; Marchant [1916] 1975, 130). Two essays on animal coloration that appeared in 1867—"Mimicry, and Other Protective Resemblances among Animals" and "On Birds' Nests and Their Plumage; or the Relation between Sexual Differences of Colour and the Mode of Nidification in Birds"—reveal the degree to which Wallace had come to differ from Darwin (Wallace [1867a] 1891a, [1867b] 1891a).

Among the more curious modifications of the coloring and external form of animals are those instances when one species resembles another unrelated species so closely as to make it difficult to distinguish between them by appearance. It was Wallace's friend and coexplorer Bates who, in 1862, first explained such imitation—which he termed mimicry—by natural selection. During his travels in South America, Bates had noticed that the brilliantly hued heliconid butterflies of the Amazon region were copied both in color and pattern by several unrelated species, including the Leptalides (Dismorphia) butterflies. The Heliconidae secrete substances (with nauseous odors) that render them unpalatable to insectivorous birds and are avoided as prey. Bates reasoned that the mimicking butterflies (which lack the offensive secretions) acquire protection merely by looking like the original. Natural selection, he argued, would favor just those variations that more closely approximated the appearance of the protected species. Bates suggested that "the selecting agents [were] insectivorous animals, which gradually destroy those sports or varieties that are not sufficiently like [the Heliconidae] to deceive them." Over time, cumulative selective pressure would result in the production of those remarkable insects that exactly resemble (externally) the mimicked species. Both Darwin and Wallace recognized Bates's explanation as providing powerful empirical support for their theory. Wallace explicitly endorsed Batesian mimicry in 1865 in his important essay "On the Phenomena of Variation and Geographical Distribution as Illustrated by the Papilionidae of the Malayan Region" (Bates 1862; Wallace 1865a, 19; Darwin wrote a favorable (but unsigned) review of Bates's article in the *Natural History Review* 3 [1863]: 219–224). Two years later, in his essay on mimicry in the *Westminster Review* (1867), Wallace extended Bates's idea to incorporate within the evolutionary framework the widespread phenomena of protective resemblances in general among animals (Wallace [1867a] 1891a).

Invoking the principle of utility, Wallace asserted that many aspects of the coloration and external appearance of animals—including traits hitherto regarded as useless or trivial by naturalists—are (or were) often of the utmost importance for survival (Wallace [1867a] 1891a, 35–36; Wallace [1889] 1975, 13–35). The diverse instances of resemblance, whether to the surrounding environment or to other animals, were evolutionary adaptations that served to conceal creatures either from their predators or from those animals they themselves prey on. Wallace cited the green-plumed groups of tropical birds (such as the parrots, barbets, and touracos), the many dusky nocturnal creatures (including rats, bats, and moles), the polar bear, the arctic fox, the alpine hare, and the "flounder and the skate, [which] are exactly the colour of the gravel or sand on which they habitually rest," as evidence of such adaptive coloration. The so-called walking-stick insects of the family Phasmidae provided Wallace with a particularly striking example. Their coloring, external form and texture, and the arrangement of the head, legs, and antennae render them identical in appearance to the twigs and branches on which they rest. Wallace claimed that all such traits—"from the mere absence of conspicuous colour or a general harmony with the prevailing tints of nature, up to such a minute and detailed resemblance to inorganic or vegetable structures as to realise the talisman of the fairy tale, and to give its possessor the power of rendering itself invisible"—were explicable by natural selection. Batesian mimicry becomes a special case of protective coloration and one, Wallace suggested, that might include not only insects but also snakes and birds (Wallace [1867a] 1891a, 36–41, 46–47, 49, 70–76).

NATURAL SELECTION OR SEXUAL SELECTION?

Wallace's 1867 essay on mimicry concluded with a brief but significant discussion of the relation of protective coloring and mimicry to the sexual differences of animals. For those species of insects (and birds) in which the sexes are dissimilar in color or marking, Wallace suggested that the generally duller and less conspicuous coloration of the females was an adaptation that served to conceal them from predators during the depositing of eggs. "In the spectre insects (Phasmidae)," he noted, "it is often the females alone that so strikingly resemble leaves, while the males show only a rude approximation." Conversely, those insects with little need for protective concealment, such as the Heliconidae and the stinging Hymenoptera (wasps, bees, ants), display no (or only slightly developed) sexual differences in color. Wallace regarded the general absence of color differentiation between the sexes in species of insects protected by "disagreeable flavour, . . . by their hard and polished coats, [or by] their rapid motions" as compelling evidence against the hypothesis of female choice. Although he did not completely abandon sexual selection—which "has often manifested itself [among insects] by structural differences, such as horns, spines, or other processes"—Wallace's analysis of the development and function of color in the animal kingdom foreshadowed his increasing commitment to the strict operation of natural selection in these instances (Wallace [1867a] 1891a, 79–80).

In his second 1867 essay, "Theory of Birds' Nests," Wallace further developed his views on the dull coloration of females in many species. He argued dull color was due not to selection by the females of more handsomely colored males but to their greater need for concealment. Wallace claimed that birds, because of their prolonged period of incubation, provided a decisive support for his own hypothesis. In the majority of cases in which male birds are more brilliantly colored, he noted that the female hatched the young in open nests. During brooding, the female would be "exposed to the attacks of enemies, and any modification of colour which rendered her more conspicuous would often lead to her destruction and that of her offspring." Natural selection, Wallace indicated, would tend to eliminate any variations in this direction. Conversely, any variations in color that tended to render the female less conspicuous by assimilating her to the surroundings would be favored by natural selection. Male birds, since they are not subject to such periods of enforced helplessness—and, hence, to selective pressure against (random) conspicuous color variations—would be capable of acquiring the brilliant plumage characterizing their sex only. Wallace's argument extended to those groups of birds, including the kingfishers, trogons, and mynahs, in which the female was as conspicuously colored as the male. With very few exceptions, these birds construct nests that are either domed or concealed in the hollows of trees or in burrows in the ground. The females of these species, since they are effectively hidden from predators during incubation of their eggs, are free to acquire "the same bright hues and strongly contrasted tints with which their partners are so often decorated." Finally, in those few cases (such as the gray phalarope) in which the female is more conspicuously colored than the male, Wallace declared "it is either positively ascertained that the latter performs the duties of incubation, or there are good reasons for believing such to be the case" (Wallace [1867b] 1891a, 129–132).

WALLACE ON DARWIN'S *DESCENT OF MAN*

While Wallace had not altogether repudiated female choice in the evolution of certain aspects of sexual dimorphism, he did relegate it to a position of minimal importance. The publication of Darwin's *Descent of Man* in 1871 ([1871] 1874) accentuated the divergence between the two naturalists on sexual selection. Aside from human origins, no issue divided Wallace and Darwin as sharply as did sexual selection (Marchant [1916] 1975, 175–78, 181–90, 210–14, 245–48; Wallace [1889] 1975, 282–88, 294–96).[6] Darwin elaborated at length on what he considered to be the widespread operation of male rivalry and female choice throughout the animal kingdom, including humans (Jann 1994). Wallace's review of Darwin's *Descent* emphasized his growing conviction that sexual selection was incompetent to account for the overwhelming majority of sexual differences Darwin had documented.

Even if one granted that female animals were capable of exercising a preference in the choice of mates, Wallace denied that the individual tastes of successive generations could produce any constant effect. "How are we to believe," he asked, "that the action of an ever varying fancy for any slight change of colour could produce and fix the definite colours and markings which actually characterize species? Successive generations of female birds choosing any little variety of colour that occurred among their suitors would necessarily lead to a speckled or piebald and unstable result, not to the beautifully definite colours and markings we see" (Wallace 1871, 182). Wallace explicitly objected to Darwin's assertion that conscious mate selection had been an important agent in determining human racial and sexual differences. Such selection would require "the very same tastes to persist in the majority of the race during a period of long and unknown duration." Wallace insisted there was no evidence for this hypothetical identity of tastes on the part of human ancestors. He further emphasized that, as Darwin's own examples demonstrated, members of "each race admire all the characteristic features of their own race, and abhor any wide departure from it; the natural effect of which would be to keep the race true, not to favour the production of new races" (Wallace 1871, 179–180). Only natural selection, Wallace insisted, by "unerringly" selecting or rejecting variations according as they are either useful or disadvantageous, could produce fixed racial or secondary sexual characteristics. Although Wallace could assign adaptive value to only certain secondary sexual characteristics, he did not doubt the "existence of some laws of development capable of differentiating the sexes other than sexual selection" (Wallace 1871, 182). Human racial differences, he suggested, were either adaptive themselves or correlated with useful variations (Wallace [1870] 1969, 178–179).

For the next two decades, Wallace continued to develop his case against sexual selection. *Darwinism* was intended, in part, to demonstrate that the varied phenomena of sexual dimorphism could be subsumed under the action of natural selection. In addition to protective coloration, Wallace declared that the need for recognition had played a decisive role in modifying the comparative coloration of the sexes. Since hybridization between members of closely related species generally results in either infertile or otherwise less fit offspring, any development that served to reduce the possibility of such crosses would be favored by natural selection. "The wonderful diversity of colour and of marking that prevails, especially in birds and insects," Wallace suggested, "may be due to the fact that one of the first needs of a new species would be, to keep separate from its nearest allies, and this could be most readily done by some easily seen external mark of difference." He emphasized that either the male or the female could be modified in color apart from the opposite sex "in the process of differentiation for the purpose of checking the intercrossing of closely allied forms" (Wallace [1889] 1975, 218, 227, 272–73).

The fundamental disagreement between Wallace and Darwin on sexual selection emerged most clearly with respect to male bird song. Darwin believed birds "to be the most aesthetic of all animals, excepting of course man, and they have nearly the same taste for the beautiful as we have" (Darwin [1871] 1874, 407–408). The ability to sing, he maintained, is a powerful means employed by male birds "to charm the females." Darwin again invoked the analogy of human breeding. Just as humans can modify domesticated birds by selecting particular variations, Darwin argued "the habitual or even occasional preference by the female of the more [melodious] males would almost certainly lead to their modification . . . augmented to almost any extent compatible with the existence of the species" (Darwin [1871] 1874, 421, 479). To Wallace, evolutionary explanations of behavioral or physical traits predicated on an aesthetic sense in lower animals were unacceptable. The discontinuity between human higher faculties and the mental processes of the rest of the animal kingdom had become axiomatic for him. Wallace declared that imputing aesthetic tastes to birds (and insects) was an anthropomorphism as unwarranted as that made by "writers who held that the bee was a good mathematician, and that the honeycomb was constructed throughout to satisfy its refined mathematical" sense. He adroitly cited Darwin to support his critique. Wallace noted that the *Origin* properly ascribed the bee's instinctual hive-making ability to a gradual accumulation by natural selection of those variations that were conducive to the construction of the best cells with the least expenditure of labor and precious wax (Wallace [1889] 1975, 336–337, 461–464). The song of a male bird, Wallace argued analogously, functions not to charm the female but as a call to indicate his presence. In addition to their value as a means of recognition between the two sexes of a given species, characteristic birdcalls also are signals that the pairing season has arrived. Wallace pointed out correctly that when the individuals of a species are widely scattered, such calls are of crucial importance in enabling pairing to take place as early as possible. This reduces the period during which the potential mates are exposed to predation and other dangers in their search for each other (Mayr 1972, 96). The "clearness, loudness, and individuality of the song," Wallace concluded, "becomes a useful character, and therefore the subject of natural selection" (Wallace [1889] 1975, 284).

Sexual Selection in Humans: A Stunning Reversal

Darwinism forcefully summarized Wallace's efforts to minimize the importance of sexual selection among animals. By 1890, however, Wallace dramatically reversed his position with respect to its efficacy in human evolution. This reversal reflects the extent to which Wallace had now integrated his biological and social theories. Bellamy's *Looking Backward* provided Wallace with more than a cogent defense of socialism. It advanced an explicit context in which human evolutionary progress could be permanently effected. In Bellamy's egalitarian future state, "sexual selection, with its tendency to preserve and transmit the better types of the race, and let the inferior types drop out, has unhindered operation." Women, free from the demands of poverty or wealth, could now choose as the fathers of their children only those men who possessed traits—"beauty, wit, eloquence, kindness, generosity, geniality, courage"—worthy of transmission to posterity, and ensure that every "generation is sifted through a little finer mesh than the last" (Bellamy [1888] 1960, 218). Wallace's rejection of neo-Larmarckism as well as capitalist variants of social biology left him with only the guidance of spiritual intelligences as a *vera causa* for further human evolution. Bellamy's sexual selection yielded an auxiliary mechanism whose explanatory potential Wallace fully appreciated. Female choice under socialism was a revelation for Wallace. It was contingent on the presence of faculties and characteristics that owed their development to spiritual intelligences. It also functioned as a naturalistic mechanism.

Female choice had the further advantage of providing a selective force and, hence, the possibility of evolution in an otherwise egalitarian society. It has been stated that "socialism and nineteenth-century evolutionism were always very uneasy bedfellows, and in the conflict between them Wallace chose socialism" (Young 1971, 224). This is inaccurate. Sexual selection provided a mechanism by which socialism and evolution could be made compatible. Wallace did not have to choose between the two. As a biologist, Wallace sought to demonstrate that sexual selection under socialism was scientifically valid. He had to confront the objection that in an egalitarian society, with its presumed absence of the Malthusian checks of war, famine, and pestilence, an inevitable overpopulation would plunge mankind into a bitter struggle for existence. But George's *Progress and Poverty* had taught Wallace that the Malthusian principle, though valid in the case of animals and plants, was not operative in human society. Wallace could now confidently argue that delayed marriage would be enshrined as one of the fundamental conditions of socialist society. Ironically citing Galton, he noted that the proportionate fertility of women decreased with increased age at marriage. Wallace also invoked Spencer's essay "A Theory of Population Deduced from the General Law of Animal Fertility" (1852). Spencer had suggested that the increasing complexity of civilization encouraged intelligence and self-discipline. It also diminished fertility. Spencer's hypothesis fit perfectly with Wallace's new espousal of female choice in human evolution (Wallace [1890] 1900c, 1:521–523).[7]

Wallace asserted that socialism would eradicate the dangerous conditions of labor under capitalism. Since men were more generally employed in dangerous occupations, this would significantly lower the rate of male mortality relative to females. Wallace knew that there was a statistically higher percentage of male births. He predicted that the observed excess of females, in capitalist society, during the ages of most frequent marriage (from twenty to thirty-five years) would be neutralized under socialism. In the monogamous socialist state that he and Bellamy envisioned, women would be in a slight minority. Female choice, based on an excess of males (or at least not a minority), would become biologically significant. Wallace believed the greater option of female celibacy under socialism, possible because of financial independence, would augment the rigor of sexual selection. Female choice, Wallace asserted, would ensure that those individuals "who are the least perfectly developed either mentally or physically . . . or who possessed any congenital deformity [or tendency to hereditary disease] would in hardly any case find partners, because it would be considered an offence against society to be the means of perpetuating such diseases or imperfections" (Wallace [1890] 1900c, 1:524–525). Such individuals, he was careful to point out, would not be deprived of the ability to lead contented lives. They would, by social consent, simply not transmit their defective traits to offspring (Wallace [1892] 1900c, 2:507).

"HUMAN PROGRESS: PAST AND FUTURE"

In "Human Progress: Past and Future" (1892), Wallace elaborated on the thesis that sexual selection under socialism afforded a potent means of effecting a permanent amelioration of human society. The advance in material civilization in historical times was undoubted. Wallace questioned whether there had been a corresponding advance in human mental and moral nature (Wallace [1892a] 1900c). He granted that "during the whole course of human history the struggle of tribe with tribe and race [with] race has inevitably caused the destruction of the weaker and the lower, leaving the stronger and the higher, whether physically or mentally stronger, to survive." He doubted whether such a process did, or ought to, operate under the conditions of modern civilization. Wallace fulminated that the system of inherited wealth— "which often gives to the weak and vicious an undue advantage both in the certainty of subsistence without labour, and in the greater opportunity for early marriage and leaving a numerous offspring"—had unfortunate consequences for human evolution. He also considered that the preservation of the weak or malformed could be construed as interference with the course of nature (Wallace [1892] 1900c, 2:496–97). Wallace noted, however, that the cultivation of sympathetic feelings "has improved us morally by the continuous development of the characteristic and crowning grace of our human, as distinguished from our animal nature" (Wallace [1890] 1900c, 1:526). The fact that some who in infancy were weak or physically deformed later exhibited superior mental qualities afforded an ethical sanction for civilization's protection of the weak.

Wallace reiterated his concern that humans under capitalist social systems were retarding evolution's "general advance." He stressed that the widespread modern trust in education and environmental reform as the main engines of human betterment was misplaced. Under socialism, the mistaken belief in the hereditary transmission "of the effects of training, of habits, and of general surroundings" would be replaced by the powerful action of sexual selection (Wallace [1892] 1900c, 2:496, 505). Wallace wryly added that Weismann's argument against the inheritance of acquired traits, whether physical or cultural, was cause for relief not despair. The debauched practices of the wealthy and the sordid habits of the oppressed workers in Victorian society could not produce any permanent degradation of humanity. But Wallace was cognizant of what were commonly perceived to be the pessimistic cultural consequences of Weismannian biology. He sought to allay the fears of friends and colleagues with Larmarkican proclivities, such as Ward and LeConte. They believed that Weismann's "germ plasm" theory of heredity doomed to failure all proposals for human betterment "except by methods which are revolting to our higher nature" (Wallace [1892] 1900c, 2:508).[8] Far from negating the influence of education and of beautiful and healthful surroundings, Wallace asserted that socialist society would treasure them. Sexual selection, when informed by an ethos of freedom and human dignity guaranteed by economic equality, necessarily entailed "that education has the greatest value for the improvement of mankind." Moreover, Wallace declared, for the first time in human history "selection of the fittest may be ensured by more powerful and effective agencies than the destruction of the weak and the helpless" (Wallace [1892] 1900c, 2:508).

Wallace deemed the principle of sexual selection under socialism to be "by far the most important of the new ideas I have given to the world" (Wallace [1905] 1969, 2:389). This principle, however, was not a new idea of Wallace's. He borrowed it almost verbatim from Bellamy. But Wallace's particular deployment of sexual selection under socialism is highly significant and was crucial for the development of his vision of human evolution. Malthus had long ago provided Wallace with one stimulus to bring various lines of thought together. Bellamy, and George, now gave Wallace an equally powerful stimulus. He was able to refashion a number of conceptual and ideological themes that had existed in a sometimes uneasy proximity in his life and career. Wallace had long embraced certain socialist ideas, dating from his youthful attendance at Owenite lectures in the working-class Halls of Science and Mechanics' Institutes. The ending of his 1864 essay "The Origin of Human Races" had echoed that Owenite vision (Wallace 1864b, clxix–clxx). But Wallace changed the ending in an 1870 version, which emphasized spiritualist rather than socialist themes. He never abandoned his belief in the guidance of spiritual intelligences as agents in human evolution. Nor did he waver in his insistence that spiritualist claims could be verified empirically and thus constituted a body of demonstrable scientific knowledge (Oppenheim 1985, 316, 320). Nonetheless, Wallace clearly appreciated the polemical advantages of the naturalistic mechanism of sexual selection within the broader framework of his evolutionary teleology (Durant 1979). Bellamy and George legitimated Wallace's ongoing effort to forge a synthesis of social progressionism with biological progressionism.

BRITISH SOCIALISTS: A MIXED CREW

In a story filled with ironies, Wallace's debt to George for his own articulation of evolutionary socialism was profound. George never permitted himself to be labeled a socialist. But Wallace was part of a broad English socialist movement that regarded George as one of its patron saints. When Sidney Webb wrote in 1889 that the "present English popular Socialist movement may be said to date entirely from the circulation here of *Progress and Poverty*," it was only a modest exaggeration. Even Marx's son-in-law Edward Aveling lectured enthusiastically to London audiences on George's theories (Rodgers 1998, 70). Thus, Wallace was now well-positioned to reject any proposals, including involuntary eugenics schemes, for social and/or biological amelioration that accepted Victorian competitive capitalism:

> They all attempt to deal at once, and by direct legislative enactment, with the most important and most vital of all human relations, regardless of the fact that our present phase of social development is not only extremely imperfect, but vicious and rotten at the core. . . . Let any one consider, on the one hand, the lives of the wealthy as portrayed in the society newspapers . . . with their endless round of pleasure and luxury, their almost inconceivable wastefulness and extravagance; . . . and, on the other hand, the terrible condition of millions of workers—men, women, and children—as detailed in the *Report on the Lords' Commission on Sweating,* on absolutely incontestable evidence, and the still more awful condition of those who seek work of any kind in vain. . . . Can any thoughtful person admit for a moment that, in a society so constituted that these overwhelming contrasts of luxury and privation are looked on as necessities, and are treated by the Legislature as matters with which it has practically nothing to do, there is the smallest probability that we can deal successfully with such tremendous social problems?
>
> (Wallace [1890] 1900c, 1:516–517)

Interestingly, Huxley—who was no foe of capitalist imperialism and who regarded Wallace's socialism as anathema—expressed similar concerns about harsh treatment of many individuals. Huxley argued that eugenic intervention would destroy the bonds of social sympathy (Paradis 1989, 47–48). Of course, this was precisely what most eugenists considered the virtue of their schemes: scientific experts would manage societal evolution. Wallace's opposition to eugenics was not shared by all socialists. Pearson saw eugenics, as did certain of the Fabians, as compatible with an "elitist socialism"—a planned socialism by middle-class experts and administrators (MacKenzie 1981, 75–79). The Fabian Society had been formed, in part, by disillusioned middle-class liberals. They rejected Marx's theory of revolution and replaced that with an ideology of evolutionary socialism. For most of the Fabians, the transformation of capitalist society would be accomplished by a reorganization of government ownership and management commencing with the municipalizing of public utilities and transit systems throughout England. The key to Fabian schemes for reform of society was the recruitment of socialist intellectuals to serve as experts and managers of the new order.

The first attempt to present a comprehensive Fabian doctrine was made in December 1889, with the publication of *Fabian Essays in Socialism*. Consisting of eight essays by leading Fabian theorists, including Shaw, Sidney Webb, and Sydney Olivier, the volume sold well. It pushed the Fabian Society into the public gaze (Fabian Society 1889). The volume played down the importance of class conflict. Its contributors suggested that the path to a socialist society lay in the hands of impartial bureaucrats, drawn from the types of clerical and managerial groups represented within Fabian ranks (Laybourn 1997, 22). This was not Wallace's nor many other British socialists' take. They stipulated the active participation of the working classes in effecting social change. Wallace's relationship to the Fabians was not straightforward. The spiritualist affiliation of a number of members of the society, particularly in its early phases, was congenial to Wallace. That characteristic countered the Fabians' increasingly elitist outlook, in Wallace's eyes, but only to a certain degree. Wallace, mindful of his youthful experiences and his Owenite lessons, was now prepared to articulate a strategy that incorporated spiritualism and theism within the framework of what he envisioned as a truly egalitarian socialist society (Barrow 1980).

An exchange of letters (1901–1906) between Wallace and Sydney C. Cockerell reveals much about the central role played by Owen in instilling that life-long passion that marked Wallace's social, and socialist, philosophy. The Wallace-Cockerell connection, which is little known, sheds important light on other thinkers who exerted a major influence on the development of Wallace's political views.[9] Cockerell served at various times as secretary of Morris's Kelmscott Press. He edited the last book issued by that press, *A Note by William Morris on His Aims in Founding the Kelmscott Press* (1898). In 1901, Cockerell sent Wallace a copy of J. W. Mackail's sketch, *William Morris: His Life and Work*. Wallace informed Cockerell that the sketch was well done and showed "what a remarkable man Morris was." But Wallace was irked by Mackail's "absurd remarks on Bellamy's work, which I expect he really never read." In Wallace's opinion, *Looking Backward* was "a better *story* than 'News from Nowhere,' and [gave] a sketch of a far truer and more *practicable*, and also more enjoyable socialist regime than that sketched by Morris." He added that Bellamy's *Equality* provided "the whole theory of Socialism" and clearly indicated the "mode of bringing it about" (Wallace to Cockerell, 9 August 1901, MS. 46442, ARWP). Cockerell then sent Wallace a copy of one of Tolstoy's books. Wallace wrote (1904) that he would "read [it] with intense interest." He informed Cockerell that he was "just now reading Robert Owen's 'Autobiography.' What a marvellous man he was! A most clear-seeing socialist and educator ages before his time, as well as one of the most wonderful *organisers* the world has seen. Both this, and his son's R. Dale Owen's 'Threading My Way' are intensely interesting. One only regrets that neither was completed." Two days later, Wallace wrote again. He told Cockerell that he would go even further in calling "Owen one of the *best* as well as the *greatest* men of the 19th century, an almost ideally perfect character but too far in advance of his time. He was my *first* introducer to mental philosophy and social reform. I heard him speak *once*"(Wallace to Cockerell, 21and 23 August 1904, MS. 46442, ARWP).

WALLACE AND KROPOTKIN

Cockerell was an eager supplier of books to Wallace. He sent him a copy of Prince Peter Kropotkin's *Memoirs of a Revolutionist* in 1905. Wallace immediately began reading it "with very great pleasure." In a letter to Cockerell thanking him for the gift, Wallace made a rare reference to his own childhood. He confided that Kropotkin's "early life—its childhood I mean—allowing for immense difference of rank, wealth and country—was, in *essentials* (education, play, &c.) not unlike my own and affords another indication of how wonderfully alike is human nature under all external changes" (Wallace to Cockerell, 17 December 1905, MS. 46442, ARWP). The comparison is apt. Wallace was never overly impressed with the titles and paraphernalia of wealth and privilege. He could easily discern the similarities between his own character and thought with that of a prince of the tottering Russian Empire. Wallace finished Kropotkin's *Memoirs* in a few weeks. He wrote Cockerell about how fascinating he found Kropotkin's account for the light it shed on the Russian character and "the horrible despotism to which [the Russian people] are still subject, equivalent to that of the days of the Bastille and the system of 'Lettres de cachet' before the great Revolution in France." Inspired by Kropotkin's fervor, Wallace ventured a prophecy of his own. "It seems to me probable," he told Cockerell, "that under happier conditions—perhaps in the not distant future, Russia may become the most advanced instead of the [most] backward in civilization—a real leader among nations, not in war and conquest but in social reform." Kropotkin's political views, as well as his cooperative interpretation of the evolutionary process, resonated with Wallace's convictions.

Wallace also possessed a copy of the 1912 revised edition of Kropotkin's ([1898] 1912) *Fields, Factories and Workshops*. In that book, Kropotkin wanted to demonstrate, from data drawn from Russia, France, Germany, and England, that modern science and technology could be combined with certain of the best features of rural agriculture and industries. Given an appropriate egalitarian societal framework, Kropotkin believed it possible to effect a benevolent "synthesis of human activities" in "modern economical evolution." The revised edition, which Wallace annotated heavily, is noteworthy because of Kropotkin's increased emphasis on economic and social conditions in the United Kingdom. Kropotkin felt the situation in the United Kingdom showed both the necessity and possibility of widespread reform. His focus on the combination of industrial with agricultural activities in a decentralized and egalitarian setting paralleled Wallace's economic philosophy. Like Wallace, Kropotkin believed that the perpetuation of outmoded political theories by the privileged classes in various countries obstructed the creation of a new world order. Wallace underlined the following passage in Kropotkin, which identified the primary obstacles to an immediate start to societal reorganization:

> They are entirely in our institutions, in our inheritances and survivals from the past—in the "Ghosts" which oppress us. But to some extent they lie also—taking society as a whole—in our phenomenal ignorance. We, civilised men and women, know everything, we have settled opinions upon everything. . . . We only know nothing about whence the bread comes which we eat . . . we do not know how it is grown, what pains it costs to those who grow it . . . what sort of men those feeders of our grand selves are . . . we are more ignorant than savages in this respect, and we prevent our children from obtaining this sort of knowledge—even those of our children who would prefer it to the heaps of useless stuff with which they are crammed at school.
>
> (Kropotkin [1898] 1912, 239–240)

Kropotkin was one of several authors whose refusal to equate evolutionary biology with competitive societal struggle impressed Wallace. Wallace possessed Anna Blackwell's *Whence and Whither? Or Correlation between Philosophic Convictions and Social Forms* (1898). Blackwell's main thesis was that late Victorian society was degenerating owing to "the rapid spread of theoretic Materialism, which denies the existence of the Spiritual element of the universe as the corollary of its denial of the existence of an Intelligent CREATOR." She saw around her "the substitution of selfish appetites and interests, in place of the nobler psychic motives of action." This, Blackwell believed, was the practical consequence of materialism's denial of an intelligent creator.

Blackwell maintained that "the Materialistic hypothesis should, therefore, be regarded as only a passing phase of the reaction of modern science" to certain outmoded beliefs and systems. She asserted that "a rational belief of a Beneficent Creator and Overruler of the universe and a rational acceptance of the all-important moral consequences inseparable from that belief" were the most potent forces capable of dislodging the "erroneous assumptions" of materialist theories. Blackwell praised all those who advocated these twin forces as "clearing the ground for the establishment of the Scientific Theism which—as the only certain guarantee of the eternal persistence of the Spiritual Principle, the only sound foundation of Physical Science, and the only safe guide to the elucidation of social questions—is the most urgent need of the present day" (Blackwell 1898, 9–10). Blackwell's views meshed neatly with those of Wallace and Kropotkin.

Wallace marked two passages in Blackwell's book with double vertical lines, underscoring his enthusiastic appraisal of her argument. The first passage claimed that "all the evils of our social state result from the substitution of individualism and antagonism in place of co-operation and mutual helpfulness, and can only be successfully dealt with by substituting *co-operation* for *individualism.*" The second passage marked by Wallace stated that various philanthropic efforts to "ameliorate what is radically bad are mainly to be rejoiced in." Blackwell welcomed them, but not for the obvious reasons. She predicted that philanthropy, no matter how generous, would always fail. Continued failures would lead to the recognition that it is impossible to diminish "the evils of our social state otherwise than by ridding ourselves of the causes to which those evils are due." Wallace shared Blackwell's view that individual acts of philanthropy attacked the symptoms but not the causes of societal inequities. Many other radical social reformers held a similar opinion. All agreed that only an egalitarian reconstruction of economic and political institutions could guarantee lasting change and improvement. Blackwell's assertion that the transformation of society would "eventually be achieved, and in the way implied in the words of Christ, viz., by *the application of the principle of cooperative helpfulness to every department of human life,*" accorded with Wallace's evolutionary teleology. Wallace found Blackwell's work compelling because it mirrored his own quest for an integrative principle for social and scientific thought (Blackwell 1898; the passages Wallace marked are 27–29; emphasis in the original).

SOCIALISM, SPIRITUALISM, AND SEXUAL SELECTION

Wallace's combination of socialism and biology was not unique. Nor was his synthesis of female sexual selection, spiritualism, and reformist social evolutionism uncommon (Owen 1990, 26–27). A number of influential feminist writers utilized sexual selection within a socialist framework. In *Women and Economics* (1898), Charlotte Perkins Gilman (1860–1935) agreed with Darwin that sexual selection was a force in human evolution. But she declared that under capitalism it served not to improve (as Darwin held) the species but to weaken it. Until women became economically independent, Perkins made it clear that female mate selection was illusory. Socialism, Gilman argued, would restore to females the evolutionary potential to make "their rightful contribution to the future of the race" (Russett 1989, 84–86). Gilman was also a Bellamy enthusiast (Love 1983, 121) and had affinities with the Fabians (Pittenger 1993, 72–79). The socialist Eliza Burt Gamble argued, in *The Sexes in Science and History: An Inquiry into the Dogma of Woman's Inferiority to Man* (1894), that under capitalism, women had become "economic and sexual slaves . . . dependent upon men for their support." This had dispossessed females of their "fundamental prerogative" of aesthetic choice. Gamble's book was considered a major nineteenth-century rebuttal of Darwinian arguments for the continuing inferiority of women. She envisioned a noncapitalist future in which women would regain their rightful power of sexual selection. Through the hereditary transmission of the "more refined instincts and ideas peculiar to the female organism" (such as altruism and sympathy), Gamble believed women would found a "new spiritual age" (cited in Richards 1983, 110n. 155). In *Social Environment and Moral Progress*, Wallace offered a similar scenario. He predicted that when woman was "conceded full political and social rights on an equality with man, she will be placed in a position of responsibility and power which will render her his superior, since the future moral progress of the race will so largely depend upon her free choice in marriage. As time goes on," Wallace wrote, "and she acquires more and more economic independence, *that* alone will give her an effective choice which she has never had before. . . . We hope and believe that [women of the future] will be fully equal to the high and responsible position which, in accordance with natural laws, they will be called upon to fulfil" (Wallace 1913b, 163–164).

Wallace's conversion to female sexual selection signaled, as perhaps no other episode in his career did, the convergence of evolutionary biology and sociopolitical reformism. His previous theoretical objections to the efficacy of sexual selection, in the human realm, were no longer valid. Wallace had never denied that human females could exercise some degree of individual choice in mating. But mate selection is not equivalent to sexual selection. For sexual selection to occur, mate choice must act as a cumulative selective force. It must bring about differential rates of reproduction favoring those individuals that bear the preferred traits—and that differ genetically in this respect from other individuals of their sex (Cronin 1991, 114, 168–174).

One of socialism's main attractions for Wallace was precisely the fact that it appeared to be the only social system under which female choice would be made a constant, biologically effective agency in evolution (Wallace 1913b, 163–165).

Wallace continued to deny, or minimize, the efficacy of sexual selection among nonhuman animal species. In a "Note on Sexual Selection" published in *Natural Science* in 1892, he asserted that among other animal species the role of female choice, even if it did exist, would be swamped by the enormously greater power of natural selection (Wallace 1892d, 749–750). This persistence of Wallace's long-held dismissal of the efficacy of sexual selection among animals accentuates the singularity of his reversal with respect to human sexual selection. It is fully consistent with his increasing tendency (since the late 1860s) to emphasize the gulf between certain human attributes and those of other animal species. Since female choice, for Wallace, is associated with the exercise of higher moral and intellectual faculties, it reinforced the role of spiritual guidance at critical stages of human evolution. Wallace explicitly linked his spiritualist convictions to his advocacy of socialism (Wallace [1898b] 1900c).

Wallace's path toward biological socialism was a complex journey. Female sexual selection was a crucial component, but that did not make him a card-carrying Victorian feminist. Like his mentor Bellamy, Wallace's advocacy on behalf of women was colored by some middle-class and patriarchal values (Strauss 1988, 80). Further, although socialists usually supported equality for women, socialism and feminism were sometimes uneasy allies. The immediate gains from women's emancipation mainly benefited middle-class women. As long as the class divisions of Victorian society remained as rigid as they were, an improvement in the status of middle-class women did not necessarily benefit women of the working classes. The "woman question" was entangled with questions of class and status as well as sex (Harrison 1991, 183). As mentioned previously, Wallace had, in 1864, spoken against the Ethnological Society's admission of women to its meetings on the grounds that "consequently many important and interesting subjects cannot possibly be discussed there" (Richards 1989a, 264). Against this, however, we must note Wallace's actions at the local natural history society in Croydon, when he was living in that town between 1878 and 1881. Wallace pushed for the admission of women to meetings of the Croydon club. His formal motion that the society's rules be amended to allow women to attend was made in early 1880. The motion was turned down "by a very large majority." Despite this procedural setback, Wallace continued his campaign for the remainder of his stay in Croydon (Sowan 2001, 17–18).

Bellamy's advocacy on behalf of women also became more pronounced.

In *Equality* (1897), the sequel to *Looking Backward,* Bellamy asserted the equal intellectual capacities between the sexes and gave women the suffrage (Strauss 1988, 88). He explicitly drew the parallel between economic and sexual inequality—and outlined their dual eradication (Bellamy 1897, 128–138). Wallace applauded the more fully developed socialism of Bellamy's sequel (Wallace [1905] 1969, 2:268–272). Wallace drew on *Equality* to provide his own sketch of how the transition to a socialist society could be brought about (Wallace 1905a).

Equality provided Wallace with detailed suggestions for effecting sociopolitical change at which Bellamy's earlier book only hinted. Wallace wrote inside the frontispiece of his copy of *Equality* that "this book contains more than twice the amount of matter of 'Looking Backward.'" He read *Equality* with great care and enthusiasm. Bellamy's insistence on the conjunction of economic equality *and* the protection of individual liberty as the bedrock of socialism resonated deeply with Wallace. Dr. Leete's assertion that "the immortal preamble of the American Declaration of Independence . . . logically contained the entire statement of the doctrine of universal economic equality guaranteed by the nation collectively to it members individually" (Bellamy 1897, 16) elicited from Wallace the comment that the "Declaration of Independence implies Socialism." Bellamy's prescriptions regarding "the equalization of the distribution of work and wealth" (306), the abolition of "private capitalism—that is to say . . . an end to the direction of the industries and commerce of the people by irresponsible persons for their own benefit" (117), and "the abolition of all war or possibility of war between nations" with the growth instead "of a fraternal sympathy and mutual good will, unconscious of any barrier of race or country" (278) all drew exclamatory comments from Wallace such as "fine!" "good!" and "a beautiful argument" (Wallace's copy of Bellamy's *Equality* in ARWL). Wallace was prepared to embrace Bellamy's comprehensive sequel to *Looking Backward.* It appeared almost simultaneously with his own catalog of the successes and failures of nineteenth-century industrial capitalism in *The Wonderful Century.*

In a 1905 essay in *The Clarion,* Wallace cited *Equality* as the clearest and most direct blueprint by which a "capitalist could be changed into a Socialist regime." He was delighted by Bellamy's blueprint. It would work "quietly and systematically, without any compulsion, and yet in such a way as to secure before long the assent and co-operation even of the non-Socialist workers and capitalists." Bellamy's book, Wallace declared, "ought to form part of every Socialist's library . . . [and its] method . . . only needs to be thoroughly understood by all Socialists who really study their subject, in order that it may be adopted when the good time comes." In England, that "good time" would occur when "we have a Socialist majority in the House of Commons and a Socialist Government in power." Working toward that end during the first decade of the twentieth century, Wallace stated that he—and those who followed his strategies for making England a socialist society—would "follow the general lines of Mr. Bellamy's forecast" (Wallace 1905a).

THE JOURNEY TO SOCIALISM COMPLETED

Wallace was, ultimately, unable to resolve all the possible obstacles inherent in the attempt to integrate Victorian socialism, evolutionary biology, feminism, and spiritualism. Who could have? But this does not detract from the historical significance of his biological socialism. Wallace's career demonstrates that the demarcation between biology and politics was (and is) one of shifting, often elusive or nonexistent, boundaries of discourse (Fichman 1997). He refused to divorce biology from ethical, sociopolitical, and theistic thought. Wallace's biological socialism must be assessed against the broader background of Victorian efforts to invoke evolutionary science on behalf of causes that spanned the political spectrum. The protean guises assumed by social evolutionism testify more to the fervor than to the validity of the varied political and moral claims educed from biology (Bowler 1993).

Wallace's soul mate Kropotkin presented his particular thesis of cooperative evolution in reply to Huxley's influential essay "The Struggle for Existence" (1888). Kropotkin set it out in a series of articles for the *Nineteenth Century* between 1890 and 1896. These were collected and published in 1902 as *Mutual Aid* (Kropotkin [1902] 1972). The affinities Kropotkin and Wallace perceived between socialism and evolution assume additional interest because Huxley's specter haunted them both. It has been argued convincingly that Wallace was the "unmentioned target" of Huxley's celebrated Romanes Lecture, "Evolution and Ethics" (1893). Huxley's lecture purported to demonstrate that social ethics could not be teased out from evolutionary biology. Ironically, "far from limiting and depoliticizing the authority of evolutionary science," Huxley's covert attack on Wallace exposed the degree to which politics and science had become entwined in the Victorian debates (Helfand 1977, 176–177). But such merging of evolution and politics does not support the oft-cited claim that evolutionary biology, especially the hypothesis of natural selection, was a "transcription," direct or indirect, of economic theory from society to biology.[10] In Wallace's case, the connection between evolutionary science and political philosophy is more accurately described as confluence rather than transcription. Biological and sociopolitical speculations and hypotheses flowed together and meshed in Wallace's vision for a more humane, less antagonistic, society than that in which he lived. He used biology, as he used social and political reformism, to fight against what he perceived as the injustices of the late Victorian era.

The power of Wallace's worldview was recognized by many of his admirers, who shared his syncretic approach to science and culture. David Maxwell, author of *Stepping Stones to Socialism* (1891), neatly characterized the appeal of Wallace's theistic evolutionary teleology. After reading *The Wonderful Century*, Maxwell sent Wallace a copy of his own small tract accompanied by a letter. Maxwell's letter demonstrates the high regard in which Wallace as scientist/social activist was held. Maxwell wrote that he had "pretty well studied 'The Wonderful Century' and would really call it a Wonderful Book, with an immense amount of matter . . . of the greatest interest." He was particularly impressed with Wallace's effortless weaving together of scientific issues with philosophical and sociocultural ones. Chapters on the glacial epoch, the antiquity of man, and "The Importance of Dust"—in which Wallace explained the significance of that seemingly trivial, if not bothersome, entity as a source of beauty, especially as giving humans "the pure blue of the sky, one of the most exquisitely beautiful colours in nature"—sidle next to chapters dealing with phrenology, vaccination, and militarism. While Maxwell did not agree with all of Wallace's specific verdicts (Maxwell did not consider phrenology as neglected a field of study as did Wallace), he applauded the book's broad scope. He was encouraged to send Wallace a copy of *Stepping Stones to Socialism* because Maxwell seemed "to think judging from many of your remarks, that my little book would not be considered by you obtrusive or beyond your regard" (Maxwell 1891).[11] Indeed, there was little that Wallace deemed beyond his regard. His definition of socialism as the "use by every one of his faculties for the common good, and the voluntary organization of labour for the equal benefit of all" (Wallace [1905] 1969, 2:274) reflects his commitment to spiritualism, mesmerism, and phrenology as well as to evolutionary biology. Wallace's journey to socialism represents another stage in his life-long effort to forge a holistic and activist philosophy of humans and/in nature.

———•◆•———

NOTES

1. While Darwin and many other scientists avoided the political limelight in which Wallace reveled, social and political concerns were never far from their thoughts, however much they may have sought to conceal or minimize them; for an analysis of Darwin's deep

but muted interest in matters sociopolitical, see Rayher (1996). Despite its comprehensive title, Rayher's dissertation uses Wallace primarily as a "reverse Darwin" to demonstrate that Darwin's science was ideologically laden.

2. Wallace's annotated copy of *Data of Ethics* is in ARWL.

3. Stigler 1969; although this article carries Stigler's name, the text of the lectures, with accompanying footnotes, was compiled by R. H. Coase; likewise, the Marshall-Wallace letters were also assembled by Coase from the originals published in the *Western Daily Press*, on March 17, 19, 23, and 24, 1883.

4. For a cautionary analysis that questions many of the historiographic categories usually invoked in discussing the relative influence of "neo-Lamarckians" and "neo-Darwinians" in late-nineteenth-century America, see Numbers (1998, 33–40).

5. Rayher (1996, 204–205) correctly concludes that Wallace had come "to see colonial mismanagement not as an anomaly, but as an inevitable outcome of capitalism, greed, and the unequal distribution of power."

6. For recent analyses, see Ghiselin (1969, 214–31) Campbell (1972), Kottler (1985, 417–419), and Cronin (1991).

7. On the ambiguities inherent in Spencer's treatment of population, and evolutionary theory generally, see Bannister (1979, 41–56).

8. On the disturbed reaction to Weismann's "neo-Darwinism," see Bannister (1979, 138–141).

9. Copies of those letters are included in ARWP MS. 46442; of the five letters from Wallace to Cockerell there, only one, dated 15 January 1906, is reprinted in Marchant ([1916] 1975).

10. Coleman (1999) is one of the more effective of recent arguments against the facile linkage, ideologically or epistemologically, of natural selection and political economy. Significantly, Coleman agrees that Wallace's mature worldview was theistic (18n. 22).

11. Maxwell's letter to Wallace, dated 21 August 1898, is placed inside the front jacket of Wallace's copy of *Stepping Stones*, in ARWL.

CHAPTER 6

Toward a Synthesis:
Wallace's Theistic Evolutionary Teleology

Writing in 1918, five years after Wallace's death, the zoologist and geneticist Lancelot T. Hogben noted that "one of the most significant traits in Wallace's character was his courageous faith in the ultimate goodness underlying the purpose of the world. . . . With eyes fixed beyond the immediate spectacle, he saw to the last the clear gleam of the Light Beautiful in the City of God. And it was . . . because he was able to cultivate and to retain a sense of the reality of the spiritual values that he succeeded in preserving his hope and his humanity throughout his long life" (Hogben 1918, 61–62). Wallace's evolutionary synthesis provides a case study for examining the role of theistic beliefs as one component in shaping the formation of Victorian scientific theories. At the outset of his career, Wallace emphasized scientific naturalism as central to his worldview. But he had also been receptive to additional metaphysical and methodological frameworks (see chap. 3). During the 1860s, Wallace first publicly broadened his position to one consonant with a theistic reading of evolution. After 1870, Wallace continued to be an eloquent and formidable defender of natural selection. But natural selection was incorporated into a broader framework. From the 1870s onward, Wallace's growing commitment to a theistic teleology helped mold his mature evolutionary synthesis. Along with deep political and ethical convictions, theism was integrated with his science.

THEISM IN CULTURAL CONTEXT

In the terminology of modern theology, Wallace's evolutionary theism treats divine activity as a complement to scientific language, not a competitor. As Ian Barbour remarks, the "cosmic drama can be interpreted as an expression of the divine purpose. God is understood to act in and through the structure and movement of nature and history." Wallace was not a classical theist. Divine transcendence is less emphasized in his evolutionary theism than in traditional Christianity. In important respects, Wallace's evolutionary theism can be regarded as a precursor of twentieth-century process theology. In the process model, the Supreme Being is a creative participant in the cosmic community, however different from all other participants. Process theology is consonant with an ecological and evolutionary understanding of nature as a dynamic and open system, characterized by emergent levels of organization, activity, and experience. It avoids rigid dualisms between mind/body, humanity/nature, and man/woman and emphasizes a holistic outlook (Barbour 1997, 231–239, 326, 331).[1] Wallace's odyssey led him to embrace holism as a key to understanding and improving the world. His evolutionary teleology also foreshadows certain contemporary notions of naturalistic theism.[2]

John Hedley Brooke has provided provocative analyses of the terms "religious," "metaphysical," and "scientific" for exploring the interplay between theism and science. Brooke has asked, pertinently, "What kinds of interactions between science and religion should we be looking for? And what is the shape of a map that most faithfully represents the diversity of historical positions adopted by specific individuals?" Such a map is multidimensional: (1) it must recognize the different levels on which theological language has impinged on and, sometimes, penetrated scientific discourse; (2) it must recognize that scientific discourse itself can be subdivided into many different types; and (3) it must recognize that statements linking nature and God have historically fulfilled a multitude of social, political, and religious functions (Brooke 1996, 1–26). What would such a map look like for Wallace? This chapter provides elements for mapping Wallace's construction of an evolutionary theism. By taking Wallace's theism seriously, one avoids the pitfalls that have marked previous attempts to characterize the interplay of metaphysics, religion, and science in Wallace's mature statement of evolutionary biology as retrograde, marginalized, or misguided (Fichman 2001b).

WALLACE'S EARLY RELIGIOUS ENVIRONMENT

Wallace grew up, as did many of his contemporaries, in a fairly traditional religious environment. His parents were both "old-fashioned religious people, belonging to the Church of England." Alfred and his siblings went to church twice each Sunday. They were periodically examined in their catechisms and were frequently read chapters from the Bible by their father in the evenings. Also, the Wallaces counted among their friends "some Dissenters, and a good many Quakers, who were very numerous in Hertford."

Wallace found his family's occasional visits to the Friends' Meeting Houses "dull and wearisome." In contrast, he delighted in the livelier atmosphere of the Dissenters' chapels. "The extempore prayers, the frequent singing, and the usually more vigourous and exciting style of preaching was," to Wallace, "far more preferable to the monotony of the Church [of England] service; and it was there only that . . . I felt something of religious fervour, derived chiefly from the more picturesque and impassioned of the hymns." Even at this early stage, however, Wallace was disappointed in the apparent lack of intellectual rigor to match the sense of religious fervor. His apprenticeship in London with his nineteen-year-old brother John in 1837, quickly—but not permanently—dampened Wallace's "religious fervor." The strident anti-Church attitudes of his brother's companions were contagious. Wallace was taken regularly to evening meetings at the Hall of Science in Tottenham Court Road, where the followers of Robert Owen lectured. Wallace termed it a "kind of club or mechanics' institute for advanced thinkers among workmen." The principles of Owenite social and political philosophy, though hardly appreciated fully by the thirteen-year-old, were to influence Wallace's own reformist views profoundly (see chaps. 2, 5). In London, there were also lectures on secularism and what would soon be labeled as agnosticism. These lectures, coupled with his reading of Thomas Paine's *The Age of Reason,* effectively challenged the elements of institutionalized religion Wallace had imbibed from his parents.

Wallace seems to have escaped the endemic Victorian crises of belief. But the hold of secular rationalism on him was to prove incomplete. At this period, however, Wallace was swept up in the fashionable anticlericalism of the London workers among whom he lived. He felt satisfied in dismissing orthodox arguments for the existence of God as both logically flawed and intellectually bankrupt. As shown in chapter 4, Wallace found the writings of Owen's eldest son, Robert Dale, compelling. Dale Owen argued that the "orthodox religion of the day was degrading and hideous, and that the only true and wholly beneficial religion was that which inculcated the service of humanity, and whose only dogma was the brotherhood of man." Wallace claimed in his autobiography that such teachings laid the "foundation of his religious scepticism." It would be more accurate, however, to interpret this early period in London as eradicating whatever remnants of the institutional structure and message of the Church of England Wallace still retained from his childhood. What he termed his "religious scepticism" did not turn him to atheism or agnosticism. Theistic sentiment remained a force in Wallace's life. But in these heady days of his first exposure to social and political radicalism, religious orthodoxy was eclipsed. When Wallace read about "the very old dilemma as to the origin of evil" in a secular pamphlet, he discussed the problem of theodicy with his father. Rather than being shocked at his son's first "acquaintance with such infidel literature," Wallace's father "merely remarked that such problems were mysteries which the wisest cannot understand." This response, not surprisingly, failed to satisfy the inquisitive youth. Wallace felt that arguments about theodicy "did not really touch the question of the existence of God." They "did seem," however, "to prove that the orthodox ideas as to His nature and powers cannot be accepted" (Wallace [1905] 1969, 1:77–79, 86–89). Religious orthodoxy, not theism, was wiped away for Wallace.

WHAT TYPE OF THEIST WAS WALLACE?

The literature on the varieties of theism is immense (Quinn and Taliaferro 1997, 197–521). The definition provided by the prolific Scotch scholar Robert Flint, in his *Theism*, was a generally accepted one in late Victorian Britain: "Theism is the doctrine that the universe owes its existence, and continuance in existence, to the reason and will of a self-existent Being, who is infinitely powerful, wise, and good" (Flint 1877, 18).[3] In adopting Flint's definition, some qualifications are necessary for clarifying Wallace's theism. Wallace believed in a God but avoided allegiance to any traditional confessional, doctrinal, or institutional position. Moreover, Wallace never asserted that God's existence is provable. Rather, he belongs to the ranks of those who, when taking "account of a sufficiently comprehensive range of data—not only the teleological character of biological evolution but also man's religious, moral, aesthetic, and cognitive experience"—argue that theism is the most probable worldview (Hick 1990, 26–28).

Designating Wallace a "spiritualist theist," a term that might appear appropriate, is problematic. Not all spiritualists in the late Victorian period were theists. Conversely, not all—indeed very few—theists were spiritualists. Although spiritualism was a major component of Wallace's worldview, he ultimately moved beyond the conventional Victorian spiritualist teachings and environment. Spiritualist beliefs and experiences reinforced Wallace's theistic conceptions. In 1874, he deemed spiritualism to be "the only sure foundation for a true philosophy and a pure religion" (Wallace 1874). But Wallace's mature evolutionary theism included elements of belief that were not among the canons of spiritualism. A number of scholars have clearly recognized the significance of spiritualism in Wallace's life and writings. Few have adequately examined Wallace's broader religious worldview, his evolutionary theism.[4] In developing a theistic framework that transcended spiritualism, Wallace was able to refine his understanding of key concepts like nature and evolutionary teleology. Theism enriched what he deemed to be the legitimate ethical and social implications of evolutionary science. Wallace's conception of the scope of natural selection functioned in ways similar to that of contemporaries such as Asa Gray, who sought also to provide a theistic yet scientifically rigorous rendering of evolutionary theory (Gray 1876). Like Wallace, Gray agreed that, in human evolution, certain forces other than natural selection had aided in the development of "the transcendent character of the superadded" mental and moral attributes. These forces were, like Wallace's, divine (Gray 1880, 44, 99–103).

Wallace's personal journey testifies eloquently to the tensions inherent in efforts to mediate between theism and science in Victorian culture. The second half of the nineteenth century was a period when attempts to demarcate between scientific and nonscientific factors in shaping the social construction of knowledge had become the subject of profound and bitter debate. As Hogben trenchantly remarked, "It required a reputation so powerful as that of Wallace to withstand the odium with which orthodox sociologists [and scientists] greeted what they were pleased to regard as a naturalist erring from his proper bent" (Hogben 1918, 58). The fact that Hogben's book was published by the Society for Promoting Christian Knowledge—which paid great attention to the work of those scientists whose views they felt accorded with Christian theism—indicates that Wallace's evolutionary teleology had come to be recognized as theistic by many. Wallace's use of theistic concepts in formulating scientific claims was a matter of deep personal conviction. Wallace's clash with George John Romanes in this sensitive area is significant.

Romanes was one of Darwin's most devoted followers. He consistently attacked those who, like Wallace, adduced what he felt were supernatural factors to explain the origin of the higher human faculties (see, e.g., Romanes 1888). Wallace was particularly bitter toward Romanes. Romanes had flirted briefly with spiritualism but kept those forays concealed from his fellow naturalists. Thus, when Romanes publicly accused the "scientist" Wallace of succumbing to the "spiritualist" Wallace, he regarded Romanes's public posture as cowardly and duplicitous. Wallace's account of this episode in his autobiography includes an exchange of letters in 1890 between Romanes and him that is further testimony to the fluidity of the categorization of "science" in the late Victorian period (Wallace [1905] 1969, 2:309–326). Carl Jung's comment on Wallace in this context is apt. Noting that even if a spiritualist interpretation of observed psychical facts be disputed, Jung asserted that Wallace—along with Myers, Crookes, and the Cambridge philosopher Sidgwick—merited praise for "having thrown the whole of [his] authority on to the side of nonmaterial facts, regardless of . . . the cheap derision of [his] contemporaries; even at a time when the intellect of the educated classes was spellbound by the new dogma of materialism, [Wallace] drew public attention to phenomena" that were contrary to accepted convictions (Jung 1921, 75–76).

From the 1880s onward, Wallace became less active publicly in spiritualist causes. With the exception of his membership in the newly formed (1882) SPR and attendance at séances during the North American lecturing tour in 1886–1887 (Wallace [1905] 1969, 2:337–349), Wallace's interest in spiritualist concerns was confined primarily to private correspondence and contributions to selected periodicals. His increasing involvement in sociopolitical debates and activities consumed a great deal of energy. But his commitment to theism also demanded more of his time. He dedicated himself to articulating an evolutionary teleology in which science and theism merged. His worldview would receive its fullest expression in the publication of *Man's Place in the Universe* ([1903] 1907) and *The World of Life* (1910a). Far from clouding his scientific acumen, Wallace's "worldview commitments" enabled theism to function positively in his evolutionary synthesis. Wallace's reconceptualization of the scope of natural selection is a potent instance of the interaction, and integration, between worldviews and science (Wykstra 1997). Entrenched historiography has interpreted Wallace's publicly announced 1869–1870 views on certain aspects of human evolution as a volte-face with respect to his previous philosophy of evolution. The analysis of Wallace's thoughts and writings from 1845 to 1870, in chapter 4, yielded a different picture.

Ethical, epistemological, and sociopolitical as well as biological interests and investigations characterized Wallace's odyssey from the outset. His caveats with respect to the scope of natural selection had their roots in Wallace's earliest evolutionary speculations. As shown in chapter 4, theism and natural selection were never viewed by Wallace as mutually exclusive components of a larger evolutionary teleology. A teleological epistemology pervaded Wallace's approach to all subjects, including science. What had been implicit in the younger Wallace's outlook became explicit in the older man. In a 1911 interview, Wallace rejected the allegation that he "had left the path of science in touching on final causes in my books." Wallace countered the charge that teleological explanations were speculative by declaring that it "is no speculation to point out that any mechanical explanation of the universe really explains nothing, and that you must have an intellect, or a Being, or a series of Beings." Targeting mechanistic materialism in particular, Wallace argued that such reductionism was antithetical to the fundamental nature of scientific explanation. He believed that, by the turn of the century, "there

[were] a greater number of scientific men now than ever before who see that the deeper we go into things the more mystery there is, and the more need for Mind rather than [mere] Force" ([Wallace] 1911).

By embedding natural selection within the framework of an evolutionary teleology, Wallace had, from the late 1850s/early 1860s, signaled that natural selection was compatible with other higher, directed powers (Wallace [1891] 1969, 213–214). These "signals" came to permeate, in an increasingly fundamental mode, his elaboration of an evolutionary theism during the last three decades of his life. Wallace remained committed to naturalism but not to the version that precluded theistic components. He saw theism as a natural complement to organic evolution. It transformed a bleak vision into one of optimism. The evolutionary progression of man, he believed, would culminate in a higher, entirely spiritual form of existence that lay beyond the individual's biological life span. Theism, for Wallace, constituted a legitimate component of a valid and inclusive system of investigating nature. As shown in chapter 4, Wallace's was scarcely a lone voice in the scientific community. Important groups existed within the ranks of professional scientists in the later Victorian period that fully endorsed the notion that there was an integral religious dimension to science. Theistic science was a powerful paradigm in Wallace's era (Lightman 2001). Wallace counted himself among those who believed that scientific theism afforded a path toward resolving many of the disturbing contradictions of Victorian culture.[5]

SCIENTIFIC THEIST

Wallace's embrace of theism did not cause his scientific productivity to suffer. Indeed, the converse is true. During the period from 1890 to 1913, Wallace contributed a steady stream of influential articles and books on technical subjects. They ranged from animal mimicry, glacial theory, the geological permanence of the great ocean basins, and biogeography to *Man's Place in the Universe: A Study of the Results of Scientific Research in Relation to the Unity or Plurality of Worlds* ([1903] 1907), *Is Mars Habitable?* (1907), and *The World of Life: A Manifestation of Creative Power, Directive Mind and Ultimate Purpose* (1910a). Some of these areas were not as directly affected by Wallace's growing commitment to theism as were others. Biogeography showed minimal theistic influence. Wallace's views on organic evolution and astronomy displayed far greater penetration by theistic convictions. His polemical assaults on starkly materialistic interpretations of nature enhanced the interest in Wallace's evolutionary pronouncements. Wallace's sociopolitical writings and activities served further to increase public awareness of his scientific work.

Wallace became one of the best-known figures of natural science among the general public in the latter decades of the Victorian period (Smith 1991, 117–118, 509–529, 533–534).

Wallace's tour of the United States and Canada provides striking testimony to the wide appeal of his efforts to integrate religious convictions and science into a theistic evolutionary framework (see chaps. 3, 5). In well-attended lecture series from Boston to San Francisco, Wallace solidified his public stance as the "greatest living champion" of evolutionary theory. Joseph LeConte, the eminent professor of geology and zoology at the recently founded University of California, introduced Wallace in just those terms "to a large and cultivated audience" on 25 May 1887. Wallace had been asked to lecture on "The Darwinian Theory, What It Is and How It Is Demonstrated" in San Francisco. LeConte was particularly well placed to introduce Wallace. In addition to his scientific credentials, LeConte had earned a distinctive place in American culture by his many books and articles designed to accommodate Christianity to modern science. LeConte was one of that group of Protestant evolutionists who, by the early 1880s, claimed to be "pioneers" in opposition to "the materialistic and irreligious [interpretations] of the doctrine of evolution." LeConte regarded himself and Wallace as kindred thinkers who sought to place theism securely within an evolutionary context (LeConte 1903, 335–337).

Wallace never referred to himself as a Christian. He was, however, expounding an explicitly theistic evolutionism in San Francisco. Wallace concluded his survey "The Darwinian Theory" with the assertion that although the human bodily structure is primarily the product of natural selection operating on lower animals, "the changes of his mental nature do not appear capable of the same explanation. . . . Holding as I do that mind is more fundamental than matter, and that the spirit or soul is the real man, of which the body is but the temporary manifestation or dwelling-place," it is the spirit, guided by higher agencies, that ultimately will develop "the noble and perfect human form."[6] Two weeks later (5 June 1887), Wallace delivered another lecture "before quite a large audience" in which he elaborated on the question of an afterlife. Wallace declared this a question "which the ancient scientists considered [an] unsolved problem, and that modern scientists had either left untouched or precisely where they found it." He argued that modern science, "having decided that all force was the result of molecular motion of matter," had hardened by the mid-nineteenth century into "this compact, fortified and nearly impregnable condition" that afforded no credence to spiritual manifestations. The growing body of evidence for the existence of spiritual phenomena, Wallace declared, had by the 1880s altered the scientific playing field. He claimed that a theistic contextualization of spiritualism could achieve what traditional science or religion alone had been unable to do. It would provide "a rational account" of the history and destiny of the human species. For Wallace, "spiritualism . . . proves that mind may exist without brain, and places a new light upon death [and the afterlife]."[7]

A BUSY OCTOGENARIAN

Wallace turned eighty years old in 1903. His last decade saw him publish five new full-length books, a two-volume "edited and condensed" version of Richard Spruce's *Notes of a Botanist on the Amazon and the Andes,* and his autobiography *My Life.* Seven books in the last ten years of an octogenarian's life might appear prodigious enough. Not so for Wallace. A steady stream of more than 120 articles, leaflets, letters to editors, and interviews issued forth from Old Orchard, his home in Broadstone (Dorset). They dealt with subjects ranging from "The Birds of Paradise in the Arabian Nights" (1904) to "A Statement of the Reasons for Opposing the Death Penalty" (1906) to "The Native Problem in South Africa and Elsewhere" (1906) to "Fertilisation of Flowers by Insects" (1907) to "Evolution and Character" (1908) to "The Present Position of Darwinism" (1908) to "Flying Machines in War" (1909) to "A Policy of Defence" (1912) to "The Origin of Life" (1912; full citations in Smith 1991, 521–536). The eclecticism that had been the hallmark of Wallace's personal and public odyssey for nearly eight decades continued to be manifest "unto his last." Barrett, one of Wallace's oldest friends, visited him in the early summer of 1913, a few months before Wallace's death. Barrett found Wallace in failing health, sitting "wrapped up before a fire in his study, though it was a warm day. . . . [But] his eyesight and hearing seemed as good as ever, and his intellectual power was undiminished." Barrett recounted that, suddenly, Wallace,

> pointing to the beautiful expanse of garden, woodland and sea which was visible from the large study windows, burst forth with vigorous gesticulation and flashing eyes: "Just think! All this wonderful beauty and diversity of nature results from the operation of a few simple laws. In my early unregenerate days I used to think that only material forces and natural laws were operative throughout the world. But these I now see are hopelessly inadequate to explain this mystery and wonder and variety of life. I am, as you know, absolutely convinced that behind and beyond all elementary processes there is a guiding and directive force; a Divine Power or hierarchy of powers, ever controlling these processes so that they are tending to more abundant and to higher types of life."

Evolutionary theism also permitted Wallace to achieve the synthesis of science and sociopolitical thought, which was one of his most cherished goals. As Barrett noted, "Then our conversation turned upon recent political events, and it was remarkable how closely [Wallace] had followed, and how heartily he approved, the legislation of the Liberal Government of the day. His admiration for Mr. Lloyd George was unfeigned . . . and he confidently awaited still larger measures which would raise the condition of the workers to a higher level; and nothing was more striking than his intense sympathy with every movement for the relief of poverty and the betterment of the wage-earning classes. The land question, we agreed, lay at the root of the matter, and land nationalisation the true solution" (Marchant [1916] 1975, 469–470).

A PRAGMATIC THEISM

The integration of scientific theism with powerful elements of sociopolitical and ethical reformism is another manifestation of the comprehensiveness of Wallace's thought. Of all the scientific theists of the late Victorian period, it is Peirce with whom Wallace shared the closest affinity (see chap. 3). Their evolutionary cosmologies are predicated on a definition of truth that rests on the existence of an objective reality. Drawing on a combination of metaphysical idealism and empirical realism, Wallace, like Peirce, conceived metaphysics as a scientific inquiry with profound human significance. The stakes were high. Were humans simply aggregates of bits of matter moving and acting according to mechanically determinant laws? Or were they integral and active parts of a living universe that is moving toward some good end? Science, particularly evolutionary science, was central to answering the question. Neither Wallace nor Peirce shied away from declaring explicitly that their scientific theism was a potent guide for investigating the human condition. For Peirce, knowing the nature of truth and reality, and humanity's relation to them, afforded powerful aid in helping realize truth and reality and the full meaning of human existence. It was this possibility of social and individual action based on knowledge that led Peirce to characterize his own synthetic evolutionary philosophy as "a highly practical and common-sense position." Peirce initially referred to this as pragmatism but later renamed it pragmaticism (Esposito 1980, 127; Parker 1998, 189, 193), but this distinction, while germane to the history of philosophy, is not relevant to the identification of Wallace as a follower of the pragmatic criterion of meaning.

Peirce dismissed traditional metaphysics. He felt it offered nothing significant for human life other than either reassuring platitudes or terrifying dogma (Parker 1998, 200–201). Wallace agreed (see chap. 3). Both Wallace and Peirce, in the 1880s and 1890s, embarked on developing their versions of evolutionary teleology. They felt confident that a rigorous antidote to materialism and physicalism was viable and necessary. Peirce used as a guide for his project Francis Ellingwood Abbot's *Scientific Theism* (1885; Anderson 1995, 154–155). Wallace possessed a copy of Abbot's *The Way Out of Agnosticism* (1890), which developed the ideas of the earlier book more fully. *Scientific Theism* had received highly favorable press notices. These were printed at the end of Wallace's copy of *The Way Out of Agnosticism*. One notice was by Peirce, in the *New York Nation*. He described *Scientific Theism* as "a strongly characterized and scholarly piece of work, doing honor to American thought." A review that appeared in the *Boston Daily Advertiser* caught Wallace's attention. The Boston reviewer noted that the educated community "are not usually much attracted by books on scientific theism. Too commonly they are attempts to make use of the general interest in science to call attention to some not very original or profound speculations about religion. The result often is a syncretism of poor science and worse theology. Such a prejudice cannot attach itself to any work from the pen of Mr. Abbot." In sending a presentation copy of *The Way Out of Agnosticism* to Wallace, Abbot enclosed a letter (dated 27 April 1890). He asked Wallace "to express your critical opinions of the new argument, grounded solely on science and philosophy, which it presents in support of theism." Abbot added that he "entertained a stronger hope of sympathy in this endeavor from [Wallace] than I do from most scientific men of the day [because] you have shown what seems to me a deeper insight than they into the indestructible nature of our great religious convictions." Abbot concluded with words that could not fail to attract Wallace: "Ignoring wholly the traditional grounds, I make my appeal [for scientific theism] solely to the modern intelligence." Wallace's extensive annotations to *The Way Out of Agnosticism* indicate that it encouraged him to undertake his own articulation of scientific theism (Wallace's annotated copy of Abbot [1890] is in ARWL). Wallace was ready to integrate fully the strands of evolutionary biology and other sciences, theism, social and political activism, and ethical goals, which had become inextricable components of his vision of a humanistic social order.

Man's Place in the Universe

Wallace's quest for synthesis was evident in his earliest forays in natural history in the 1840s and 1850s. It had grown steadily in power and scope in the following half century. His 1903 book on the implications of recent astronomical findings for the question of the origin of life shows that quest in undiminished force. *Man's Place in the Universe: A Study of the Results of*

Scientific Research in Relation to the Unity or Plurality of Worlds is vintage Wallace. It is controversial, idiosyncratic, and highly readable (Wallace [1903] 1907). Wallace had entered into a field of lively scientific and popular interest and debate. The book went through seven editions in just five years. The appearance of "cheap" editions in 1912 and 1914 testifies to the interest generated by Wallace's pronouncements on the controversy concerning the plurality of worlds. There was a long history of debate as to whether life is restricted to Earth alone or existed (in some form) elsewhere in the universe (Dick 1982). The debate was cast in a new mold in the second half of the nineteenth century. From 1877 onward, the Italian astronomer and statesman Giovanni Schiaparelli observed about a hundred systems of rectilinear marking on Mars's surface. He dubbed them *canali*, Italian for "channels" or "canals." The American astronomer Percival Lowell became the leader of those who believed the markings to be bands of vegetation, bordering irrigation ditches dug by intelligent beings to carry water from the polar caps. Most astronomers could see no canals, and many doubted their reality. Experiments, such as photography through the earth's atmosphere, were not feasible. The lines were near the limit of resolution of the human eye and of existing telescopic cameras. The controversy was resolved only when pictures were taken from the Mariner spacecraft in 1969. These demonstrated that the canals are illusions caused by chance alignment of large craters and other features of the Martian surface. It was not just the disputed appearance of the canals that made the controversy into which Wallace leaped so vehement. Also at stake were "disputed ideas of the proper route from observation to theory." Had the observational data been more definitive, they "might have settled the controversy themselves by precluding intelligence" (Dick 1996, 79).

Debates on extraterrestrial life and intelligence raised crucial philosophical and religious as well as scientific issues. Scientists and laypersons were confronted with the perplexing question of whether there might be other cultures apart from those on Earth (R. W. Smith 1999, 238). Such debates were part of a broader process in which the often-messy interactions between scientists in and between specialized disciplines became more openly discussed. Victorian popularizers of science since midcentury had become crucial mediators between the professional elites and the rapidly increasing mass market of educated, middle- and working-class readers eager for scientific information. John G. Wood (who had encouraged Wallace to go on the lecture tour of North America), Richard Proctor, and Agnes Clerke were three of the most influential science popularizers in the latter half of the nineteenth-century. Significantly, all three saw their works as tools to refashion and revitalize the natural theology tradition in light of the most recent findings of modern science. They were explicitly aiming to keep natural theology vibrant at a time when Darwinian theory and scientific naturalism posed serious threats to the public's belief in divine wonder in nature (Lightman 2000). Wallace could not remain aloof from the fray. *Man's Place in the Universe* was intended to integrate the most recent findings of astronomy and of evolutionary biology with theological considerations. Wallace wanted to convince both the scientific community and the general reader of the uniqueness of life on Earth (Wallace [1903] 1907, vi). Wallace's arguments for the earth's uniqueness set him at odds with Proctor, an advocate for a pluralistic theory of life in the universe. But Wallace and Proctor shared the overriding motivation to demonstrate the hand of God in nature to elite and popular audiences.

Astronomy had long fascinated Wallace. An entire chapter of *Island Life* was devoted to astronomical causes of changes of climate (Wallace [1880] 1892, chap. 8). The evolutionary implications of astronomical factors, especially those bearing on the geographical distribution of animals and plants, were key elements in Wallace's cosmology at least since 1880. But astronomy now played an even more crucial role in Wallace's speculations. It permitted him to buttress his evolutionary theism with modern findings of other scientific disciplines. For Wallace, the question of whether life on Earth was unique bore directly on broader questions of religion and culture. "During the last quarter of the past century," he declared,

> the rapidly increasing body of facts and observations leading to a more detailed and accurate knowledge of stars and stellar systems have thrown a new and somewhat unexpected light on this very interesting problem of our relation to the universe. . . . They do tend to show that our position in the material universe is special and probably unique, and that it is such as to lend support to the view, held by many great thinkers and writers to-day, that the supreme end and purpose of this vast universe was the production and development of the living soul in the perishable body of man.
>
> (Wallace 1903a)

Wallace's goal was to construct an irresistible argument against the possibility of life elsewhere in the universe.

Wallace drew six major conclusions from his interpretation of astronomers' findings. Three of these he regarded as incontestable: (1) The universe "though of enormous extent, is yet finite, and its extent determinable." (2) The solar system is situated near the center of the Milky Way, and the earth "therefore nearly in the centre of the stellar universe." (3) The universe "consists throughout of the same kinds of matter, and is subjected to the same physical and chemical laws." Three additional conclusions, Wallace claimed, had high probabilities in their favor: (4) No "other planet in the solar system than the earth is inhabited or habitable." (5) The "probabilities are almost as great against any other sun possessing inhabited planets." (6) The "nearly central position of our sun is probably a permanent one, and has been specially favourable, perhaps absolutely essential, to life-development on the earth." Few of his contemporaries would dispute Wallace's first trio of conclusions. It was the latter three assertions that catapulted *Man's Place* into turbulent waters. Wallace claimed that his synthesis led to "one great and definite conclusion—that man, the culmination of conscious organic life, has been developed here [on Earth] only in the whole vast material universe we see around us." Wallace assured his readers that this conclusion held nothing "that need alarm either the scientific or the religious mind" (Wallace [1903] 1907, 317–318). *Man's Place* was a deftly constructed argument for evolutionary theism and teleology. It was destined for controversy.

RECEPTION OF *MAN'S PLACE*

The book "set all the world talking." It created "quite a sensation" among both the scientific community and the general public. Interviewing Wallace shortly after the publication of *Man's Place*, Albert Dawson suggested that a major reason for Wallace's extensive popularity among Victorian readers was his willingness to enter into controversial areas and express his opinions with "candour, open-mindedness and high courage." Dawson's perception indicates—once again—that the professionalization and specialization of science, so precious to Huxley and many other scientists of the late Victorian period, was not one of Wallace's own driving goals or talents. Dawson captured Wallace's personality well in calling him "a pioneer. The beaten track has no attraction for him; his adventurous spirit and quenchless enthusiasm [even in his eighty-first year] sometimes carry him into regions that are under the ban of the orthodox scientist. . . . That Dr. Wallace has not shrunk from incurring the odium scientiae is one of the reasons of his popularity with the general public." Numerous reviews of *Man's Place* quickly appeared (Dawson 1903). Wallace gained the support of a number of important astronomers in advancing his views as to the position of the earth and the solar system in the universe. What sparked debate was Wallace's assertion that the intricate web of conditions and forces that resulted in human evolution on Earth was unique. He declared it "in the highest degree improbable that they can *all* be found again combined in the solar system or even in the stellar universe" (Wallace [1903] 1907, 274–275, 310–317). To think otherwise, Wallace insisted, "would imply that to produce the living soul in the marvellous and glorious body of man—man with his faculties, his aspirations, his powers for good and evil—[was] an easy matter which could be brought about anywhere, in any world. It would imply that man is an animal and nothing more, is of no importance in the universe, needed no great preparations for his advent, only, perhaps, a second-rate demon, and a third or fourth-rate earth" (Wallace [1903] 1907, 321–322).

Wallace hoped that *Man's Place* might provide additional support for Christians (and other religious groups) in their efforts to defend their faith against skeptical onslaughts. But that would be a secondary bonus. His main objective in *Man's Place* was to buttress scientific theism further. Wallace's theism rested not on faith but on what he considered the fact that the "law of the universe seems to be growth by evolution—from the lower to the higher, smaller to greater, worse to better." Theistic evolutionary teleology, Wallace maintained, "may actually govern the action of God Himself. The old idea that God is omnipotent in the sense that He can do anything, even make two and two add up into five, is not a working theory. Limitation, pain, struggle are evidently essential factors in the development of spiritual beings, and if we believe in a Supreme Being with faculties at all similar to those with which He has endowed us, we cannot help also believing that His purpose is the perpetuation of the greatest happiness of the greatest number" (Dawson 1903).

Wallace's optimism that evolution afforded a philosophical as well as scientific basis for belief in human progress—subject, always, to prerequisite socioeconomic and political reforms—struck some of his contemporaries as extreme. Such a verdict ignores the realistic and experiential basis of Wallace's optimism. Marchant asked Wallace how he had endured the loss of his valuable Amazonian specimens and notes when the vessel transporting them back to England burned and sank. Wallace replied that he thought that loss was "the most fortunate thing that happened to me. . . . As the result of this accident I went to the Malay Archipelago, a perfectly virgin country, which hardly any naturalist had then properly explored. My experiences in the Far East were of singular interest to me, and I look back upon them as standing for probably the best part of my life" (Marchant 1905). Admittedly, Wallace is characterizing his feelings more than half a century after the event occurred. It is difficult not to imagine that when the ship actually sank, so too did Wallace's spirit—but only temporarily. He took the loss in stride, went on to explore the Malay Archipelago for eight years, and made scientific history. Wallace's optimism was rooted from early youth until old age in a deeply practical, as well as metaphysical, conviction that adversity could be made to serve a beneficial purpose in human life.

Wallace knew that his astronomical argumentation might be viewed with skepticism by some members of the specialized scientific community. In the sixth edition of *Man's Place*, he added an appendix titled, authoritatively, "An Additional Argument Dependent on the Theory of Evolution." Wallace marshaled the results of his life-long studies as one of the world's preeminent biologists. One of his original and enduring hypotheses was that the "rigidity of natural selection and the severity of the struggle for existence . . . [proves] that no species has ever arisen independently in different places or at different times" (Wallace [1903] 1907, 329). There is an irony in Wallace's self-reference to the famous 1855 essay "On the Law Which Has Regulated the Introduction of New Species." One of the key motives for writing that essay was to oppose biblical creationists. But Wallace had also wanted to deflate those evolutionists who incorporated traditional, and what he regarded as ineffective, teleological arguments drawn from orthodox theology and outmoded science (McKinney 1972, 45; Ospovat 1978, 35, 49–52). In *Man's Place*, Wallace now used a scientifically updated evolutionary teleology to demonstrate that

> the ultimate development of man has . . . depended on something like a million distinct modifications. . . . The chances against such an enormously long series of definite modifications having occurred twice over, even in [*sic*] the same planet but in different isolated portions of it . . . are almost infinite. . . . But if such long-continued identity of the whole course of evolution is hardly conceivable on different parts of the *same* planet, where all the . . . essential conditions are equally fulfilled, how infinitely improbable it becomes that such an identity should have arisen . . . on other planets of other suns, where the whole series of fundamental conditions which I have shown to be essential for *any* high development of life, though they might in rare cases approximate those of the earth, could certainly never have been quite identical. And without absolute identity to the smallest details, any identity of development, resulting after millions of ages in the *same* forms of the higher animals, is manifestly impossible.

For Wallace, this impossibility of identical physical conditions was supplemented by the conviction that the "very definite and peculiar mental and moral nature" of mankind is still more unique and providentially designed. The improbabilities of intelligent life elsewhere in the universe were, to him, "so great as to approach very closely, if not quite to attain, the actually impossible" (Wallace [1903] 1907, 332–335).

Proof of the overarching argument of *Man's Place* required, Wallace admitted, another volume. This would appear in 1910 as *The World of Life: A*

Manifestation of Creative Power, Directive Mind and Ultimate Purpose (Wallace 1910a). But Wallace had a more immediate challenge. Percival Lowell's *Mars and Its Canals* was published in 1906. Lowell contended that extraterrestrial life was probable but more so closer to Earth than in the remote reaches of the universe. Few professional astronomers were convinced by Lowell's arguments that the "canals" of Mars implied intelligent beings as their architects. But his ideas exerted considerable influence on the general public. In addition to his position as one of the United States' leading astronomers, Lowell was a brilliant speaker and always in great demand for lecture tours (Marsden 1973). Lowell's career provides highly useful insights into the ways in which scientific images, both visual and rhetorical, travel through different cultural spheres (Strauss 1998). Wallace thus had an opponent who enjoyed a scholarly/popular repute similar to his own.

Wallace had first sought to counter Lowell and other critics of *Man's Place* in an article he published in 1903 (Wallace 1903b). The publication of *Mars and Its Canals* demanded an extended refutation. Wallace had to show that Lowell failed to invalidate the central thesis in *Man's Place*, that neither Mars nor any other planet was habitable. The result was *Is Mars Habitable?* published in 1907 (Wallace 1907). Wallace considered his book "as furnishing [to the general reader] a quite natural explanation of features of the planet which have been termed 'non-natural' by Mr. Lowell" (Wallace 1907, vii). He believed complete repudiation of all of Lowell's claims was requisite because the latter's works, though "not very well written" and, in places, actually "twisting" the evidence, would "no doubt [impress] the newspaper men [who] think that as he is such a great astronomer he must know what it all means" (Marchant [1916] 1975, 408). Wallace, as always, was aiming for public as well as professional recognition of his ideas. In this case, he was on the side of the angels in two respects. Most professional astronomers were unreceptive to Lowell's speculations. Equally encouraging was the fact that religious thinkers and many laypersons found Wallace's *Man's Place* highly congenial. According to Marchant, Wallace's astronomical writings from 1903 to 1907 were efforts to buttress his deep conviction "that the near future would show the strong tendency of scientists to become more religious or spiritual" (Marchant [1916] 1975, 405).

THEISM AND POLITICS

Most of Wallace's writings in the final years of his life were devoted to the advancement of evolutionary theism, which also permeated his sociopolitical convictions. Wallace was an ardent reader of and contributor to the *Clarion*, one of England's most prominent socialist newspapers. He regretted that its editor, Robert Blatchford, had embarked on a public campaign against Christianity. Wallace regarded Blatchford's attacks on Christianity as ill founded and personally distasteful. Although Wallace was a theist and not a traditional Christian, he recognized common ethical goals between theistic evolutionism and Christianity. That Blatchford was also the author of the best-selling work *Merrie England* further perturbed Wallace. He was concerned that an influential socialist was posing a dichotomy between the moral teachings of socialism and religious ethics (Wallace 1912c). Wallace's holistic worldview had made him wary of efforts to divorce not only politics from science but also politics and science from religious ethics. Some years earlier, Wallace had corresponded with the Reverend H. Price Hughes. In September 1898, Hughes wrote Wallace to say that Hughes had just been elected president of the Wesleyan Methodist Conference. He regarded Wallace as an important voice in the campaign to link politics to theistic ethics. Hughes told Wallace that as president of the Methodist Conference, he had "special opportunities . . . of propagating Social Christianity, which in fact, and to a great extent in form, is what you yourself are doing" (Marchant [1916] 1975, 394).

Wallace's conception of religious ethics was nondoctrinaire. In *World of Life,* he directed his readers to two books that he regarded as exemplifying the ethics of theism: *Psychic Philosophy, as the Foundations of a Religion of Natural Law* (1901) by V. C. Desertis and *Spirit Teachings* (1894) by William Stainton Moses (Wallace 1910a, 398n. 1; Desertis was the pseudonym often used by Stanley De Brath). Germane to understanding Wallace's evolutionary theism is what Moses called the "theology of the spirit," which emphasized that human ability to grasp truth, like change in the material world, is evolutionary. Equally pertinent is Moses's claim that reason must be the final court of appeal in human perceptions of the divine. Wallace asserted that "we are forced to the assumption of an infinite God by the fact that our earth *has* developed life, and mind, and ourselves . . . I can imagine the supreme, the Infinite being, foreseeing and determining the broad outlines of a universe which would, in due course, and with efficient guidance, produce the required result, . . . the life-world of man." Wallace believed the "vast whole" to be a manifestation of God's power, "perhaps of his very self—but by the agency of his ministering angels through many descending grades of intelligence and power" (Wallace 1910a, 393–396). In 1911, Wallace's close friend Barrett published *Creative Thought and the Problem of Evil.* Barrett discussed evolution and the impossibility of explaining the phenomena of life without a supreme directing force. Wallace wrote Barrett, on 15 February 1911, that "it is very curious that even the religious reviewers [of *Creative Thought*] seem horrified and pained at the idea that the Infinite Being does not actually do every detail himself, apparently leaving his angels, and archangels, his seraphs and his messengers, which seem to exist in myriads, according to the Bible, to have no function whatsoever!" Wallace declared that Barrett's concepts concerning the relation between science and religion "closely . . . coincide[d]" with his own (Marchant [1916] 1975, 439).[8] Like Wallace, Barrett had moved beyond conventional spiritualism to an explicitly theistic conception of the universe (see chap. 4). Barrett's theism was further exemplified in *On the Threshold of the Unseen*. Published in 1918, five years after Wallace's death, *Threshold* shows that Barrett and Wallace held close views on the meaning of scientific theism. Moreover, Barrett's comments on the distinction between spiritualism and theism accord with Wallace's insistence that spiritualism has limits that can only be transcended by theism (Barrett 1918, 28–29). Barrett's analysis of science and religion is, like Wallace's evolutionary epistemology (see chap. 3), a reassessment of such categories and of the utility of any sharp demarcation between them.

RECONCILING SCIENCE AND RELIGION

Wallace differed from the exponents of traditional natural theology. Writing at the start of the twentieth century, he was acutely aware of the vast sociopolitical, environmental, and metaphysical transformations wrought by advances in Victorian science and technology. The confidently comforting message that had characterized much of eighteenth- and nineteenth-century natural theology was no longer viable for Wallace. Nor could it be. The relationships between science and religion have changed over time and are complex and highly diverse even within a given historical community (Brooke and Cantor 1998, 15–69). Wallace's mature evolutionary teleology was also optimistic. But unlike most traditional natural theology, his optimism was qualified by the stringent proviso that radical social and political reforms be implemented. Wallace's evolutionary theism was a response to the challenges posed to questions of human values as science emerged as an increasingly potent and professionalized cultural institution. Theism completed Wallace's evolutionary worldview. He saw theism, in terms of intelligent design, as providing an account of the emergence of those human traits he deemed inexplicable by natural selection and necessary for the possibility of future human progress. Wallace came to regard intelligent design as guiding certain aspects of the development of the nonhuman organic world as well. This reenvisioned evolutionary teleology informed the thought and writings of his later years. Wallace described this period as " 'the third chapter of my life'; [just as] *Man's Place in the Universe*—a totally new subject for me—may well be termed the 'third chapter of my book,' that is, of my literary work" (Wallace [1905] 1969, 2:382, 399).

As noted previously (see chaps. 3, 4), there are clear parallels between Oliver Lodge's and Wallace's writings at the turn of the century regarding theism. Lodge considered progressive evolution an empirically established scientific theory. He also felt that scientific naturalism served, erroneously, to preclude any incorporation of divine and/or spiritual agency in the course of human evolution. Lodge believed that "a ministry of benevolences surrounds us—a cloud of witnesses—not witnesses only but helpers, agents like ourselves of the immanent God" (Lodge 1910, 34, 155).[9] Finally, in Lodge's scientific theism, science was more than an empirical knowledge of nature, and religion was more than a biblical knowledge of God (Laudan 1993, 3, 21; Wilson 1996, 34).

In *Man and the Universe* (1909), Lodge proposed a conception of the universe that he hoped would reconcile science and religion. It lay "open to all manner of spiritual influences, permeated through and through with a Divine spirit, guided and watched by living minds, acting through the medium of law indeed, but with intelligence and love behind the law" (Lodge 1909, 22–23). Wallace was entirely comfortable with this attempt to define the compatibility of science and religion. Earlier, in a letter to Arabella Buckley Fisher (9 April 1897), Wallace stated that he admired Lodge's recent address to the Spiritualists' Association on similar matters. It is worth noting that Wallace's incorporation of spiritualism into a broader theistic framework was in stark contrast to his highly critical views on theosophy and ideas of reincarnation In that same letter to Fisher, Wallace admitted that "I have tried several Reincarnation and Theosophical books, but *cannot* read them or take any interest in them. They are so purely imaginative, and do not seem to me rational. Many people are captivated by it—I think most people who like a grand, strange, complex theory of man and nature, given with authority—people who if religious would be Roman Catholics" (Marchant [1916] 1975, 432–433).

In October 1895, Wallace had agreed to write an introductory note to Stanley De Brath's *Psychic Philosophy*. He considered De Brath's book a work of "great lucidity, a philosophy of the universe and of human nature in its threefold aspect of body, soul, and spirit." Wallace also wrote a prefatory note to the expanded second edition in 1908, stating that he fully agreed with all of De Brath's changes and additions. Wallace endorsed De Brath's view that late-nineteenth- and early-twentieth-century scientific developments, most notably in evolutionary theory, were potent grounds for a conciliation between the findings of modern science and the basic teachings of Christianity. The basic teachings of Christianity, for both De Brath and Wallace, were, however, something quite specific: they were religious lessons divorced from dogma and institutionalized churches. According to De Brath, the "new mode of thought recognises fully that the valid test to us of the existence of spiritual force is its material effect, but that all spiritual causation can only be expressed by metaphor, simile, and trope, straining the resources of language to express the higher verity, and not by scientific terms having only one sense. To literalise is to degrade the whole broad and grand treatment of God and human life which characterises the teaching of Jesus, into formula, making it no longer truth to be known but dogma to be assented to." De Brath continued by declaring that if "we know we are spirits veiled in flesh, for whom there is no death; having within ourselves infinite possibilities of health and growth; having faculty to receive strength and guidance from the very Creative Spirit Himself . . . then how differently would the world look to each one of us. We should see it as it is—the garden of God, wherein He brings flowers from corrupt and dead matter; as His undeveloped Kingdom wherein we may be His agencies whereby shall be made the new heaven and the new earth." Wallace's own evolutionary theism is echoed by De Brath's belief that there "is also a future sense to the individual man, when, leaving the body, his true self is manifest by his entrance on spirit-conditions. It is to this aspect that Jesus alludes when He says the righteous shall shine forth as the sun; shall inherit the Kingdom, prepared indeed from the foundation of the world, for it belongs to conditions where Time has no place" (De Brath 1921, v–vi, 9, 30, 269).

Wallace was sufficiently impressed with De Brath's book to inform Arabella Buckley (Mrs. Fisher) that he found it "a really fine and original work." Buckley was then reading *A Scientific Demonstration of the Future Life* by Thomson Jay Hudson (1896). Wallace told Buckley (4 January 1896) that he found Hudson's volume "so pretentious, so unscientific . . . that I do not feel inclined to read more of the same author's work." He hoped Buckley would read De Brath as a rigorous corrective to Hudson (Marchant [1916] 1975, 431). Given the length and depth of Wallace's friendship with Buckley, his recommendation of De Brath to her is significant. He felt that she was immersing herself uncritically in the still vast spiritualist literature. This was the literature Wallace himself once devoured but now regarded as partially superseded by more penetrating attempts to reconcile science and religion. Wallace admired De Brath for demonstrating that the solution to the pressing need to establish a basis for a modern "religion of natural law" lay in grounding such a religion in "the most advanced conclusions of science." This was one of Wallace's prime goals in writing *Man's Place in the Universe* and *The World of Life*. He regarded De Brath's approach as kindred to his own.

De Brath, for his part, cited Wallace's dictum that the term "miracle"— used disparagingly by critics of theism and spiritualism—was generally misunderstood. For Wallace, so-called miracles were simply natural events whose explanation "known laws are inadequate" to provide. De Brath shared Wallace's conviction that such events were increasingly being shown to be explicable by the findings of modern science. Wallace, in turn, applauded De Brath's emphasis that the primary virtue of a "religion rooted and grounded not only in Love but in scientific Law" was to provide humanity with a "scientific touchstone . . . to solve those problems of health, political action, and personal conduct where we now see but darkly." Wallace had found yet another ally in De Brath. He felt that De Brath's religion of natural law, "when thoroughly realised, becomes a sure guide to right action both for individuals and communities, and often affords a clue to the solution of the most vital political and social problems." Wallace had a twofold reason for his high estimation of De Brath's treatise. It aimed to demonstrate that theism was fully consistent with "the world of sequence and sensation, which is to us the ultimate basis of all our real knowledge." Metaphysics would become "an experimental science." But equally crucial was De Brath's "sympathetic and elevated tone" and "high moral teachings." His work, Wallace declared, "was well calculated to raise the ethical standard of public life, and thus assist in the development of a higher civilisation" (De Brath 1921, v–vi, 29–31).[10]

SPOKESMAN FOR THEISTIC EVOLUTION

Wallace's zeal in promoting theistic evolutionary teleology was obvious in his North American tour (see chaps. 3, 5). Two events that occurred at that time are helpful in documenting Wallace's path to becoming a spokesman for evolutionary theism. One was his visit to Cincinnati for twelve days. There, he again lectured successfully on evolutionary theory, including the phenomena and causes of animal coloration, before the Natural History Society. Wallace also met several people who had become religion skeptics through their reading of Spencer and Darwin. These individuals confided to him that they were regaining their religious beliefs through spiritualism (Wallace [1905] 1969, 2:145). Spiritualism was a divisive issue in Cincinnati at the time of Wallace's visit. A letter to the editor in the *Cincinnati Weekly Times* is revealing. The writer complained "that there are hundreds of people in Cincinnati who have been entangled in the web of modern spiritualism. But the fact is more surprising still when it is considered that they have been drawn into the belief by men and women who cannot produce a single intelligent reason for the faith they profess." Interestingly, spiritualism was not dismissed for the usual reasons, fraudulent mediums and so on. The writer's main charge was that spiritualism failed to achieve what its advocates assert: "There are thousands of doubters in the world—men who would rejoice if it were possible to grasp with unshaken faith the gracious declarations of Holy Writ. Spiritualism . . .

possesses the power to scatter all doubts like chaff before a driving wind. Then why not do it? There is an easy answer: Spiritualism is not in harmony with the Bible, and any attempt to reveal the hidden things of God must of necessity prove a disastrous failure."[11] Wallace encountered a sizable community in Cincinnati receptive to spiritualism. He also encountered hostility from fundamentalist Christians, who rejected even the possibility of a conciliation between biblical literalism and the "scientific" claims of spiritualists. Wallace never courted biblical literalists. He had dismissed their position since his youth. However, the bitter debates over spiritualism he was again witnessing were a further prod for him to move beyond the confines of spiritualism to a more encompassing evolutionary theism. Newspaper accounts from Boston, New York, Baltimore, San Francisco, Washington D.C., Toronto, Kingston, and Montreal, in addition to those of Cincinnati, document the widespread public interest in Wallace's increasingly overt theistic evolutionism.

Support for Wallace's theistic views was important. Criticisms of his publicly articulated theism are equally revealing. As noted in chapter 4, overtly theistic evolutionists in the United States were less common in the last two decades of the nineteenth century than they had been in the two decades prior to Wallace's North American tour. Theistic evolution was undergoing "privatization." Many American evolutionists retained their religious beliefs, but explicit references to the divine became less visible in the scientific literature. Specialization and professionalization fostered a separation of religion from science among the scientific elite in their public statements and writings (Numbers 1998, 40). Wallace, as a spokesman for evolutionary science, was bucking this trend. Why? Several factors were at play in his decision to go public with theistic evolution. Wallace's ambivalent status as a card-carrying professional scientist contributed to his freedom to express views that other professional scientists may have held but regarded as inappropriate in scientific discourse. Wallace's refusal to be pigeonholed as a specialist entailed that he make public his worldview commitments. Theism was one crucial commitment. His theistic convictions became more overt than they had been in the period prior to 1870 because they had become more precisely formulated since then. Theism informed Wallace's science. The converse was also true. For Wallace, evidence from biology and astronomy, in particular, supported theistic conclusions. Since he had never subscribed to any traditional institutionalized religion, Wallace's theism did not carry the burden of defending orthodox creeds that he found oppressive.

The second event, actually just prior to Wallace's arrival in America, occurred in New York. Noah Porter, president of Yale, read a lengthy paper entitled "Some Thoughts upon Evolution" at the Nineteenth Century Club in New York City on 25 May 1886. Porter's main aim was to show, like many others, that in his opinion there was no conflict between evolutionary science and theism. Porter cited first "the well-known fact that Darwin himself asserted his belief in an intelligent Creator." He then embarked on a torturous explication of what Herbert Spencer's teachings "really mean." Darwin's commitment to theism, however, is a source of continuing debate among scholars. And Spencer's concept of the "Unknowable"—that some things simply could not be known—was, and remains, largely inscrutable. When Porter finally concluded that evolutionary theory becomes "luminous with thought when projected against the bright background of the living God," he is no longer dealing with Darwin or Spencer. The closing paragraphs of his address proclaim that evolutionary science makes it easier, and more necessary, to assert that the very laws of universal, progressive development must have arisen in the mind of a supreme being. "Man," Porter declared, "also is conceived as the culmination of the history of the Universe hitherto, and as the brightest and most consummate product of all its progressive movements. Why, then, may he not be worthy of the constant care and the fatherly love of Him who has had him in his thought from the beginning till now, and toward whom his plans and movements have ever been tending" (Porter 1886). This is Wallaceism—though Porter did not call it that—not Darwinism or Spencerism. Wallace began expounding the evolutionary theism Porter was promoting when he arrived in New York at the commencement of his North American tour. American audiences, if not all American scientists, were primed to hear theistic versions of evolution.

Wallace was in his element. His visit to the United States, despite—or because of—his controversial views, made him something of a sage for North Americans. As late as 1909, he told an interviewer that hardly a week passed without a request for his autograph, especially from America. "I always send it, particularly," he added with a smile, "if a stamp is included for return." Wallace relished his transatlantic fame. "I am always being asked to write," he confessed to his interviewer with a humorous twinkle, "even if I know nothing about the subject, and American editors think nothing of having my articles cabled across. In some cases I ignore the request, but when I can I oblige." Even the polymathic Wallace set limits to his own areas of expertise. "I wish it were more widely recognised that this is an age of specialisation," Wallace noted, "and that even scientists ought not to be expected to know everything" (Rann 1909). But Wallace did not always conform to that restriction. He was a generalist who wandered among specialized fields, trying to pull things together. Wallace always considered himself as pursuing his varied social and cultural concerns with the rigor and methodology of science. Those areas that most deeply meshed with his underlying commitments became part of his scientific worldview. Works such as *The World of Life* demonstrated the degree to which theism flowed into Wallace's evolutionary cosmology.

Wallaceism

Wallace's article "The Present Position of Darwinism" appeared in the August 1908 issue of the *Contemporary Review*. It was not simply a survey of the various biological theories that, by 1900, offered alternatives to natural selection as the mechanism of evolutionary change. Wallace had composed a calculated and absolutely timely counterattack against what he perceived to be an onslaught, particularly from the ranks of the "Neo-Lamarckians, the Mutationists, and the Mendelians. The general public," Wallace complained, "are being told to-day that Darwinism is played out; that as a means of explaining the origin of *species* and the general development of the organic world, it is entirely superseded by newer and more scientific views. Of course the public, ever ready to accept new things in science, believes these statements, which are put forward with so much confidence and, apparently, on such good authority; while the theologians are especially glad to seize upon this weapon against what they have long considered to be their most formidable enemy" (Wallace 1908a, 129). That Darwinism was under severe attack was not the paranoid imaginings of the aged cofounder of the theory of natural selection. At the 1894 meeting of the BAAS, Huxley once again had been called on to defend Darwinism. This time, however, Huxley was not responding to a religious offensive, as he had in his much earlier confrontation with Bishop Wilberforce at Oxford. His mission was to counter the charge of the Marquis of Salisbury, a former prime minister and president of the BAAS that year, that natural selection was wholly inadequate as a mechanism for evolution. Salisbury's charge was itself not new, since natural selection had endured significant and persistent opposition from 1858 onward. But it was a harbinger of what grew to be a loud chorus of attacks from physicists as well as biologists.

Books such as Eberhart Dennert's *At the Deathbed of Darwinism* (1903) were not coming from the fringes of evolutionary thought. Dennert's work was indicative of a powerful surge of scientific opinion. Dennert, a neo-Lamarckian, purported to demonstrate that selection was at best only a secondary force in evolution. Mutation theory, a term popularized by Hugo De Vries, was another influential alternative. De Vries and his followers argued that evolution occurred not by the slow and gradual accumulation of almost imperceptible variations as natural selection demanded but by the sudden appearance of major variations that were both random and (initially) nonadaptive. A third important challenge to natural selection emerged from the rediscovery of Mendel's laws in 1900. Mendelian genetics, although ultimately reconciled with natural selection to create the modern synthesis of contemporary evolutionary theory, initially was regarded by many scientists as fueling the anti-Darwinian attacks. Wallace shared this assessment of Mendelism; but his own views on the actual relationship between natural selection and genetics were more complex and far less dogmatically negative than his statements in the *Contemporary Review* suggested (Marchant [1916] 1975, 333–334). Finally, William Bateson and his school advocated a discontinuous and antiselectionist theory of evolution (Bowler 1983, 3–10, 186–197). The critics singled out by Wallace were powerful and numerous. They created a crisis significant enough to warrant the term, coined by Julian Huxley, of an "eclipse of Darwinism" in the early 1900s (Huxley 1942, 22–28).

Wallace undertook a formidable task. He sought to minimize the claims of Lamarckians, while conceding that Darwin himself had "actually accepted [the inheritance of acquired characteristics] though he always maintained that it had very little or no effect in producing modification of species." One of Wallace's main neo-Lamarckian targets was the American paleontologist Edward D. Cope. He characterized Cope's *The Primary Factors of Organic Evolution* (1896) as repudiated by recent and "very valuable experiments and observations." The mutationist hypothesis of De Vries and his advocates was dismissed by Wallace as "a mountain of theory reared upon such an almost infinitesimal basis of fact!" He deemed the criticisms of eminent antimutationists such as Thistleton Dyer and E. B. Poulton conclusive. Wallace's attacks on mutationism may also have been prompted in part by his equally persistent advocacy of gradualism, in contrast to sudden change, in effecting sociopolitical transformations. Mendel's experiments and so-called laws were regarded by Wallace as inconclusive at best. He thought them scarcely capable of "playing any essential part in the scheme of organic development. . . . They arise out of what are essentially abnormalities, whether called varieties, 'mutations,' or sports. These abnormalities are very rare in a state of nature, as compared with the ever-present individual variability ample in amount and affecting every part and organ which furnishes the material for both man's and for nature's selection." Wallace concluded his article on a triumphant note:

> To anyone who has devoted a considerable portion of his life to the
> study of nature, both in field and in cabinet, both at home and in
> distant regions, the vast complex of phenomena presented by the
> organic world . . . is almost overwhelming in its grandeur and its
> beauty. Almost all such loving students of nature have found in the

theory of Darwin, in his many stimulating works and in those of his friends and followers, the only intelligible clue to the mighty labyrinthe of nature. To such students of nature the claims of the Mutationists and the Mendelians, as made by many of their ill-informed supporters, are ludicrous in their exaggeration and total misapprehension of the problem they profess to have solved. To set upon a pinnacle this mere side-issue of biological research, as if it comprised within itself all the phenomena and problems presented by the organic cosmos, is calculated to bring ridicule upon what, in its place may be an interesting and perhaps useful line of study. To myself these monstrous claims suggest a comparison with those of the perhaps equally enthusiastic and equally ill-informed admirers of the immortal Pickwick, who believed his "Speculations on the Source of the Hampstead ponds with some Observations on the Theory of Tittlebats," to have been a most important contribution to the science of that period.

(Wallace 1908a)

Wallace's motives in writing this particular defense of Darwinism were surely complex. The state of evolutionary theory in 1908 was itself in tremendous flux, if not in outright crisis. And while Wallace was correct in pointing out certain of the more exaggerated claims of the opponents he targeted, he was in a precarious position to act as *the* spokesman of the Darwinian camp. First, the Darwinists themselves were unable to maintain a united front against the mounting opposition (Bowler 1983, 4, 10–15). Second, Wallace had distanced himself from Darwin on a number of fundamental issues in the decades since the publication of the *Origin*. In addition to their radical separation on the question of human evolution, Wallace and Darwin disagreed on the origin of cross- and hybrid sterility and the origin of sexual dimorphism (Kottler 1985, 367–432). Wallace had also developed an increasingly stringent adaptationist view of natural selection. He emerged as the archadaptationist, arguing that all of the myriad and, often, minute differences that distinguished even the most closely related species had adaptive significance. All such differences, according to Wallace, fell securely within the province of natural selection alone—with the exception, of course, of human evolution (Wallace 1896b). Wallace's stance on coadaptation and other aspects of the concept of variation were sufficiently contentious to cause some Darwinians to regard him as not the most authentic exponent of Darwinism at the start of the twentieth century (Bowler 1976; Ridley 1982, 56–61). Samuel Butler, that astute and acerbic observer of the Darwinian scene, had suggested that Wallace's differences with Darwin necessitated coining a new term in the lexicon of late Victorian evolutionary thought: "Wallaceism" (Butler [1890] 1970, 236). Why Wallace himself so long resisted adopting that far more appropriate term for the evolutionary synthesis he was, in fact, defending is puzzling.

Other commentators showed no such hesitation. In the November 1908 issue of the *Contemporary Review*, A. A. W. Hubrecht declared that he and many "continental naturalists" had for quite some time been distinguishing "that 'variety' of Darwinism which may be termed 'Wallaceism' and the real foundation of Darwinism, which is the Selection Theory as it was formulated fifty years ago by Darwin and Wallace simultaneously" (Hubrecht 1908, 629). In the December issue of the *Review*, Wallace protested *"very strongly"* against Hubrecht's allegation "of any such divergence of opinion between Darwin and myself as he states to have existed, without, so far as I can see, one particle of evidence to support it." Hubrecht's arrow had pierced its target. Wallace accused Hubrecht of distorting Darwin's theories (Wallace 1908b).

Wallace's accusation is, simply, wrong. His contemporaries recognized the significant differences between his and Darwin's conceptions on a number of fundamental aspects of evolutionary theory, including their divergent views on variation. Recent scholarship has only served to clarify and document further the Wallace-Darwin divide (Provine 1985, 825–842; Hodge 1989, 163–182; Asma 1996). Wallace's increasingly strident selectionist/adaptationist views had, by 1900, become clear enough to both his supporters and critics to warrant using the term "Wallaceism." Why did Wallace reject that term and why did he demand that Hubrecht "make such an apology as seems to him proper for having so prominently asserted an antagonism between Darwin and myself which had no existence whatever" (Wallace 1908b, 717)?

The answer to this question relates, in part, to Wallace's espousal of evolutionary theism. During the 1860s and 1870s, Wallace publicly—albeit not in private correspondence and conversation—minimized the significance of his adherence to spiritualism for his conception of human evolutionary history. At that period, he couched his arguments against the total sufficiency of natural selection in scientific terminology. Could the aged Wallace, now more openly theistic, have sought—consciously or not—to refrain from exacerbating criticisms of his evolutionary teleology by denying opponents the symbolic weapon of the term "Wallaceism"? Evidence for this exists in one further ambiguity embedded in "The Present Position of Darwinism." Wallace's opening paragraph chides those theologians who were especially eager to seize on the rival hypotheses to natural selection for the purpose of extirpating "what they have long considered to be their most formidable enemy." Though no theologian himself, Wallace had—as this study has demonstrated—developed by 1900 an all-encompassing theistic evolutionary teleology. Many critics argued that Wallace's evolutionary teleology did as much to alter the very fabric of natural selection theory as Lamarckism, mutationism, and Mendelism combined. The great challenge of Wallace's life, especially during his last years, was centered on the question of how to reconcile his commitment to natural selection with his increasingly overt theistic and teleological worldview. Darwin has been characterized as a "tormented evolutionist" (Desmond and Moore 1991). Wallace was not tormented. But he was a perplexed, and perplexing, evolutionist. Wallace sought to alleviate this perplexity, both for himself and for others, in *The World of Life*.

THE WORLD OF LIFE

The convergence of theism and science in Wallace reached a peak in *The World of Life*. Published in 1910, this was not his final book. In the year of his death (1913), two more books—*The Revolt of Democracy* and *Social Environment and Moral Progress*—appeared. In both, Wallace dealt primarily with sociopolitical matters (Wallace 1913a, 1913b). The full title of the 1910 work reveals its polemical goal: *The World of Life: A Manifestation of Creative Power, Directive Mind and Ultimate Purpose*. Wallace's preface announced that he planned to "summarise and complete my half-century of thought and work on the Darwinian theory of evolution." But he immediately added that he was going to extend the scope of that theory "in several directions [to show] that it is capable of explaining many of the phenomena of living things hitherto thought to be beyond its range." Wallace amplified and updated already familiar topics, including biogeography, the geological record, the pervasive adaptation of animal and plant species to their environments, and the factors governing heredity and variation. But the most prominent and novel feature of *World of Life*, Wallace informed his readers, was a "popular yet critical examination . . . of the nature and causes of Life itself; and more especially of its most fundamental and mysterious powers—[cellular] growth and reproduction." Such a strategy, Wallace believed, would convince his readership that the diverse phenomena of the natural world demanded a "Creative Power," a "directive Mind," and an "ultimate Purpose." This purpose, Wallace contended, was "the development of Man, the one crowning product of the whole cosmic process of life-development; the only being which can to some extent comprehend nature . . . , appreciate the hidden forces and motions everywhere at work, and can deduce from them a supreme and over-ruling Mind as their necessary cause." Finally, Wallace pointed out in the preface "that, however strange and heretical some of my beliefs and suggestions may appear to be, I claim that they have only been arrived at by a careful study of the facts." This claim was required because many "critics of *Man's' Place in the*

Universe (to which this [work] may be considered supplementary)—treated the conclusions there arrived at as if they were wholly matters of opinion or imagination, and founded (as were their own) on personal likes or dislikes, without any appeal to evidence or to reasoning. This is not a method I have adopted in any of my works" (Wallace 1910a, v–viii).

Wallace's preface was polemical because *Man's Place* had been subjected to harsh criticism from some of his scientific colleagues. Their displeasure was directed not toward his astronomical arguments on the centrality of life on Earth, with which a number of critics concurred, but toward Wallace's theistic and teleological envelope. Wallace could no longer separate, however, his scientific argumentation from his religious convictions. The two formed, by this stage in his career, an indissoluble unity. He was fully "convinced that at one period in the earth's history, there was a definite act of creation, [and] that from that moment evolution has been at work, guidance has been exercised." Wallace characterized "Materialism [as] a most gigantic foolishness," which would soon be banished by a deeper understanding of the evolutionary process as he conceived it. He conceded that there was some excuse for the embrace of materialism initially. The findings of Victorian science threw a "bomb of the most deadly power . . . into the authoritative nonsense and superstitions of Clericalism." But Wallace regarded "those whose intelligence had been outraged and irritated by this absurd priestcraft" as throwing the baby out with the bathwater. Materialists "rushed to the conclusion that religion was destroyed, . . . that in mud was the origin of mind, and in dust its end." Wallace declared that the embrace of materialism was illogical and a denial of what evolution and other sciences actually implied. "There are laws of nature," Wallace asserted, "but they are purposeful. Everywhere we look we are confronted by power and intelligence. The future will be full of wonder, reverence, and a calm faith worthy of our place in the scheme of things." Wallace's theistic evolutionism led him to predict that "materialism is as dead as priestcraft for all intelligent minds" ([Wallace] 1910b).

Wallace was still sensitive to the criticism from some close colleagues that while his science was excellent, his theistic teleology detracted from the technical merit of his later works. When the expert botanist Thistleton-Dyer pronounced Wallace's exposition of "the more evolutionary part" of *The World of Life* accurate in its botanical details, Wallace was delighted. He gratefully thanked Thistleton-Dyer in a letter sent on 8 February 1911. Wallace's delight was short-lived. Thisleton-Dyer replied, almost immediately (12 February 1911), that "we are in agreement as to Natural Selection being capable of explaining evolution 'from amoeba to man.'" But he quickly added that he referred to natural selection as "a mechanical or scientific explanation. That is to say, it invokes nothing but intelligible actions and causes. . . . But if we admit that it is scientific, then we are precluded from admitting a 'directive power.' " Thistleton-Dyer further noted that he entirely sympathized "with anyone who seeks an answer [to the riddle of the Universe] from some other non-scientific source. But I keep scientific explanations and spiritual craving wholly distinct" (Marchant [1916] 1975, 341–344). Darwin had made a more blunt but similar point a half-century earlier. His target then was Lyell, not Wallace. In a letter to Lyell, Darwin responded to Lyell's criticisms after the latter had received an advance copy of the first edition of the *Origin*. Darwin fulminated that he "entirely reject[ed] as in my judgment quite unnecessary any subsequent addition of 'new powers & attributes & forces;' or of any 'principles of improvement'. . . . If I were convinced that I required additions to the theory of natural selection, I would reject it as rubbish" (Darwin to Lyell, 11 October 1859, in Burkhardt and Smith, 1985–, 7:345). At the time of the *Origin*'s appearance, Darwin could not have known that Wallace was sailing in the same waters as Lyell. That bombshell would come only in 1869, with Wallace's review of the tenth edition of Lyell's *Principles of Geology* (see chap. 4).

Wallace had never viewed science and theism as mutually exclusive universes of discourse. His position had become more assured with each succeeding decade after 1870. His reason for writing *The World of Life* was to obliterate the demarcation between certain aspects of science and religion. Wallace regarded *Man's Place* and *World of Life* as forming "together a very elaborate, and I think, conclusive, scientific argument in favour of the view that the whole material universe exists and is designed for the production of Immortal Spirits." He found comfort in scientific colleagues like Lodge. In one of the last letters Wallace wrote (9 October 1913), he expressed his admiration for Lodge's recent address to the BAAS. He told Lodge that he thought the address "especially notable for your clear and positive statements as to the evidence in all life-process of a 'guiding' Mind" (Marchant [1916] 1975, 410–411).

WALLACE CONTRA MECHANISTIC MATERIALISM

The World of Life presents an interpretation of biological phenomena, including the "nature and causes of Life itself," which renders explicit the theistic framework that had come to permeate Wallace's evolutionary synthesis. The vast evidence drawn from the study of plant and animal life, supplemented by data and concepts from chemistry, physiology, geology, and astronomy, indicated "a prevision and definite preparation of the earth for Man." Wallace asserted that the ancient doctrine "that the universe is not a chance product," far from being "exploded" by late Victorian science, is substantiated within "the realm of scientific inquiry." Evolutionary science, in particular, represented for him the clear manifestation of an "Infinite and Eternal Being" who nonetheless requires the "continuous coordinated agency of myriads of [spiritual] intelligences." As Wallace himself stated, directly and unambiguously, in one of the last letters he wrote (to his close friend and editor Marchant), the "materialistic mind of my youth and early manhood has been slowly moulded into the socialistic, spiritualistic, and theistic mind I now exhibit." He added that the whole cumulative argument of *World of Life* accorded with "the teaching of the Bible, of Swedenborg, and of Milton" (Wallace 1910a, v–vi, 394–400; Marchant [1916] 1975, 413). An interview of Wallace, shortly after the publication of *The World of Life,* gives further testimony to the role now played by theism in his worldview (Wallace 1912e).

Wallace was asked about his response to a paper on the origin of life given by the chemist Edward A. Schaefer at the eighty-second annual BAAS meeting in Dundee in September 1892. Criticizing the mechanistic assertions made by Schaefer concerning the chemical origins of life, Wallace declared: "I maintain that you cannot explain the smallest portion of dead [chemical] matter without a series of forces which imply mind, which imply direction. . . . If you assume that the directing power is essentially a spiritual power, then you can understand all this, but without it you cannot understand it." In his refutation of Schaefer's mechanistic premise that life is a consequence of the organization of chemical elements into compounds and molecules of increasingly complex structures, Wallace's theism is fundamental. His conception of the nature of matter makes this clear. For Wallace, "living matter"—the substratum of evolutionary development—requires the intervention of directive, spiritual forces. But what he terms "dead matter" (the chemical elements) has been shown to be almost as complex as organic beings. Late-nineteenth- and very early twentieth-century experimental and theoretical studies on the structure of the atom proved, for Wallace, that atoms themselves are composed of yet smaller constituents that are imbued with force. He declared that mechanistic chemists, like Schaefer, failed to deal with the "ultimate cause," the "directing power" that has created the forces that then act on inert matter to produce both chemical and biological activity. The nature of the relationship between force and matter was (and is) a perennial problem in the history of science. Wallace's argument that spiritual mind was a necessary component of the explanation of chemical, as well as biological, phenomena demonstrates the cognitive dimension of theism in his science (Wallace 1912d, 1912e).

The World of Life was Wallace's last comprehensive attempt to resolve, for himself, any potential conflict between the demands of science and the demands of theistic belief. As a public document, it was also intended to convince many others that there was no conflict. This was not only a conceptual debate. It had emotional consequences as well for its participants. These profound conceptual and emotional dilemmas were reflected in language. Writings in the mid- and late Victorian periods were rife with an obvious tension generated by competing linguistic choices. When was it appropriate to use the language of ostensibly value-neutral naturalistic science? And when was it appropriate to employ the value-laden discourse that figured prominently in the ethical, sociopolitical, and religious works of the time? Many subjects seemed to straddle convenient but arbitrary demarcation boundaries. Evolutionary biology was among the most notorious of the fields in which linguistic choices were never clear-cut. Spencer, Mill, Romanes, Huxley, Tyndall, Lodge, and many of their contemporaries had wrestled with the issue (Mandelbaum 1971; Beer 1999). Could Wallace resolve the conceptual, semantic, and evidentiary challenges posed in his polemical treatise?

Much has been written about the evolutionary challenge to theology. Despite the wide variety of scholarly views on the impacts of such a challenge, it is now clear that evolutionary theory forced a reevaluation of the nature of the relationship between science and religion in two crucial respects. First, there was a collapse of consensus as to how scientific and religious beliefs were to be related and integrated. Some writers seemed simply to abandon the attempt, at least in their published works. Others, such as Gray, articulated worldviews that tended to relegate science and religion to separate realms of discourse. Still others, such as James McCosh at Princeton and Henry Drummond, became advocates of theistic evolution. At one level, Wallace belonged to the ranks of McCosh and Drummond. But their views were directed primarily to religious, not scientific, audiences. Wallace, in contrast, aimed at both audiences. He was confronted by a rhetorical dilemma by the second major impact of evolutionism with respect to the science/religion interface: a pronounced shift in the nature of scientific texts. By the end of the nineteenth century it had become uncommon to find references to divine design, direction, or control in scientific texts on evolution. This was in sharp contrast to the situation at midcentury, when such references were abundant (Brooke and Cantor 1998, 161–162). Wallace's *World of Life* is noteworthy precisely because he was attempting to elaborate his theistic and teleological convictions within the structure and language of a scientific text. That he made such an attempt is crucial for understanding Wallace's evolution.

WALLACE'S RHETORICAL STRATEGY

Wallace refused to accept the comfortable responses to the evolutionary challenge to theology. This refusal can be seen as yet another example of the alleged vitiated scientific acuity of the aged Wallace, as some of his critics charged. Or it can be seen as indicative of his deeply rooted belief that a comprehensive evolutionary theism, overtly teleological in structure and language, afforded the surest avenue for confronting the concerns he had raised in *Wonderful Century*. In that book, Wallace sketched the potentially detrimental cultural as well as environmental effects linked to the increasingly powerful and globalized science (and technology) of the close of the nineteenth century. It has been argued that "despite the recent renaissance of interest in natural theology, the decline of the design argument in the later decades of the nineteenth century has left a major lacuna that had not been adequately filled. So much of modern science is directed to the solution of either theoretical problems or practical ones arising form the requirements of business, industry and the military. Hence science is now usually justified to the public either in terms of expanding the [endless] frontiers of human knowledge or as the goose that lays the golden egg." One consequence of this twentieth-century attitude toward science has been to dismiss the design argument as either wrong or trivial and easily undermined by the philosophically sophisticated (Brooke and Cantor 1998, 200). From this perspective, *The World of Life* was a major effort by Wallace to modernize the design argument.

The World of Life, despite its broader objectives, must first be read as a scientific text. Wallace was clear on this point. "During the fifty years that have elapsed since the Darwinian [*sic*] theory was first adequately, though not exhaustively, set forth," Wallace asserted that "it has been subject to more than the usual amount of objection and misapprehension both by old-fashioned field-naturalists, and by the new schools of physiological specialists." In his opinion, most such objections had been shown to be "fallacious by some of the most eminent students of evolution both here and on the continent." But there remained "stumbling blocks [that] are continually adduced as being serious difficulties to the acceptance of natural selection as a [major] explanation of the origin of species." Wallace's first, though not ultimate, task was to address the scientific community and demonstrate that their objections were ill founded. In addition to his own critique of the neo-Lamarckians, the De Vriesian mutationists, and the Mendelian geneticists, Wallace sought to demolish objections to natural selection by presenting the work of August Weismann. Weismann is known primarily for his theory of the continuity of the "germ plasm." This was his term for the substance of heredity that was transmitted from generation to generation. Weismann's ideas were highly influential contributions to biological thought. In 1908, he was honored by both the Linnean Society (at the Darwin-Wallace celebration) and the Royal Society. By citing Weismann in *World of Life* Wallace was enlisting one of the most celebrated biological experimental theorists—whose researches were affording new support for natural selection at the cellular level—to convince his early-twentieth-century scientific audience (Wallace 1910a, 252, 271–277; Robinson 1976; Johnston 1995).

The bulk of the 400-page volume (thirteen out of twenty chapters) was devoted to a detailed and sophisticated analysis of the most recent findings of biogeography, heredity, adaptationism, geology, and geophysics. Wallace incorporated lengthy citations from the research of scientists ranging from the British zoologist Ray Lankester to Hooker to the American vertebrate paleontologist Othniel Marsh (who established the fossil pedigree of the modern horse) to A. E. Shipley, president of the BAAS in 1909 (Wallace 1910a, 84–86, 204, 221–223, 245). Wallace's extensive scientific documentation was a strategic prelude to his ultimate goal. *World of Life* was written to demonstrate that the most recent scientific researches rendered natural theology (in sharp contrast to revealed theology) both reinvigorated and essential for the twentieth century.

In chapters 14–20, Wallace argued that the profound complexity of the natural world as demonstrated by modern science requires, "to afford any rational explanation of its phenomena, . . . postulat[ing] the continuous action and guidance of higher intelligences; and further, that these have probably been working toward a single end, the development of intellectual, moral, and spiritual beings." This was scarcely a new stance for Wallace. He pointed out to his audiences, that he first presented his objections to the complete sufficiency of natural selection in the 1870 *Contributions to the Theory of Natural Selection*. Wallace's suggestion that Earth had been designed for the advent of man was, he noted, immediately attacked by "Haeckel and the whole school of Monists, as well as [by] most of the followers of Spencer and Darwin [as] being unscientific or . . . priest-ridden." Wallace pointedly added that "several critics accused [Wallace] of 'appealing to first causes' in order to get over difficulties; of maintaining that 'our brains are made by God and our lungs by natural selection'; and that, in point of fact, 'man is God's domestic animal.'" Objections by such formidable opponents, friends and enemies alike, induced Wallace—as has been previously shown—either to couch his reservations as to the scope of natural selection in terms of the principle of utility or to keep his convergence of science and teleological theism muted in much of his published work in the period from 1865 to 1880.

Wallace could, however, "recur to the subject after forty years of further reflection." By the start of the twentieth century, developments in evolution and other scientific disciplines now afforded Wallace evidence that he interpreted as scientific validation of his prior speculations. *The World of Life* was, for Wallace, a vindicating synthesis of his long-held conviction that science and theistic teleology were not merely compatible. They were both epistemologically necessary for a comprehensive account of nature. Wallace emphasized that, although *The World of Life* would have its critics in 1910 just as *Contributions* had its detractors in 1870, the playing field had altered dramatically. In the mid-Victorian era, the "opposition was between science and religion, or, perhaps more correctly, between the enthusiastic students of the facts and theories of physical science in the full tide of its efforts to penetrate the inmost secrets of nature and the more or less ignorant adherents of dogmatic theology. Now, the case is wholly different. Speaking for myself I claim to be as whole-heartedly devoted to modern science as any of my critics" (Wallace 1910a, 315–317, 332).

A RENEWED NATURAL THEOLOGY

Wallace's claim is valid. Natural theology, although besieged, was still common as a rhetorical strategy at the time of the publication of the *Origin*. Darwin used such rhetoric. But Wallace was correct in pointing out that a rift had developed in the mid-nineteenth century that rendered natural theology far less respectable as a vehicle for scientific expression than it had been in its heyday, during the eighteenth and early nineteenth centuries (Lindberg and Numbers 1986). In a supreme irony, it was Wallace himself who, in 1866, criticized Darwin's very use of the term "natural selection." Wallace thought it tended to encourage the belief that Darwin's metaphor implied "the constant watching of an intelligent 'chooser' like man's selection to which you so often compare it." Wallace suggested, instead, the use of Spencer's phrase "survival of the fittest." Spencer's phrase, he hoped, as "the plain expression of the *fact*" would avoid any misunderstandings arising from an anthropomorphic or divine implication of the metaphor "natural selection" (Wallace to Darwin, 2 July 1866, in Marchant [1916] 1975, 140–143). By 1910, Wallace no longer feared an anthropomorphic or theistic interpretation of evolution. He openly and confidently embraced such interpretations. *The World of Life* is a manifesto of evolutionary teleology. Its narrative form and rhetorical strategy show Wallace as a powerful advocate for a renewed natural theology. Natural theology was, and is, far from monolithic. It has fulfilled many functions in the history of science, one of which has been as a form of rhetoric. The latter function moves historiographic analyses into new and suggestive directions (Drees 1995; Shortland 1996, 187–205, 287–300; Brooke and Cantor 1998, 176). If natural theology is studied rhetorically, not only *The World of Life* but also the entire work of Wallace's last decade assume both enhanced significance and coherence.

The concluding four chapters of *World of Life* displayed Wallace's rhetorical deployment of natural theology at its best. They forcefully present his conviction that science and theism are integral and mutually reinforcing parts of a broader, holistic worldview. Wallace considered recent studies in the biochemistry of living cells as providing the newest scientific evidence for supporting his evolutionary cosmology. He cited the research of Weismann, Lloyd Morgan (particularly his *Animal Life and Intelligence*), and the cytobotanist A. Kerner, among others. Wallace concluded from their investigations that the extraordinarily complex structures and functions "of these minute unit-masses of living matter, the cells . . . must, I think, convince the reader that the persistent attempts made by Haeckel and Verworn to minimise their marvellous powers as mere results of their complex chemical constitution, are wholly unavailing. They are mere verbal assertions which prove nothing; while they afford no enlightenment whatever as to the actual *causes* at work in the cells leading to nutrition, to growth, and to reproduction." Wallace condemned the biological reductionists precisely because all "questions of antecedent purpose, of design in the course of development, or of any organising, directive, or creative mind as the fundamental *cause* of life and organisation, are altogether ignored, or, if referred to, are usually discussed as altogether unscientific." He accused the reductionists of "showing a deplorable want of confidence in the powers of the human mind to solve all terrestrial problems."

Wallace constructed an argument that would appeal to his readers' imagination and emotions as well as to their reason. This is one of the cornerstones of rhetorical strategy. "We are asked," he thundered, "to believe that these cells and all their marvellous outcome are the result of the fortuitous clash of atoms!" This was a dilemma that also troubled Huxley and Spencer acutely. For Wallace, the dilemma was resolved by the increasing realization that modern science actually required an "immanent directive and organising MIND, acting on and in every living cell of every living organism, during every moment of its existence." Wallace's deity was no mere "God of the gaps." He "venture[d] to hope and to believe that such of my readers as have accompanied me . . . through the present volume, and have had their memory refreshed as to the countless marvels of the world of life; culminating in the two great mysteries—that of the human intellect with all its powers and capacities as its outcome, that of the organic cell with all its complexity of structure and of hidden powers as its earliest traceable origin—will not accept the loud assertion, that everything exists because it is eternal, as a sufficient or convincing explanation" (Wallace 1910a, 349–354).

Wallace's purpose and technique in writing had shifted during the course of nearly a half-century. The younger man excelled, as Darwin, Hooker, and others so often commented, in devastatingly specific argumentation. He ferreted out the weaknesses of opponents of evolutionary theory and demolished the premises of many of their objections. The older Wallace wanted to probe the "great questions." He was now in a position, buttressed not only by evolutionists but by chemists and physiologists as well, to pronounce that further advances "in our knowledge of the universe" will afford man "more and more adequate conceptions of the power, and perhaps to some extent of the nature, of the author of that universe; will furnish him with the materials for a religion founded on knowledge, in the place of all existing religions, based largely on the wholly inadequate conceptions and beliefs of bygone ages." Most crucially, the old doctrine that maintained a "prevision and definite preparation of the earth for Man," rather than being exploded as the materialist reductionists would have it, "will, I hope, no longer seem to be outside the realm of scientific inquiry." Wallace ended his rhetorical tour de force with the assertion that he had found a way of expressing Spencer's comprehensive, if not always comprehensible, ideas of the "Unknown Reality which underlies both Spirit and Matter . . . in a more concrete and intelligible manner" (Wallace 1910a, 390–391, 399). *The World of Life* was Wallace's most sustained effort to synthesize the vast array of facts, theories, speculations, and experiences with which nine decades had provided him. He believed he had gained at least a partial insight into the mysteries of nature. Holistic principles, Wallace maintained, united the specialized domains of science as well as the diverse domains of human political, ethical, religious, and socioeconomic reality. He was not naive enough to expect that his synthesis would be the final one. Nor did he think it would be received without criticism. "Truth," Wallace recognized, "is born into this world only with pangs and tribulations, and every fresh truth is received unwillingly. To expect the world to receive a new truth, or even an old truth, without challenging it, is to look for one of those miracles which do not occur" (Northrup 1913).

Of the many responses to Wallace's synthesis, one of the most pertinent for assessing the impact of his theistic teleology is that of the Reverend J. M. Mello. In his 1911 *The Mystery of Life and Mind*, Mello used Wallace's *World of Life* to support the thesis that "the whole universe, in all its aspects, is the work of a Designing Mind, and is, as it has been strikingly expressed, 'the Externalization of the Thought of God.' " Mello also relied on arguments from other leading scientists such as Lord Kelvin, the chemist Lionel Beale, and St. George Mivart. But Wallace was his main source for claiming that scientific evidence and argumentation lead to the conclusion "that the only adequate cause of all that we see around us is a Ruling and Creative Power to which the Universe is due." Mello, as a minister, regarded science as powerfully supporting a theistic and teleological interpretation of nature. He saw Wallace as one of the foremost thinkers to use scientific expertise to discredit atheism, materialism, and the denial of design in the natural world (Mello 1911, 1–20).[12] Wallace's synthesis was by no means universally accepted. But Mello's tract testifies to the potency and wide appeal of Wallace's theistic evolutionary teleology. Wallace's odyssey had ended. What he found at the end of his quest fulfilled him. Wallace's passion and expertise in pursuing a holistic and humane philosophy of nature remain a crucial legacy from the Victorian era.

————•◆•————

NOTES

1. On "classical theism," see Brian Leftow (1998, 98–100).

2. For a suggestive account of naturalistic theism and its compatibility with evolutionary science at a constitutive level, see Griffin (1989), chap. 5, "Evolution and Postmodern Theism," 69–82.

3. *Theism* reached its eleventh edition by 1905. Flint was also the author of the article "Theism" in *Encyclopaedia Britannica,* 9th ed. (1888).

4. Hick (1990, 125–130), offers an insightful treatment of the differences between theism and spiritualism.

5. See chap. 5 for a discussion of Wallace's endorsement of Anna Blackwell's statement that "the establishment of Scientific Theism . . . is the most urgent need of the present day" (Blackwell 1898, 9–10).

6. "Man and Monkey. Dr. Wallace Expounds the Human Pedigree." *San Francisco Chronicle* 45, 26 May 1887, 6.

7. Wallace's comments appeared in the newspaper article "The Life Hereafter. Future State Considered; Interesting Address by Dr. Alfred Russel Wallace" (*San Francisco Chronicle* 45, 6 June 1887, 8).

8. Barrett's ideas had originally been delivered as a lecture before the Quest Society in London but were quickly published as *Creative Thought and the Problem of Evil* (1911).

9. See also Wallace 1912d and 1912e.

10. For the significance of the notion that theists emphasize their conviction that belief in God involves a profound sense of moral obligation to do good, see Swinburne (1993, 184–216).

11. "A Blow to Spiritualism," *Cincinnati Weekly Times,* 7 April 1887, 2, cols. 5–6.

12. Wallace's copy of Mello's book, which he expressed "many thanks" for receiving, is in ARWL.

CHAPTER 7

Epilogue

1913: THE LAST TWO BOOKS

Just as *World of Life* was intended as a sequel to *Man's Place in the Universe*, Wallace's last two books were, in part, sequels to *World of Life*. *Social Environment and Moral Progress* and *The Revolt of Democracy* were both published in the year of his death (1913a, 1913b). The guiding vision of Wallace's life was to link science and the broader culture to ensure a benevolent future for all humanity. Forging that benevolent link, he knew, would be an uphill battle. In the midst of his analysis of the historical development of life as revealed by the geological record in *World of Life*, Wallace had launched into an attack on the great urban cities of the early twentieth century. He called them "the 'wens', the disease-products of humanity." The geological record indicated that species exist only when environmental conditions are appropriate for their health and survival. The price paid for maladaptation to past environments was displayed in the fossil record of extinct species. Such, Wallace believed, was the possible fate of humans in modern society.

"The teaching[s] of that true and far-seeing child of nature, William Cobbett, [of] all our greatest sanitarians, [and] of Nature herself in the comparative rural and urban death-rates," had left their mark on Wallace. He declared that until industrial wens were abolished, or brought under control, "there can be no approach to a true or rational civilization." Because he saw no legislators or ministers willing or able to put an end to the continued growth of modern cities—"which are wholly and absolutely evil"—Wallace, in *World of Life*, took "this opportunity of showing *how* it can be done":

> There is much talk now of what *will* and *must* be the growth of London during the next twenty or fifty years; and of the *necessity* of bring-

ing water from Wales to supply the increased population. But where
is the necessity? Why provide for a population which need never have
existed, and whose coming into existence will be an evil and of no
possible use to any human beings except the landowners and speculators who will make money by the certain injury of their fellow citizens? If the House of Commons and the London County Council
are not the bond-slaves of the landowners and speculators, they have
only to refuse to allow any further water-supply to be provided for
London except what now exists, and London will cease to grow. Let
every speculator have to provide water for and on his own estate, and
the thing will be done—to the enormous benefit of humanity.

                                                    (Wallace 1910a, 285)

The professionalization and specialization of science made the juxtaposition
of sociopolitical and moral/theistic concepts with scientific ones, in discourse
on science, suspect or at least decidedly unfashionable by the time Wallace
wrote *World of Life*. Wallace had never been a slave to fashion, scientific or otherwise. The old man at Old Orchard was simply more aggressive in asserting
his holistic vision.

Wallace demanded, both as scientist and social reformer, that the governing body of any growing town or city should announce to its citizenry
that when "you have not a gallon of polluted water in your town, and when
its death-rate is brought down to the average standard of rural areas, we
will reconsider the question of your future growth." Wallace had come to
rely increasingly on Malthusian preventive checks (as opposed to positive
checks such as famine, disease, or wars) to remedy what he perceived as
the pernicious effects of industrialized urbanization (Wallace 1910a, 285). In
1899, Wallace had written an essay on the best way to improve the general
conditions of workers in an increasingly industrialized society. He rejected
strikes as an inefficient and inappropriate strategy. Wallace argued that although the "strike may have been an essential weapon in the past—perhaps
the only weapon the worker possessed—now, however, all the higher grades
of workers are better educated, better organised, and have higher ideals." He
regarded strikes as retrograde because they "effect nothing of a permanent
nature." Strikes also usually cost the strikers more in lost wages than they
could ever regain. Wallace's solution to labor unrest, as he had advocated in
his many other writings and lectures on land nationalization and socialism,
was cooperative action that stopped short of the disruptive and controversial
functions of strikes. Federations of workers, he believed, could then take
over control of shops and factories by gradually amassing, through union
dues and other methods of financial accumulation such as savings, the funds

to purchase the means of production. Workers would beat employers at their own game (Wallace 1899, 105).

Wallace's views on strikes echoed those of some Fabian theorists, particularly Beatrice and Sidney Webb. Like Wallace, the Webbs stressed mutual improvement and self-help activities by laborers rather than outright confrontation. The first decade of the twentieth century witnessed an explosion of labor unrest, which accelerated in the years immediately preceding World War I. At issue was a fundamental disagreement, at times erupting into open antagonism, between rank-and-file workers and their union leaderships. Many trade union officials took their cue from middle-class strategists (like the Fabians) and members of Parliament. In 1896, an act of Parliament had offered conciliation facilities to employers and union negotiators. After 1900, when George Asquith became the Board of Trade's "trouble-shooter," these facilities were expanded. Union officials were appointed as "labour correspondents," sending information to a central statistical office. Militant union activity in the period 1900–1914 was very much due to pressure from below, in reaction against the more conciliatory approach of union leaders and their middle-class supporters.

In addition to the Webbs, some of Britain's more influential left-wing political groups, including certain factions of the emerging Labour Party, were ideologically as well as tactically opposed to strikes. The Fabian executive had written to the press before the threatened railway strike of 1907 that "in the case of the nation's principal means of land transport, resort to the characteristic trade-union weapon of the strike is . . . a national calamity." The *Clarion*, to which Wallace was a regular contributor, declared on 16 September 1910 that "Socialist teaching has consistently pointed out the futility of the strike, and advocated the better way of arbitration." Even the attitude of the revolutionary Social Democratic Federation, which renamed itself the Social Democratic Party in 1907, was surprisingly similar to that of the more mainstream left. H. M. Hyndman wrote in the 10 August 1907 issue of *Justice* that "we of the Social Democratic Party and *Justice* are opposed to strikes on principle" (Ward 1998, 88–90). Wallace was opposed to strikes as a permanent feature of political life. But he conceded that until socialist reforms were truly implemented, occasional major strikes—such as the Great Strike of 1913—did serve a valuable function. They brought the plight of the workers forcibly to national attention (Wallace 1912b; 1913a, chap. 3).

SOCIAL ENVIRONMENT AND MORAL PROGRESS

Despite his rejection of strikes as an appropriate strategy for effecting reform, Wallace trumpeted his increasing dissatisfaction with existing social and political conditions in *Social Environment and Moral Progress*. According to Marchant, the conclusions Wallace drew in that work were "startling" (Marchant [1916] 1975, 383). In the final chapter, "How to Initiate an Era of Moral Progress," Wallace's generally temperate language was replaced by outrage. "It is not too much to say," he thundered, "that our whole system of society is rotten from top to bottom, and that the Social Environment as a whole, in relation to our possibilities and our claims, is the worst that the world has ever seen" (Wallace 1913b, 169). Marchant noted that this "terrible indictment was doubly underscored in his MS" (Marchant [1916] 1975, 383). Wallace's condemnation was prefaced with a brief summary of his well-known views on the origin of the higher human faculties. His purpose was to demonstrate that "what is commonly termed morality is not wholly due to any inherent perception of what is right and wrong conduct, but that it is to some extent and often very largely a matter of convention. . . . The actual morality of a community is largely a product of the environment, but is local and temporary, not permanently affecting the character."

The strategy of *Social Environment* was to bring together evidence, drawn from the record of human evolution as well as contemporary history, "to distinguish between what is permanent and inherited and what is superficial and not inherited, and to trace out some of the consequences as regards what we term 'morality.' " Wallace zeroed in on the elusive term "character— in individuals, in societies, and especially in those more ancient and more fundamental divisions of mankind which we term 'races.' " Character, he defined, "as the aggregate of mental faculties and emotions which constitute personal or national individuality. It is very strongly hereditary, yet is probably subject to more inherent variation than is the form and structure of the body." Wallace drew on his long-held conviction (both on evolutionary and theistic grounds) that the higher faculties of the human species were the product not of natural selection but of divine provision. He sided with those historians and anthropologists who regarded the ancients as in no respects inferior, and in many respects superior, to modern man. Wallace cited the spectacular achievements of the ancient Babylonians, Indians, and Egyptians to demonstrate that the higher human faculties were present in full force in early history. His conclusion was unambiguous: "Now it is this inher[itable] character itself that tends to be transmitted to offspring, and this being the case, there can be no progressive improvement in character without some selective agency tending to such improvement. . . . There is no proof of any real advance in it during the whole historical period." The only selective agency that Wallace regarded as capable of effecting a permanent, and progressive, elevation in human morality was the principle of sexual selection under socialism (Wallace 1913b, 9–13, 22–35, 147–152, 163–165).

This was the Bellamy-inspired principle that Wallace had first presented in the 1890 essay "Human Selection" ([1890] 1900c) and the 1892 essay "Human Progress" ([1892] 1900c). Both were reprinted in his *Studies Scientific and Social* of 1900. Wallace had also contributed a lengthy essay titled "Evolution and Character" to the volume *Character and Life* that resulted from the 1912 symposium of the same name (Parker 1912, 3–50). That essay parallels the argument of *Social Environment*. It concluded with Wallace's judgment on nineteenth-century attempts at improving the human character. "We shall, perhaps, realise, before it is too late," he wrote, "that we have begun at the wrong end. Improvement of social conditions must precede improvement of Character; and only when we have so reorganised society as to abolish the cruel and debasing struggle for existence and for wealth that now prevails, shall we be enabled to liberate those beneficent natural forces which can alone elevate Character." The preeminent natural force was female sexual selection in marriage in a socialist society. Wallace had been asked specifically to participate in the symposium because he shared "with Darwin the honour of conceiving the theory of Evolution." It was as a scientist, not as a social reformer, that his views on the bearing of evolution on character— particularly on "what characteristics in men and women does Evolution seem most to approve?"—were solicited (Parker 1912, vii, 50). Wallace, of course, never regarded science and sociopolitical issues as separate. They were parts of a unified, holistic evolutionism. But his contribution to the symposium was toned down to conform to the format. In *Social Environment and Moral Progress* there were no such constraints. Wallace offered his "mature judgement" (the phrase is Marchant's) on Britain's "moral position as a nation, [its] social environment, how it came to be what it is, and what lessons we may learn from it" (Wallace 1913b, 48).

The causes of the sorry state of Britain's national morality included those Wallace had enunciated since the 1880s: a systemic economic antagonism under competitive capitalism, monopolistic control of land and capital, and existing laws that fostered the inheritance of wealth by the few. For Wallace, as for many late Victorian socialists, the sole permanently effective solution to social ills lay in radically reversing the course of existing economic and sociopolitical practices (Laybourn 1997). Wallace welcomed certain government reforms of the late nineteenth and early twentieth centuries. But they were merely initial and provisional steps toward redressing societal ills. Wallace argued that only a fundamental commitment to socialism could "change our existing immoral environment into a moral one, and initiate a new era of Moral Progress" (Wallace 1913b, 166–174). *Social Environment* received a wide and (not surprisingly) mixed reception. Leading articles and illustrated reviews appeared in most of London's newspapers and periodicals, which covered the full range of responses to so polemical a work (Marchant [1916] 1975, 471).

The Revolt of Democracy

Wallace's final published work, *The Revolt of Democracy*, was written a few months prior to *Social Environment and Moral Progress*. The manuscript, however, was given to Marchant only a few months before Wallace's death and was thus published after the later book (Marchant [1916] 1975, 384). Little, if any, scholarly attention has been paid to the close relationship between Wallace and Marchant and to the crucial role played by the latter in the publication of Wallace's last writings. There is no evidence to suggest that Marchant edited Wallace's final books in any other than slight stylistic senses. The two shared identical social and cultural convictions. Marchant's biographical preface to *The Revolt of Democracy* is a reverential but accurate précis of Wallace's life and career. He called Wallace "the Grand Old Man of British Science, a true Revealer and Prophet, in the real sense of being a forthteller of the truth spoken to him." Marchant's sketch affords a valuable insight into Wallace's state of mind as he neared death. He remained fully in command of his mental and polemical powers. In a letter to Marchant, written in May 1913, Wallace discussed the possibility of living organisms being someday produced in the chemist's laboratory from inorganic matter. He declared such a possibility "impossible, because unthinkable, while even were it supposable that it should happen, it could not in any way explain Life, with all its inherent forces, powers and laws. . . . Recent discoveries demonstrate the need of coordinating power even in the very nature and origin of matter; and something far more than this in the origin and development of mind. The whole cumulative argument of my 'World of Life' is that, *in its every detail* it calls for the agency of a mind or minds so enormously above and beyond any human minds, as to compel us to look upon it, or them, as 'God or Gods,' and so-called 'Laws of Nature' as the action by will-power or otherwise of such superhuman or infinite beings. 'Laws of Nature' apart from the existence and agency of some such Being or Beings, are mere words, that explain nothing—are, in fact, unthinkable. That is my position!" (Wallace 1913a, xxxiv–xxxix).

The Revolt of Democracy was Wallace's parting contribution to the rapidly changing political landscape of Britain prior to the Great War. He intended it as a primer for his "readers, and especially [those of] the Labour Party." Wallace hoped *Revolt* would provide the Labour Party with informed guidance, "in order that it may have a definite program to work for, and may be able to enforce its claims upon the Government with all the weight of its combined and determined action." He sought to expose, clearly and succinctly, "the series of economic fallacies which alone prevent them from claiming and obtaining, for the workers of the whole country, a continuously increasing share of the entire product of their labour." Wallace argued that the "forces of Labour, if united in the demand for this one *primary* object, must and will succeed." He predicted that the inevitable result of implementing the social, political, economic, and moral reforms he had been advocating publicly for three decades, since he became president of the Land Nationalization Society in 1881, would be twofold. First, there would be a general rise of wages for the entire population at the expense of the unprecedented individual wealth that industrial capitalism had bestowed on the relatively few. Second, that most miserable sector of society, the unemployed workers, would be reintegrated into self-supporting rural cooperatives. Wallace regarded the destruction of the "old rural populations which were for centuries the pride and strength of Britain" as among the greatest evils perpetrated by industrial capitalism. He considered the forced migration of agricultural workers into the rapidly growing cities and towns of nineteenth-century industrial society as "the most disastrous policy . . . ever pursued by" a so-called civilized nation. Wallace pointed to the success in other European states, notably Denmark, of salubrious and profitable agricultural cooperatives. He reiterated his claim of thirty years that nationalization of "a large portion of the agricultural land of England, which has been so misused by its owners, must be acquired by the Government in trust for the nation." Wallace concluded with the exhortation that the creation of a humane and equitable society based on the twin pillars of agriculture and industry "must be the great and noble work of our statesmen of to-day and of to-morrow. May they prove themselves equal to the great opportunity which the justifiable revolt of Labour has now afforded them" (Wallace 1913a, 54–56, 66–67, 74–78).

A Socialist's Critique of the New Liberals

The analytical framework of *Revolt* shows Wallace's enduring debt to his relationship with Spencer and George, his reading of Bellamy, and his Owenite roots. What is novel, for Wallace, was his certainty that social and political conditions by the close of the first decade of the twentieth century made Britain ripe for major reform. It was the appointment of two successive reformist Liberal prime ministers, Sir Henry Campbell-Bannerman (1905–1908) and Herbert Asquith (1908–1915), which precipitated what

Wallace termed the revolt of democracy. Wallace maintained that previous Victorian prime ministers, most notably William Gladstone who served as Liberal leader four separate times from 1868 to 1894, were constrained in their ability to implement effective legislation to better the conditions of the working classes. They were crippled, Wallace declared, by their commitment to the doctrine that "wages were kept down by the 'iron law' of supply and demand; and that any attempts to find a remedy by Acts of Parliament only aggravated the disease" of poverty. Wallace credited Campbell-Bannerman with changing "this attitude of negation of all his predecessors," that poverty was due to economic causes over which governments had no power. In numerous speeches both in an out of Parliament, Campbell-Bannerman "boldly declared . . . that he held it to be the duty of a government to deal with the great problems of unemployment and poverty, and especially to attack the increasingly injurious land monopoly, and so to legislate as to make our native soil ever more and more 'a treasure-house for the poor rather than a mere pleasure-house for the rich.' " Campbell-Bannerman's choice of Herbert Asquith as chancellor of the Exchequer and his recruitment of Lloyd George into his ministry were signs to Wallace that the prime minister fully intended to make his an activist government on behalf of labor reforms. When Asquith became prime minister in 1908 and appointed Lloyd George his successor as chancellor of the Exchequer, Wallace was delighted. He admired Lloyd George greatly. Wallace especially applauded his successes in securing old age pensions and other measures calculated to benefit the working classes (Wallace 1913a, 7–8).

Wallace endorsed the New Liberals (as they had come to be known) for another reason. Many of their leading theorists, such as L. T. Hobhouse, regarded biology as providing a scientific basis for political action (Freeden 1978).[1] But Wallace was a socialist, not a Liberal. Despite his praise for the new initiatives of the Liberal governments, Wallace felt they did not go far enough. He charged that Liberal good intentions were vitiated by the absence of any truly adequate remedial legislation to eradicate the causes and not just the symptoms of societal inequities. This Liberal "failure," as Wallace termed it, coupled with the rising expectations of an increasingly well-organized and politically educated army of workers, was a recipe for disaster. It led, he suggested, to the series of major strikes by dockers, seamen, miners, and members of the railway and other transport unions during the period 1911–1912. Wallace's generic antipathy toward strikes, as damaging to workers' interests in both the short and long run, caused him to turn to the Labour Party. Labour appeared to him as the most promising Parliamentary group able to bring about the socialist remedies he considered fundamental for any true reform of British society (Wallace 1913a, 7–13, 47).

Wallace's critique of New Liberalism was shared by many of his contemporaries, particularly the Fabians. It has been estimated that approximately two thousand progressives changed their parliamentary party allegiance from Liberal to Labour. Some few, like Wallace, did this between 1908 and 1912. Most of the Liberal hemorrhage occurred during and shortly after the First World War (Blaazer 1992, 70–71, 104–109). Wallace's support first of Liberals then of Labour was, however, only provisional. He, like most of his socialist contemporaries, recognized that there was little chance at the time of achieving an official socialist presence in the form of an established party in Parliament. Wallace and others lent their support to Liberals and Labour if they believed that individuals in either of those two parties could advance socialist ideals and causes. But he and a number of socialists had certain reservations about this strategy. To them, even the Labour Party was not the ideal vehicle for bringing about a socialist Britain. Labour's initiatives often appeared to threaten individual and democratic rights with the overtly statist views it adopted in certain policy areas. This potential conflict with the voluntaristic socialism so dear to Wallace and his allies tempered their enthusiasm (Minkin 1991). But the Labour Party had made enormous political strides between 1900 and 1914. And its frequent socialist rhetoric suggested a genuine socialist core in at least certain critical areas. These factors were sufficient to attract Wallace and like-minded thinkers and activists to Labour's fold in the years prior to the First World War.

Many members of the Labour Party were evolutionary socialists and reformists at heart (Laybourn 1997, 40, 64). Wallace shared Sidney Webb's view that in Britain "important organic changes" could only be achieved democratically and gradually, constitutionally and peacefully. Although this specific statement by Webb was not published until after Wallace's death, Wallace had long endorsed the Fabian position on this matter (Shaw 1920, 34–35). In 1903, in his preface to William Morris's *Communism*, Shaw had written that the Fabians' "socialism could be adopted either as a whole or by instalments by any ordinary respectable citizen without committing himself to any revolutionary association or detaching himself in anyway from the normal course of English life" (Morris 1903, 3).[2] Wallace's search for the most appropriate political vehicle for achieving the goals of socialism, as he interpreted them, was not mere opportunism or vacillation. His political evolution reflected the challenge faced by most British socialists at the turn of the century. The efforts to forge a specifically British socialism, different from continental varieties, gave rise to a bewildering variety of options—a veritable "pilgrims' progress"—during the crucial period of socialist expansion in Britain between 1890 and 1918 (Blaazer 1992, 98).

LOOKING BACK TO LOOK FORWARD

The Revolt of Democracy is fitting as the final published work of Wallace. Far less comprehensive in scope than *The World of Life* or *Man's Place in the Universe*, it nonetheless encapsulates the essential integrative themes of Wallace's worldview. The most poignant and telling sections of the book, however, are those that evoke Wallace's childhood roots in the pastoral Welsh border country. The aged man never forgot the beauties of rural Britain to which he was first exposed. His allusions to the old rural populations, which were for centuries the pride and strength of Britain, are not simply nostalgic. Agricultural cooperatives and land reform, about which he learned early in life from Owen's lectures, remained central to Wallace's final cultural pronouncement. His ideas reflected and contributed to the programs of those late Victorian socialists who no longer accepted industrialization as the inevitable paradigm for future social organization.

Wallace's socialist vision can best be compared with those other advocates of British radicalism who fought for a new moral as well as economic/political order (Wallace [1905] 1969, 2:271–272). They saw socialism not merely as a new economic order but also as an entirely new system of life and belief. Such a sweeping concept of an ethically as well as sociopolitically reformed society characterized many figures crucial to the establishment of British socialism from the early 1880s onward (Yeo 1977). Wallace and Morris, Ruskin, H. M. Hyndman, Keir Hardie, and Blatchford used their metaphorical and historical images of preindustrial Britain to prefigure a new socialist society that would divest itself of the worst features of both industrialism and capitalism. All visions of a new socialist community involved a rejection of industry under capitalism. But most socialists believed that industrialization was the key to societal advance. Wallace and like-minded contemporaries had a different vision. They hoped that a socialist society could be built, in part, on the rural foundations of pre-nineteenth-century Britain.

The *Clarion* on 14 September 1895 published a verse "To the Farmers of England by One of Them" (Ward 1998, chap. 2). It is worth recalling, in the context of the strong rural component in Wallace's socialism, that among his first attempts at writing for publication was an 1843 sketch titled "The South-Wales Farmer."[3] Drawing on his personal observations during his surveying work in Brecknockshire and Glamorganshire, Wallace depicted the manners and customs of the Welsh peasantry. The resulting portrait was far from idyllic. The main significance of this article is the clear display of Wallace's concern, from the age of twenty, in righting the wrongs he saw inflicted by an industrializing Britain, which was condemning many sectors of the rural as well as urban population to lives of misery and penury (Wallace [1905] 1969, 1:206–222; Moore 1997, 300–303). A significant number of socialists took rural idealism and "back to the land" notions quite literally. Many London Fabians and members of other socialist organizations bought country houses if they could afford dual residences. Others, like Wallace, simply removed themselves from London altogether and sought their own socialist-inspired "new Edens" in affordable rural locations (MacKenzie and MacKenzie 1977; Marsh 1982; Gould 1988). Wallace's move from London to Grays in 1872 was the first of several changes of residence that took him and his family farther and farther from the urban metropolis. They sought a succession of increasingly more isolated, though never very distant, rural seats. This quest, with its political as well as psychological overtones, culminated in Wallace's beloved final home at Old Orchard in Broadstone.

WHERE TO BURY WALLACE:
OLD ORCHARD OR WESTMINSTER ABBEY?

By the summer of 1913, Wallace's health began to decline rapidly. He could no longer walk about his beloved gardens. He did have some of the rarer primroses and other favorite plants brought to a small area in front of the windows of his study so that he might gaze on them. His mental faculties remained strong. Wallace entered into a contract to write yet another large volume on social issues. An agreement with the publishers was signed, but by the time the plan of the proposed book was to be discussed Wallace had grown progressively weaker. On 7 November 1913, he died peacefully in his sleep at 9:25 a.m. Three days later, Wallace was buried with a simple ceremony in the small cemetery of Broadstone on a pine-clad hill swept by sea breezes. The funeral was attended by his son and daughter, his wife's sister, and Marchant. Wallace's wife Annie, by then an invalid, was unable to come. The funeral service was conducted by the Bishop of Salisbury, Dr. Ridgeway. Among the representatives of the varied facets of Wallace's life and career were Meldola and Poulton (representing the Royal Society), Dr. Scott (representing the Linnean Society), and Joseph Hyder of the Land Nationalisation Society. A monument consisting of a fossil tree-trunk from the Portland beds was erected over his grave on a base of Purbeck stone, a durable limestone. It bore the following modest inscription:

ALFRED RUSSEL WALLACE, O.M.

Born Jan. 8th, 1823, Died Nov. 7th, 1913

It had been suggested by some that Wallace merited burial in Westminster Abbey, beside Darwin. Annie Wallace and the rest of the family declined.

They stated that it ran counter to their own, and Wallace's, wishes for a modest and isolated resting place. That Wallace was buried not at Westminster Abbey but in the quiet hillside cemetery at Broadstone is itself a fitting memorial, as well as conclusion, to his long, complex, and contentious career. Would Wallace, whose entire life had been dedicated to social reform and the eradication of class and economic privilege, in addition to the pursuit of scientific knowledge, have been comfortable amid the pomp and glory of Westminster Abbey? A group of leading members of the scientific community felt that the lack of some more public memorial than the grave at Broadstone would be egregious and, indeed, scandalous. A committee was formed shortly after Wallace's death, with Poulton as chairman and Meldola as treasurer, to create a suitable national memorial. The committee presented a letter on the morning of 2 December 1913 to the dean of the abbey, Herbert Ryle. They requested permission to place a medallion there, "believing that no position would be so appropriate as Westminster Abbey." The signatories of the letter included Crookes, Archibald Geikie, Ray Lankester, Meldola, Lodge, Poulton, and William Ramsay (recipient of the 1904 Nobel Prize in Chemistry). Dean Ryle replied later that same day that their petition was unanimously granted by the chapter of the abbey. He added that "nothing could have been more satisfactory or impressive than the document with which you furnished me this morning. I hope to get it specially framed."

Wallace's medallion was unveiled on All Souls' Day, 1 November 1915, together with medallions to the memory of two of his knighted contemporaries, Sir Joseph Hooker and Lord Lister. Dean Ryle's sermon noted that the three men whose memorials were then being "uncovered to the public view, in the North Aisle of the Choir . . . will always be ranked among the most eminent scientists of the last century. . . . They were all men of singularly modest character. As is so often observable in true greatness, there was in them an entire absence of that vanity and self-advertisement which are not infrequent with smaller minds. It is the little men who push themselves into prominence through dread of being overlooked. It is the great men who work for the work's sake without regard to recognition, and who, as we might say, achieve greatness in spite of themselves." Ryle added, with circumspection and perhaps unintended irony given Wallace's highly controversial and antiestablishment writings of his later years, that "Wallace's life was spent in the pursuit of various objects of intellectual and philosophical interest, over which I need not linger" (Marchant [1916] 1975, 473–474). An interviewer who visited Wallace in 1909 offered an image that was far closer to the mark. Wallace appeared, the interviewer stated, like an "old warrior—a venerable figure crowned with white, standing, as it were, midway between the centuries. Behind him lies the one that he has done so much to mould and alter and convince; before him that mysterious future on which he gazes with unshadowed faith in the ultimate triumph of his views" (Rann 1909).

———•◆•———

NOTES

1. Freeden (1978, chap. 3) gives a detailed account of the New Liberals' use of biological arguments.

2. Ironically, Morris's 1903 pamphlet was itself an attack on the "make-shift alleviations" of reformist socialism, which he felt would come to be "looked upon as ends in themselves" and thus blunt the move toward the communism that he embraced (11).

3. The essay did not get published at the time. Wallace, however, thought it of particular interest in demonstrating his early preoccupation with sociopolitical concerns and included the full text in his autobiography (Wallace [1905] 1969, 1:206–222).

BIBLIOGRAPHY

Archival Sources

I have consulted the following archival/manuscript materials in the preparation of this study:

ARWL = Alfred Russel Wallace Library, Special Collections, Edinburgh University Library (formerly housed in the Hope Library, Oxford University Museum). This is a rich collection of more than four hundred books from Wallace's personal library. Most are heavily annotated by Wallace. This collection is of special importance to my analysis since it consists mainly of works on social and political affairs, religion, and spiritualism purchased by Wallace between 1870 and 1913.

ARWP = Alfred Russel Wallace Papers, MSS. Add 46434–46442, British Library: approximately three thousand folio pages of letters to and from Wallace, dated 1848–1913. This is an invaluable source for Wallace's correspondence on all subjects. Especially pertinent to this book are Add 46439 (correspondence on spiritualism and related subjects); Add 46440 (correspondence on land nationalization, socialism, and related subjects); and Add 46434 (general letters).

British Museum (Natural History): twenty letters; two notebooks.

Cambridge University Libraries: letters to and from Wallace, Balfour Library.

Imperial College of Science, Technology and Medicine, London: eight letters from Wallace to T. H. Huxley (1863–1891).

Linnean Society of London: letters; four notebooks from the Malay Archipelago; five journals, including "American Journal: 1886–1887"; annotated books from Wallace's personal library (dealing mainly with scientific matters and including Darwin's *Descent of Man*).

Royal Botanic Gardens, Kew: 139 letters; mostly between Wallace and the directors of Kew, including J. D. Hooker and W. T. Thistleton-Dyer.

Royal Geographical Society of London: fifteen letters from Wallace to various recipients (1853–1878).

University of London: correspondence between Wallace and Herbert Spencer, listed under Wallace letters in the Spencer Collection.

WFBP = W. F. Barrett Papers, box 2, A2, no. 11, Society for Psychical Research, London.

PRINTED MATERIALS

Abbot, Francis Ellingwood. 1885. *Scientific Theism.* Boston: Little, Brown, & Co.

————. 1890. *The Way out of Agnosticism; or, The Philosophy of Free Religion.* Boston: Little, Brown, & Co.

Allen, David. 1994. *The Naturalist in Britain: A Social History.* Princeton, N.J.: Princeton University Press.

Alter, Peter. 1987. *The Reluctant Patron: Science and the State in Britain, 1850–1920.* Oxford: Berg.

American Historical Review. 1899. "Minor Notices." *American Historical Review* 4, no. 2 [January]: 389.

Amundson, Ron. 1998. "Typology Reconsidered: Two Doctrines in the History of Evolutionary Biology." *Biology and Philosophy* 13:153–177.

Andelson, Robert V. 1993. "Henry George and the Reconstruction of Capitalism: An Address." *American Journal of Economics and Sociology* 52:493–501.

Anderson, Douglas R. 1995. *Strands of System: The Philosophy of Charles Peirce.* West Lafayette, Ind.: Purdue University Press.

Armstrong, Patrick. 1998. "Crossing the Line: Alfred Russel Wallace in Bali and Lombok, 1856." *Journal of Indian Ocean Studies* 5:267–286.

Asma, Stephen T. 1996. "Darwin's Causal Pluralism." *Biology and Philosophy* 11:1–20.

Baber, Zaheer. 1996. *The Science of Empire: Scientific Knowledge, Civilization, and Colonial Rule in India.* Albany: State University of New York Press.

Baigrie, Brian S., ed. 1996. *Picturing Knowledge: Historical and Philosophical Problems concerning the Use of Art in Science.* Toronto: University of Toronto Press.

Bannister, Robert C. 1979. *Social Darwinism: Science and Myth in Anglo-American Social Thought.* Philadelphia: Temple University Press.

Barbour, Ian. 1997. *Religion and Science: Historical and Contemporary Issues.* San Francisco: Harper.

Barbour, Thomas. 1950. *Naturalist at Large.* London: Scientific Book Club.

Barker, Charles A. 1955. *Henry George.* New York: Oxford University Press.

Barnes, Barry, David Bloor, and John Henry. 1996. *Scientific Knowledge: A Sociological Analysis.* Chicago: University of Chicago Press.

Barr, A., ed. 1997. *Thomas Henry Huxley's Place in Science and Letters: Centenary Essays.* Athens: University of Georgia Press.

Barrett, William Fletcher. 1908. *On the Threshold of a New World of Thought: An Examination of the Phenomena of Spiritualism.* London: Kegan, Paul, Trench, Trubner & Co.

————. 1911. *Creative Thought and the Problem of Evil.* London: J. M. Watkins.

————. 1918. *On the Threshold of the Unseen.* New York: E. P. Dutton & Co.

————. Papers. Box 2, A2, no. 11. Society for Psychical Research, London.

Barrington, M. R., K. M. Goldney, and R. G. Medhurst, eds. 1972. *Crookes and the Spirit World: A Collection of Writing by or concerning the Work of Sir William Crookes, O.M., F.R.S., in the Field of Psychical Research.* London: Souvenir Press.

Barrow, Logie. 1980. "Socialism in Eternity: Plebeian Spiritualism, 1853–1913." *History Workshop* 9:37–69.

————. 1986. *Independent Spirits: Spiritualism and English Plebeians, 1850–1910.* London and New York: Routledge & Kegan Paul.

Bartholomew, J. G., W. Eagle Clarke, and Percy H. Grimshaw. 1911. *Atlas of Zoogeography.* London: John Bartholomew for the Royal Geographical Society.

Bartholomew, Michael. 1973. "Lyell and Evolution: An Account of Lyell's Response to the Prospect of an Evolutionary Ancestry for Man." *British Journal for the History of Science* 6:261–303.

Barton, Ruth. 1987. "John Tyndall, Pantheist: A Rereading of the Belfast Address." *Osiris* 3:111–134.

———. 1990. " 'An Influential Set of Chaps': The X-Club and Royal Society Politics, 1864–85." *British Journal for the History of Science* 23:53–81.

———. 1998a. " 'Huxley, Lubbock, and Half a Dozen Others': Professionals and Gentlemen in the Formation of the X Club, 1851–64." *Isis* 89:410–444.

———. 1998b. "Just before *Nature:* The Purposes of Science and the Purposes of Popularization in Some English Popular Science Journals of the 1860s." *Annals of Science* 55:1–33.

Bates, Henry Walter. 1862. "Contributions to an Insect Fauna of the Amazon Valley." *Transactions of the Linnean Society of London* 23:512–513.

———. 1863. *The Naturalist on the River Amazons.* 2 vols. London: John Murray.

Beddall, Barbara G. 1968. "Wallace, Darwin, and the Theory of Natural Selection: A Study in the Development of Ideas and Attitudes." *Journal of the History of Biology* 1:261–323.

———. 1972. "Wallace, Darwin, and Edward Blyth: Further Notes on the Development of Evolution Theory." *Journal of the History of Biology* 5:153–158.

———. 1988a. "Darwin and Divergence: The Wallace Connection." *Journal of the History of Biology* 21:1–68.

———. 1988b. "Wallace's Annotated Copy of Darwin's *Origin of Species.*" *Journal of the History of Biology* 21:265–289.

———. 1998. "Wallace, Alfred Russel (1823–1913)." In *Routledge Encyclopedia of Philosophy,* ed. Edward Craig, 9:678–679. London and New York: Routledge.

Beer, Gillian. 1983. *Darwin's Plots: Evolutionary Narrative in Darwin, George Eliot and Nineteenth-Century Fiction.* London: Routlege & Kegan Paul.

———. 1999. *Open Fields: Science in Cultural Encounter.* Oxford: Oxford University Press.

Bellamy, Edward. [1888] 1960. *Looking Backward.* New York: New American Library/ Signet.

———. 1897. *Equality.* New York: Appleton.

Bennett, J. A. 1987. *The Divided Circle: A History of Instruments for Astronomy, Navigation, and Surveying.* Oxford: Phaidon-Christie's.

Benton, Ted. 1995. "Science, Ideology and Culture: Malthus and the Origin Of Species." In *Charles Darwin's "The Origin of Species": New Interdisciplinary Essays,* ed. David Amigoni and Jeff Wallace, 68–94. Manchester: Manchester University Press.

Berry, Andrew, ed. 2002. *Infinite Tropics: An Alfred Russel Wallace Anthology.* New York and London: Verso.

Blaazer, David. 1992. *The Popular Front and the Progressive Tradition: Socialists, Liberals, and the Quest for Unity, 1884–1939.* Cambridge: Cambridge University Press.

Blackwell, Anna. 1898. *Whence and Whither? Or Correlation between Philosophic Convictions and Social Forms.* London: George Redway.

Blaisdell, Muriel L. 1992. *Darwinism and Its Data: The Adaptive Coloration of Animals.* New York and London: Garland.

Bland, T. A. 1905. *In the World Celestial.* 4th ed. Chicago: T. A. Bland & Co.

———. 1906. *Pioneers of Progress.* Chicago: T. A. Bland & Co.

Blanford, W. T. 1892. "The Permanence of Oceans and Continents." *Natural Science* 1:639–640.

Blinderman, Charles. 1998. "Alfred Russel Wallace." In *British Reform Writers, 1832– 1914,* ed. Gary Kelly and E. Applegate, 325–333. Dictionary of Literary Biography, vol. 190. Detroit: Gale Research.

Block, E. 1986. "T. H. Huxley's Rhetoric and the Popularization of Victorian Scientific Ideas: 1854–1874." *Victorian Studies* 29:363–386.

Blount, Godfrey. 1911. *The Blood of the Poor: An Introduction to Christian Social Economics.* London: A. C. Fifield.

Bohlin, Ingmar. 1991. "Robert M. Young and Darwin Historiography." *Social Studies of Science* 21:597–648.

Bono, James J. 1990. "Science, Discourse, and Literature: The Role/Rule of Metaphor in Science." In *Literature and Science: Theory and Practice,* ed. Stuart Peterfreund, 59–89. Boston: Northeastern University Press.

Bowler, Peter, J. 1976. "Alfred Russel Wallace's Concepts of Variation." *Journal of the History of Medicine* 31:17–29.

———. 1983. *The Eclipse of Darwinism: Anti-Darwinian Evolution Theories in the Decades around 1900.* Baltimore and London: Johns Hopkins University Press.

———. 1993. *Biology and Social Thought: 1850–1914.* Berkeley: University of California at Berkeley, Office for History of Science and Technology.

———. 1996. *Life's Splendid Drama: Evolutionary Biology and the Reconstruction of Life's Ancestry, 1860–1940.* Chicago and London: University of Chicago Press.

Brackman, Arnold C. 1980. *A Delicate Arrangement: The Strange Case of Charles Darwin and Alfred Russel Wallace.* New York: Times Books.

Brailsford, Henry N. 1948. *The Life-Work of J. A. Hobson.* London: Oxford University Press.

Brockway, A. Fenner. 1909. "The Next Step to Socialism; Dr. A. R. Wallace's Remedy for Unemployment." *Christian Commonwealth,* 27 January, 299a–299c.

Brooke, John and Geoffrey Cantor. 1998. *Reconstructing Nature: The Engagement of Science and Religion.* Edinburgh: T. & T. Clarke.

Brooke, John Hedley. 1996. "Religious Belief and the Natural Sciences: Mapping the Historical Landscape." In *Facets of Faith & Science.* Vol. 1, *Historiography and Modes of Interaction,* ed. Jitse M. van der Meer, 1–26. Lanham, Md.: University Press of America.

———. 1998. "The Historiography of Religion and Science Interaction." Paper delivered at the Science in Theistic Contexts: Cognitive Dimensions Conference, Redeemer College, Ancaster, Ontario, July 21–25.

Brooke, John Hedley, and R. Hooykaas. 1974. *New Interactions between Theology and Natural Science.* Milton Keynes: Open University Press.

Brooks, John Langdon. 1984. *Just Before the Origin: Alfred Russel Wallace's Theory of Evolution.* New York: Columbia University Press.

Brotman, C. 2001. "Alfred Russel Wallace and the Anthropology of Sound in Victorian Culture." *Endeavour* 25(4):144–147.

Brown, Frank Burch. 1986. "The Evolution of Darwin's Theism." *Journal of the History of Biology* 19:1–45.

Browne, Janet. 1983. *The Secular Ark: Studies in the History of Biogeography.* New Haven, Conn.: Yale University Press.

————. 1992. "A Science of Empire: British Biogeography before Darwin." *Revue d'histoire des Sciences* 45:453–475.

————. 1995. *Charles Darwin: Voyaging.* Princeton, N.J.: Princeton University Press.

————. 1998. "I Could Have Retched All Night: Darwin and His Body." In *Science Incarnate: Historical Embodiments of Natural Knowledge,* ed. Christopher Lawrence and Steven Shapin, 240–287. Chicago: University of Chicago Press.

————. 2003. *Charles Darwin: The Power of Place.* Princeton, N.J.: Princeton University Press.

Buckley, Arabella B. 1876. *A Short History of Natural Science, and of the Progress of Discovery from the Time of the Greeks to the Present Day, for the Use of Schools and Young Persons.* London: John Murray.

Burchfield, Joe D. 1975. *Lord Kelvin and the Age of the Earth.* New York: Science History Publications.

Burkhardt, Frederick. 1994. *A Calendar of the Correspondence of Charles Darwin, 1821–1882, with Supplement.* Cambridge: Cambridge University Press.

Burkhardt, Frederick, Sydney Smith, et al., eds. 1985–. *The Correspondence of Charles Darwin.* Cambridge: Cambridge University Press.

Burrow, J. W. 1970. *Evolution and Society: A Study in Victorian Social Theory.* Cambridge: Cambridge University Press.

Butler, Samuel. 1872. *Erewhon.* 4th ed. London: Trubner & Co. Wallace's annotated copy in ARWL.

————. 1879. *Evolution Old and New.* London: Hardwicke & Bogue.

————. [1890] 1970. "The Deadlock in Darwinism." *Universal Review.* Reprinted in Samuel Butler, *Essays on Life, Art and Science.* Edited by R. A. Streatfield. Port Washington, N.Y.: Kennikat Press.

Butts, Robert E. 1993. *Historical Pragmatics: Philosophical Essays.* Dordrecht: Kluwer Academic Publishers.

Call, Lewis. 1998. "Anti-Darwin, Anti-Spencer: Friedrich Nietzsche's Critique of Darwin and 'Darwinism.'" *History of Science* 36:1–22.

Camerini, Jane R. 1993. "Evolution, Biogeography, and Maps: An Early History of Wallace's Line." *Isis* 84:700–727.

————. 1996. "Wallace in the Field." *Osiris* ("Science in the Field," ed. Henrika Kuklick and Robert E. Kohler) 11:44–65.

————. 1997. "Remains of the Day: Early Victorians in the Field." In Lightman 1997, 354–377.

————, ed. 2002. *The Alfred Russel Wallace Reader: A Selection of Writings from the Field.* Baltimore and London: Johns Hopkins University Press.

Campbell, Bernard, ed. 1972. *Sexual Selection and the Descent of Man, 1871–1971.* London: Heinemann.

Campbell, R. J. 1916. "The Optimist in Science and in Life." Review of *Alfred Russel Wallace: Letters and Reminiscences,* by James Marchant. *Evening Standard and St. James's Gazette,* 28 April.

Cannon, Susan F. 1978. *Science in Culture: The Early Victorian Period.* New York: Dawson.

Carpenter, Edward, ed. 1897. *Forecasts of the Coming Century by a Decade of Writers.* Manchester: Labour Press, Ltd.

Carpenter, William Benjamin. 1871. "Spiritualism and Its Recent Converts." *Quarterly Review* 131:301–353.

————. 1877a. *Mesmerism, Spiritualism, &c., Historically and Scientifically Considered: Being Two Lectures Delivered at the London Institution.* New York: D. Appleton.

————. 1877b. "Psychological Curiosities of Spiritualism." *Fraser's Magazine*, n.s., 16:541–564.

Cerullo, John J. 1982. *The Secularization of the Soul: Psychical Research in Modern Britain.* Philadelphia: Institute for the Study of Human Issues.

Chesterton, G. K. 1904. "Alfred Russel Wallace." *English Illustrated Magazine* (London), n.s., 30:420–422.

Chipman, Robert A. 1973. "Henry Charles Fleeming Jenkin." In Gillispie 1970–1980, 8:93.

Claeys, Gregory. 2000. "The 'Survival of the Fittest' and the Origins of Social Darwinism." *Journal of the History of Ideas* 61:223–240.

Clements, Harry. 1983. *Alfred Russel Wallace: Biologist and Social Reformer.* London: Hutchinson.

Clerke, Agnes M. 1903. "Man's Place in the Universe." *Knowledge* 26 (211):108.

Clode, D., and R. O'Brien. 2001. "Why Wallace Drew the Line: A Re-Analysis of Wallace's Bird Collections in the Malay Archipelago and the Origins of Biogeography." In *Faunal and Floral Migrations and Evolution in SE Asia-Australia*, ed. Ian Metcalfe et al., 113–121. Lisse, Netherlands: A. A. Balkema.

Cockerell, T. D. A. 1890. "Some Notes on Dr. A. R. Wallace's 'Darwinism.'" *Nature* 41 (1061):393–394.

Cohen, I. Bernard, ed. 1994. *The Natural Sciences and the Social Sciences: Some Critical and Historical Perspectives.* Dordrecht: Kluwer Academic.

Coleman, William. 1999. *The Strange "Laissez Faire" of Alfred Russel Wallace: The Connection between Natural Selection and Political Economy Reconsidered.* Discussion Paper 1999–04. Hobart: University of Tasmania School of Economics. Reprinted in *Darwinism and Evolutionary Economics*, ed. John Laurent and John Nightingale. Cheltenham and Northampton, Mass.: E. Elgar, 2001.

Collini, Stefan. 1991. *Public Moralists: Political Thought and Intellectual Life in Britain, 1850–1930.* Oxford: Clarendon Press.

Colp, Ralph, Jr. 1976. "The Contacts of Charles Darwin with Edward Aveling and Karl Marx." *Annals of Science* 33:387–394.

————. 1982. "The Myth of the Darwin-Marx Letter." *History of Political Economy* 14:461–482.

————. 1992. "'I Will Gladly Do My Best': How Charles Darwin Obtained a Civil List Pension for Alfred Russel Wallace." *Isis* 83:3–26.

Commager, Henry Steele, ed. 1967. *Lester Ward and the Welfare State.* Indianapolis and New York: Bobbs-Merrill Co.

Cooter, Roger. 1984. *The Cultural Meaning of Popular Science: Phrenology and the Organization of Consent in Nineteenth-Century Britain.* Cambridge: Cambridge University Press.

Cooter, Roger, and S. Pumfrey. 1994. "Separate Spheres and Public Places: Reflections on the History of Science Popularization and Science in Public Culture." *History of Science* 32:237–267.

Corsi, Pietro. 1988. *Science and Religion: Baden Powell and the Anglican Debate, 1800–1860.* Cambridge: Cambridge University Press.

Croce, Paul Jerome. 1995. "William James's Scientific Education." *History of the Human Sciences* 8:9–27.

Croll, James. 1875. *Climate and Time in Their Geological Relations.* London: Daldy, Ibister.

Cronin, Helena. 1991. *The Ant and the Peacock: Altruism and Sexual Selection from Darwin to Today.* Cambridge: Cambridge University Press.

Crookes, William. 1877a. "Another Lesson from the Radiometer." *Nineteenth Century* 1:879–887.

———. 1877b. "Letters to the Editor." *Nature* 17:7–8.

d'A Jones, Peter. 1968. *The Christian Socialist Revival, 1877–1914: Religion, Class, and Social Conscience in Late-Victorian England.* Princeton, N.J.: Princeton University Press.

D'Albe, E. E. Fournier. 1923. *The Life of Sir William Crookes, O.M., F.R.S.* London: T. Fisher Unwin.

Dana, James Dwight. 1863. "Evidence as to Man's Place in Nature." *American Journal of Science* 35:451–454.

Darwin, Charles. [1859] 1964. *On the Origin of Species by Means of Natural Selection, or the Preservation of Favoured Races in the Struggle for Life.* Facsimile reprint ed. Cambridge, Mass.: Harvard University Press.

———. [1871] 1874. *The Descent of Man, and Selection in Relation to Sex.* 2d ed. New York: A. L. Burt.

Darwin, Charles, and Alfred Russel Wallace. 1858. "On the Tendency of Species to Form Varieties; and on the Perpetuation of Varieties and Species by Natural Means of Selection." *Journal of the Linnean Society of London, Zoology* 3:45–62.

Darwin, Francis, ed. 1887. *The Life and Letters of Charles Darwin, including an Autobiographical Chapter.* 3 vols. London: John Murray.

Darwin, Francis, and A. C. Seward, eds. 1903. *More Letters of Charles Darwin: A Record of His Work in a Series of Hitherto Unpublished Letters.* 2 vols. London: John Murray.

Daston, Lorraine, and Peter Galison. 1992. "The Image of Objectivity." *Representations*, no. 40, 81–128.

Daston, Lorraine, and Katharine Park. 1998. *Wonder and the Order of Nature, 1150–1750.* New York: Zone Books.

Davies, Gordon L. 1969. *The Earth in Decay: A History of British Geomorphology, 1578–1878.* New York: American Elsevier Publishing.

Daws, Gavan, and Marty Fujita. 1999. *Archipelago: The Islands of Indonesia, from the Nineteenth-Century Discoveries of Alfred Russel Wallace to the Fate of Forests and Reefs in the Twenty-First Century.* Berkeley: University of California Press and Nature Conservancy.

Dawson, Albert. 1903. "A Visit to Dr. Alfred Russel Wallace." *Christian Commonwealth* 23, no. 1156 (10 December 1903): 176a–177d.

De Brath, Stanley. 1921. *Psychic Philosophy: With Introductory Note by Alfred Russel Wallace.* 3d ed. Huddersfield: Spiritualists' National Union, Ltd.

DeCarvalho, Roy J. 1988–1989. "Methods and Manifestations: The Wallace-Carpenter Debate over Spiritualism." Pts. 1, 2. *Journal of Religion and Psychical Research* 11 (4):183–194, 12 (1):20–25.

De Morgan, Sophia E. 1863. *From Matter to Spirit: The Result of Ten Years' Experience in Spirit Manifestations.* By C. D.; preface by A. B. [Augustus De Morgan]. London: Longman, Green, Longman, Roberts, & Green.

Desertis, V. C. 1901. *Psychic Philosophy as the Foundation of a Religion of Natural Law.* 2d ed. London: P. Wellby.

Desmond, Adrian. 1989. *The Politics of Evolution: Morphology, Medicine, and Reform in Radical London.* Chicago: University of Chicago Press.

———. 1998. *Huxley: From Devil's Disciple to Evolution's High Priest.* New York: Penguin Books.

———. 2001. "Redefining the X Axis: 'Professionals,' 'Amateurs,' and the Making of Mid-Victorian Biology—a Progress Report." *Journal of the History of Biology* 34:3–50.

Desmond, Adrian, and James Moore. 1991. *Darwin: The Life of a Tormented Evolutionist.* New York: Warner Books.

Deynoux, M., ed. 1994. *Earth's Glacial Record.* Cambridge: Cambridge University Press.

Dick, Steven J. 1982. *Plurality of Worlds: The Origins of the Extraterrestrial Life Debate from Democritus to Kant.* Cambridge: Cambridge University Press.

———. 1996. *The Biological Universe: The Twentieth-Century Extraterrestrial Life Debate and the Limits of Science.* Cambridge: Cambridge University Press.

Dickenson, John. 1996. "Bates, Wallace and Economic Botany in Mid-Nineteenth-Century Amazonia." In *Richard Spruce (1817–1893): Botanist and Explorer,* ed. M. R. D. Seaward and S. M. D. Fitzgerald, 65–80. Kew: Royal Botanic Gardens.

Doyle, Arthur Conan. [1926] 1975. *The History of Spiritualism.* 2 vols. Facsimile ed. New York: Arno Press.

Drees, W. 1995. *Religion, Science and Naturalism.* Cambridge: Cambridge University Press.

Dunbar, G. S. 1975. "Elisee Reclus." In Gillispie 1970–1980, 11:337–338.

Durant, John R. 1979. "Scientific Naturalism and Social Reform in the Thought of Alfred Russel Wallace." *British Journal for the History of Science* 12:31–58.

———. 1985. "The Ascent of Nature in Darwin's *Descent of Man.*" In Kohn 1985, 283–306.

Eisele, Carolyn. 1974. "Charles Sanders Peirce." In Gillispie 1970–1980, 10:482–486.

Ellegard, Alvar. 1958. *Darwin and the General Reader: The Reception of Darwin's Theory of Evolution in the British Periodical Press, 1859–1872.* Goteborg: Elanders.

England, Richard. 1997. "Natural Selection before the *Origin:* Public Reactions of Some Naturalists to the Darwin-Wallace Papers (Thomas Boyd, Arthur Hussey, and Henry Baker Tristram)." *Journal of the History of Biology* 30:267–290.

Esposito, Joseph. 1980. *Evolutionary Metaphysics.* Athens: Ohio University Press.

Fabian, A. C., ed. 1998. *Evolution: Society, Science and the Universe.* Darwin College Lectures. Cambridge: Cambridge University Press.

Fabian Society. 1889. *Fabian Essays in Socialism.* London: Fabian Society.

Fancher, Raymond. 1998. "Biography and Psychodynamic Theory: Some Lessons from the Life of Francis Galton." *History of Psychology* 1:99–115.

Farber, Paul L. 1976. "The Type-Concept in Zoology during the First Half of the Nineteenth Century." *Journal of the History of Biology* 9 (1):93–119.

———. 1994. *The Temptations of Evolutionary Ethics.* Berkeley: University of California Press.

Ferri, Enrico. 1905. *Socialism and Positive Science (Darwin-Spencer-Marx).* London: Independent Labour Party.

Fichman, Martin. 1977. "Wallace: Zoogeography and the Problem of Land Bridges." *Journal of the History of Biology* 10:45–63.

———. 1981. *Alfred Russel Wallace.* Boston: Twayne Publishers.

———. 1984. "Ideological Factors in the Dissemination of Darwinism in England, 1860–1900." In *Transformation and Tradition in the Sciences: Essays in Honor of*

I. Bernard Cohen, ed. Everett Mendelsohn, 471–485. Cambridge: Cambridge University Press.

————. 1997. "Biology and Politics: Defining the Boundaries." In Lightman 1997, 94–118.

————. 2001a. "Alfred Russel Wallace's North American Tour: Transatlantic Evolutionism." *Endeavour* 25 (2):74–78.

————. 2001b. "Science in Theistic Contexts: A Case Study of Alfred Russel Wallace." *Osiris* 16:227–250.

————. 2002. *Evolutionary Theory and Victorian Culture.* Amherst, N.Y.: Humanity Books/Prometheus Books.

Fisch, Max H. 1964. "Was There a Metaphysical Club in Cambridge?" In *Studies in the Philosophy of Charles Sanders Peirce,* ed. Edward Moore and Richard Robin. Amherst: University of Massachusetts Press.

Flint, Robert. 1877. *Theism: Being the Baird Lecture for 1876.* Edinburgh: W. Blackwood.

————. 1888. "Theism." In *Encyclopedia Britannica,* vol. 23, 9th ed.

Foley, Robert. 1995. *Humans before Humanity.* Oxford and Cambridge, Mass.: Blackwell Publishers.

Forbes, Edward, 1846. "On the Connection between the Distribution of the Existing Fauna and Flora of the British Isles, and the Geological Changes Which Have Affected Their Area, Especially during he Epoch of the Northern Drift." *Memoirs of the Geological Survey of Great Britain and of the Museum of Economic Geology in London* 1:336–432.

————. 1851–1854. "On the Manifestation of Polarity in the Distribution of Organized Beings in Time." *Notices of the Proceedings at the Meetings of the Members of the Royal Institution* 1: 428–433.

Forsdyke, Donald R. 1999. "The Origin of Species Revisited." *Queen's Quarterly* 106 (1):112–132.

Frankel, Henry. 1981. "The Paleobiogeographical Debate over the Problem of Disjunctively Distributed Life Forms." *Studies in the History and Philosophy of Science* 12:221–259.

————. 1984. "Biogeography before and after the Rise of Seafloor Spreading." *Studies in the History and Philosophy of Science* 15:141–167.

————. 1987. "The Continental Drift Debate." In *Scientific Controversies: Case Studies in the Resolution and Closure of Disputes in Science and Technology,* ed. H. Tristram Engelhardt, Jr., and Arthur L. Caplan, 203–248. Cambridge: Cambridge University Press.

Freeden, M. 1978. *The New Liberalism.* Oxford: Clarendon Press.

Fuller, Steve. 1995. "A Tale of Two Cultures and Other Higher Superstitions." *History of the Human Sciences* 8:115–125.

Gaffney, Mason. 1997. "Alfred Russel Wallace's Campaign to Nationalize Land: How Darwin's Peer Learned from John Stuart Mill and Became Henry George's Ally." *American Journal of Economics and Sociology* 56:609–615.

Galison, Peter, and David J. Stump, eds. 1996. *The Disunity of Science: Boundaries, Contexts, and Power.* Stanford, Calif.: Stanford University Press.

Gardiner, Brian G. 1995. "The Joint Essay of Darwin and Wallace." *Linnean* 11:13–24.

————. 2000. "Wallace and Land Nationalization." *Linnean* 16:15–18.

Garwood, Christine, 2001. "Alfred Russel Wallace and the Flat Earth Controversy." *Endeavour* 25:139–143.

Gates, Barbara T., and Ann B. Shteir, eds. 1997. *Natural Eloquence: Women Reinscribe Science.* Madison: University of Wisconsin Press.

Gauld, Alan. 1968. *The Founders of Psychical Research.* London: Routledge & Kegan Paul.

Gavin, William Joseph, 1992. *William James and the Reinstatement of the Vague.* Philadelphia: Temple University Press.

George, Henry. [1879] 1938. *Progress and Poverty.* New York: Modern Library.

George, Henry, Jr. 1906–1911. *The Life of Henry George.* In *The Complete Works of Henry George.* 10 vols. Garden City, N.Y.: Fels Fund Library Edition.

George, Wilma. 1964. *Biologist Philosopher: A Study of the Life and Writings of Alfred Russel Wallace.* London: Abelard-Schuman.

———. 1971. "The Reaction of A. R. Wallace to the Work of Gregor Mendel." *Folio Mendeliana,* no. 6, 173–177.

———. 1979. "Alfred Wallace, the Gentle Trader: Collecting the Amazonia and the Malay Archipelago, 1848–1862." *Journal of the Society for the Bibliography of Natural History* 9:503–514.

———. 1981. "Wallace and His Line." In *Wallace's Line and Plate Tectonics,* ed. T. C. Whitmore. Oxford: Oxford University Press.

Ghiselin, Michael T. 1969. *The Triumph of the Darwinian Method.* Berkeley and Los Angeles: University Of California Press.

———. 1993. "Alfred Russel Wallace and the Birth of Biogeography: The Living Evidence for Evolution." *Pacific Discovery* 46 (2):17–23.

Gieryn, Thomas F. 1999. *Cultural Boundaries of Science: Credibility on the Line.* Chicago: University of Chicago Press.

Gillispie, Charles Coulston. [1951] 1959. *Genesis and Geology: A Study in the Relations of Scientific Thought, Natural Theology, and Social Opinion in Great Britain, 1790–1850.* Cambridge, Mass.: Harvard University Press; reprint, New York: Harper Torchbooks.

———, ed. 1970–1980. *Dictionary of Scientific Biography.* 16 vols. New York: Charles Scribner's Sons.

Golinski, Jan. 1990. "The Theory of Practice and the Practice of Theory: Sociological Approaches in the History of Science." *Isis* 81:492–505.

———. 1992. *Science as Public Culture: Chemistry and the Enlightenment in Britain, 1760–1820.* Cambridge: Cambridge University Press.

———. 1998. *Making Natural Knowledge: Constructivism and the History of Science.* Cambridge: Cambridge University Press.

Gordon, Scott. 1989. "Darwin and Political Economy: The Connection Reconsidered." *Journal of the History of Biology* 22 (3):437–459.

Gotthelf, Allan. 1999. "Darwin on Aristotle." *Journal of the History of Biology* 32:3–30.

Gould, Peter C. 1988. *Early Green Politics: Back to Nature, Back to the Land and Socialism in Britain, 1880–1900.* Brighton: Harvester.

Gray, Asa. 1876. *Darwiniana: Essays and Reviews Pertaining to Darwinism.* New York: D. Appleton.

———. 1880. *Natural Science and Religion: Two Lectures Delivered to the Theological School of Yale College.* New York: Charles Scribner's Sons.

Green, Len. 1995. *Alfred Russel Wallace: His Life and Work.* Occasional Paper No. 4. Hertford: Hertford and Ware Local History Society.

Greene, Mott. T. 1982. *Geology in the Nineteenth Century: Changing Views of a Changing World.* Ithaca, N.Y.: Cornell University Press.

Griffin, David Ray. 1989. *God and Religion in the Postmodern World: Essays in Postmodern Theology.* Albany: State University of New York Press.

Groeben, Christiane, ed. 1982. *Charles Darwin—Anton Dohrn Correspondence.* Naples: Macchiaroli.

Hall, Trevor H. 1963. *The Spiritualists: The Story of Florence Cook and William Crookes.* New York: Helix Press, Garrett Publications.

Hallam, A. 1983. *Great Geological Controversies.* Oxford: Oxford University Press.

Haller, John S., Jr. 1971. *Outcasts from Evolution: Scientific Attitudes of Racial Inferiority, 1859–1900.* Urbana: University of Illinois Press.

Harding, Sandra. 1998. *Is Science Multicultural? Postcolonialisms, Feminisms, and Epistemologies.* Bloomington: Indiana University Press.

Harrison, Frederic. 1889. "The Future of Agnosticism." *Fortnightly Review* 45:144–156.

Harrison, John F. C. 1969. *Robert Owen and the Owenites in Britain and America: The Quest for the New Moral World.* London: Routledge & Kegan Paul.

———. 1991. *Late Victorian Britain, 1875–1901.* London: Routledge.

Hartman, H. 1990. "The Evolution of Natural Selection: Darwin versus Wallace." *Perspectives in Biology and Medicine* 34 (1):78–88.

Hartshorne, Charles, and Paul Weiss, eds. 1958–1965. *Collected Papers of Charles Sanders Peirce.* 8 vols. Cambridge, Mass.: Harvard University Press.

Hausman, Carl. R. 1993. *Charles S. Peirce's Evolutionary Philosophy.* Cambridge: Cambridge University Press.

Hawkins, Mike. 1997. *Social Darwinism in European and American Thought, 1860–1945: Nature as Model and Nature as Threat.* Cambridge: Cambridge University Press.

Haynes, Renee. 1982. *The Society for Psychical Research, 1882–1982: A History.* London: Macdonald.

Heffernan, William C. 1978. "The Singularity of Our Inhabited World: William Whewell and A. R. Wallace in Dissent." *Journal of the History of Ideas* 39 (1):81–100.

———. 1981. "Percival Lowell and the Debate over Extraterrestrial Life." *Journal of the History of Ideas* 42 (3):527–530.

Helfand, Michael, S. 1977. "T. H. Huxley's 'Evolution and Ethics': The Politics of Evolution and the Evolution of Politics." *Victorian Studies* 20:159–177.

Henderson, Archibald, ed. 1911. *George Bernard Shaw: His Life and Works.* London: Hurst & Blackett.

Hennis, Wilhelm. 1998. "The Spiritualist Foundation of Max Weber's 'Interpretative Sociology': Ernst Troeltsch, Max Weber, and William James' *Varieties of Religious Experience.*" *History of the Human Sciences* 11:83–106.

Herschell, Farrer Herschell, chairman. 1890. *Third Report of the Royal Commission Appointed to Inquire into the Subject of Vaccination: With Minutes of Evidence and Appendices.* 24 December. London: Printed for Her Majesty's Stationary Office by Eyre & Spottiswoode.

Hick, John H. 1990. *Philosophy of Religion.* 4th ed. Englewood, N.J.: Prentice-Hall.

Hobhouse, L. T. 1911. *Social Evolution and Political Theory.* New York: Columbia University Press.

Hobson, John A. 1891. *Problems of Poverty: An Inquiry into the Industrial Condition of the Poor.* London: Methuen & Co.

———. 1897. "The Influence of Henry George in England." *Fortnightly Review* 68 (December): 835–844.

————. [1904] 1988. "Herbert Spencer." In *J. A. Hobson: A Reader*. Edited by M. Freeden. London: Unwin Hymen.

Hodge, M. J. S. 1972. "The Universal Gestation of Nature: Chambers' *Vestiges* and *Explanations*." *Journal of the History of Biology* 5:127–151.

————. 1989. "Darwin's Theory and Darwin's Argument." In *What the Philosophy of Biology Is*, ed. Michael Ruse, 163–182. Dordrecht: Kluwer.

Hogben, Lancelot T. 1918. *Alfred Russel Wallace: The Story of a Great Discoverer*. London: Society for Promoting Christian Knowledge.

Home, Mme. Dunglas. [1888] 1976. *D. D. Home: His Life and Mission*. New York: Arno Press.

Hooker, Joseph Dalton. 1853. *The Botany of the Antarctic Voyage of H.M. Discovery Ships "Erebus" and "Terror" in the Years 1839–1843*. Vol. 2, *Flora Novae-Zelandiae*. Pt. 1, *Flowering Plants*. London: Lovell Reeve.

————. 1867. "Insular Floras." *Gardeners' Chronicle and Agricultural Gazette* (January 5), 1–76.

Houston, Amy. 1997. "Conrad and Alfred Russel Wallace." In *Conrad: Intertexts and Appropriations: Essays in Memory of Yves Hervouet*, ed. Gene M. Moore et al., 29–48. Amsterdam and Atlanta: Rodopi.

Hubrecht, A. A. W. 1908. "Darwinism *versus* Wallaceism." *Contemporary Review* 94:629–634.

Hudson, Thomson Jay. 1896. *A Scientific Demonstration of the Future Life*. London: Putnam.

Hughes, R. Elwyn. 1989. "Alfred Russel Wallace: Some Notes on the Welsh Connection." *British Journal for the History of Science* 22 (4):401–418.

————. 1991. "Alfred Russel Wallace (1823–1913): The Making of a Scientific Nonconformist." *Proceedings of the Royal Institution of Great Britain* 63:175–183.

Hull, David L. 1973. *Darwin and His Critics: The Reception of Darwin's Theory of Evolution by the Scientific Community*. Cambridge, Mass.: Harvard University Press.

Huxley, Julian S. 1942. *Evolution: The Modern Synthesis*. London: Allen & Unwin.

Huxley, Leonard, ed. 1918. *Life and Letters of Sir Joseph Dalton Hooker*. 2 vols. London: J. Murray.

Huxley, Thomas Henry, 1863. *Evidence as to Man's Place in Nature*. London: Williams & Norgate.

————. 1877. *American Addresses*. London: Macmillan & Co.

————. 1888. "The Struggle for Existence: A Programme." *Nineteenth Century* 23:161–180.

————. 1889a. "Mr. Spencer on the Land Question." *Times* (London), 12 November 1889.

————. 1889b. "The Ownership of the Land." *Times* (London), 21 November 1889.

————. 1891. *Social Diseases and Worse Remedies*. London: Macmillan & Co.

Hyman, Ray. 1989. *The Elusive Quarry: A Scientific Appraisal of Psychical Research*. Buffalo: Prometheus Books.

Inkster, I., and J. Morrell, eds. 1983. *Metropolis and Province: Science in British Culture, 1780–1850*. London: Hutchinson.

[James, William]. 1865. "Alfred Russel Wallace's 'The Origin of Human Races and the Antiquity of Man Deduced from the Theory of 'Natural Selection.'" *North American Review* 101:261–263.

————. 1893. Review of "Frederick W. H. Myers, *Science and a Future Life.*" *Nation* 57:176–177.

————. 1986. *Essays in Psychical Research.* Edited by Frederick H. Burkhardt et al. Cambridge, Mass.: Harvard University Press.

Jann, Rosemary. 1994. "Darwin and the Anthropologists: Sexual Selection and Its Discontents." *Victorian Studies* 37:287–306.

Jardine, Nicholas, James Secord, and Emma Spary, eds. 1996. *Cultures of Natural History.* Cambridge: Cambridge University Press.

Jenkins, Roy. 1986. *Asquith.* 3d ed. London: Collins.

Jensen, J. Vernon. 1988. "Return to the Wilberforce-Huxley Debate." *British Journal for the History of Science* 21:161–179.

Jevons, William Stanley. 1878. *Political Economy.* London: Macmillan & Co.

Johnston, Timothy D. 1995. "The Influence of Weismann's Germ-Plasm Theory on the Distinction between Learned and Innate Behavior." *Journal of the History of the Behavioral Sciences* 31:115–128.

Jones, Greta. 1980. *Social Darwinism and English Thought: The Interaction between Biological and Social Theory.* Sussex: Harvester Press.

————. 2002. "Alfred Russel Wallace, Robert Owen and the Theory of Natural Selection." *British Journal for the History of Science* 35:73–96.

Jones, Lamar B. 1994. "T. H. Huxley's Critique of Henry George: An Expanded Perspective." *American Journal of Economics and Sociology* 53 (2):245–255.

Jordanova, Ludmilla. 1989. *Sexual Visions: Images of Gender in Science and Medicine between the Eighteenth and Twentieth Centuries.* New York: Harvester Wheatsheaf.

Jukes-Brown, A. J. 1892. "Evolution of Oceans and Continents." *Natural Science* 1:508–513.

Jung, Carl G. 1921. "The Psychological Foundations of Belief in Spirits." *Proceedings of the Society for Psychical Research* 31:75–93.

Kain, Roger J. P., and Hugh C. Prince. 1985. *The Tithe Surveys of England and Wales.* Cambridge: Cambridge University Press.

Keller, Evelyn Fox, and Elisabeth A. Lloyd, eds. 1992. *Keywords in Evolutionary Biology.* Cambridge, Mass.: Harvard University Press.

Kevin, James J., Jr. 1985. " 'Man's Place in the Universe': Alfred Russel Wallace, Teleological Evolution, and the Question of Extraterrestrial Life." M.A. thesis. University of Notre Dame.

Keynes, R. D., ed. 1988. *Charles Darwin's "Beagle Diary."* Cambridge: Cambridge University Press.

Kleiner, Scott A. 1985. "Darwin's and Wallace's Revolutionary Research Programme." *British Journal for the Philosophy of Science* 36 (4):367–392.

Knapp, Sandra. 1999. *Footsteps in the Forest: Alfred Russel Wallace in the Amazon.* London: Natural History Museum.

Knapp, Sandra, Lynn Sanders, and William Baker. 2002. "Alfred Russel Wallace and the Palms of the Amazon." *Palms* 46 (3):109–119.

Koch-Weser, Suzanne. 1977. "Alfred Russel Wallace on Man." *Synthesis* (Harvard University) 3 (4):5–32.

Kohn, David, ed. 1985. *The Darwinian Heritage.* Princeton, N.J.: Princeton University Press.

————. 1989. "Darwin's Ambiguity: The Secularization of Biological Meaning." *British Journal for the History of Science* 22:215–239.

Kottler, Malcolm Jay. 1974. "Alfred Russel Wallace, the Origins of Man, and Spiritualism." *Isis* 65:145–192.

———. 1980. "Darwin, Wallace, and the Origin of Sexual Dimorphism." *Proceedings of the American Philosophical Society* 124 (3):203–226.

———. 1985. "Charles Darwin and Alfred Russel Wallace: Two Decades of Debate over Natural Selection." In Kohn 1985, 367–432.

Kropotkin, Peter. [1898] 1912. *Fields, Factories and Workshops; or, Industry Combined with Agriculture and Brain Work with Manual Work.* London: Thomas Nelson & Sons.

———. [1902] 1972. *Mutual Aid: A Factor in Evolution.* Edited by Paul Avrich. New York: New York University Press.

Kuklick, Bruce. 1977. *The Rise of American Philosophy: Cambridge, Massachusetts, 1860–1930.* New Haven, Conn.: Yale University Press.

Kuklick, Henrika, and Robert E. Kohler, eds. 1996. *Science in the Field.* Osiris, vol. 11. Chicago: University of Chicago Press.

Lankester, E. Ray. 1890. *The Advancement of Science.* London: Macmillan & Co.

Larson, James. 1986. "Not without a Plan: Geography and Natural History in the Late Eighteenth Century." *Journal of the History of Biology* 19:447–488.

Latour, Bruno. 1992. "One More Turn after the Social Turn . . ." In *The Social Dimensions of Science,* ed. Ernan McMullin, 272–294. Notre Dame, Ind.: University of Notre Dame Press.

Laudan, Rachel. 1987. *From Mineralogy to Geology: The Foundations of a Science. 1650–1830.* Chicago: University of Chicago Press.

———. 1993. "Histories of the Sciences and Their Uses: A Review to 1913." *History of Science* 31:1–34.

Laybourn, Keith. 1997. *The Rise of Socialism in Britain, c. 1881–1951.* Thrupp, Gloucestershire: Sutton Publishing.

LeConte, Joseph. 1882. *Elements of Geology.* Rev. ed. New York: D. Appleton & Co.

———. 1903. *The Autobiography of Joseph LeConte.* Edited by William Dallam Armes. New York: D. Appleton & Co.

Leftow, Brian. 1998. "God, Concepts of." In *Routledge Encyclopedia of Philosophy,* ed. Edward Craig, 4:98–100. London and New York: Routledge.

Le Grand, H. E. 1988. *Drifting Continents and Shifting Theories.* Cambridge: Cambridge University Press.

Lenoir, Timothy, ed. 1998. *Inscribing Science: Scientific Texts and the Materiality of Communication.* Stanford, Calif.: Stanford University Press.

Lesch, John E. 1975. "George John Romanes." In Gillispie 1970–1980, 11:516–520.

Levinson, Henry Samuel. 1981. *The Religious Investigations of William James.* Chapel Hill: University of North Carolina Press.

Lightman, Bernard. 1987. *The Origins of Agnosticism: Victorian Unbelief and the Limits of Knowledge.* Baltimore: Johns Hopkins University Press.

———, ed. 1997. *Victorian Science in Context.* Chicago: University of Chicago Press.

———. 2000. "The Visual Theology of Victorian Popularizers of Science: From Reverent Eye to Chemical Retina." *Isis* 91:651–680.

———. 2001. "Victorian Sciences and Religions: Discordant Harmonies." *Osiris* 16:343–366.

Lindberg, David C., and Ronald L. Numbers, eds. 1986. *God and Nature: Historical Essays on the Encounter between Christianity and Natural Science.* Berkeley: University of California Press.

Linnean Society of London. 1908. *The Darwin-Wallace Celebration Held on Thursday, 1ˢᵗ July, 1908, by the Linnean Society of London.* London: Burlington House, printed for the Linnean Society.

Livingstone, David. N. 1987. *Darwin's Forgotten Defenders: The Encounter between Evangelical Theology and Evolutionary Thought.* Edinburgh: Scottish Academic Press.

Lodge, Oliver J. 1909. *Man and the Universe: A Study of the Influence of the Advance in Scientific Knowledge Upon Our Understanding of Christianity.* 5ᵗʰ ed. London: Methuen & Co.

————. 1910. *Reason and Belief.* New York: Moffat, Yard, & Co.

Lorimer, Douglas A. 1997. "Science and the Secularization of Victorian Images of Race." In Lightman 1997, 212–235.

Love, Rosaleen. 1983. "Darwinism and Feminism: The 'Woman Question' in the Life and Work of Olive Schreiner and Charlotte Perkins Gilman." In Oldroyd and Langham 1983, 113–131.

Lowell, Percival. 1906. *Mars and Its Canals.* New York: Macmillan.

Lyell, Sir Charles. 1830–1833. *Principles of Geology: Being an Attempt to Explain the Former Changes of the Earth's Surface, by Reference to Causes Now in Operation.* 3 vols. London: John Murray.

————. 1867–1868. *Principles of Geology.* 10th ed. London: John Murray, 2 vols.

————. 1873. *The Geological Evidences of the Antiquity of Man, with Remarks on Theories of the Origin of Species by Variation.* 4th ed., rev. London: John Murray.

————. 1881. *Life, Letters and Journals of Sir Charles Lyell, Bart.* Edited by Mrs. [Katherine M.] Lyell. 2 vols. London: John Murray.

Lynch, Michael, and Steve Woolgar, eds. 1990. *Representation in Scientific Practice.* Cambridge, Mass.: MIT Press.

MacKenzie, Donald A. 1981. *Statistics in Britain, 1865–1930: The Social Construction of Scientific Knowledge.* Edingburgh: Edinburgh University Press.

MacKenzie, Norman, and Jeanne MacKenzie. 1977. *The First Fabians.* London: Weidenfeld & Nicolson.

MacLeod, Roy. 1970. "Science and the Civil List, 1824–1914." *Technology and Society* 6:47–55.

Malinchak, Michele Daryl. 1987. "Spiritualism and the Philosophy of Alfred Russel Wallace." Ph.D. diss., Drew University.

Mandelbaum, Maurice. 1971. *History, Man, and Reason: A Study in Nineteenth-Century Thought.* Baltimore: Johns Hopkins University Press.

Mandler, Peter, Alex Owen, Seth Koven, and Susan Pedersen. 1997. "Cultural Histories Old and New: Rereading the Work of Janet Oppenheim." *Victorian Studies* 41:69–105.

Marchant, James. 1905. "A Man of the Time: Dr. Alfred Russel Wallace and His Coming Autobiography." *Book Monthly* 2:545–549.

————. [1916] 1975. *Alfred Russel Wallace: Letters and Reminiscences.* New York and London: Harper & Brothers. Reprint, New York: Arno Press.

Marsden, Brian G. 1973. "Percival Lowell." In Gillispie 1970–1980, 8:520–523.

Marsh, Jan. 1982. *Back to the Land: The Pastoral Impulse in England, from 1880 to 1914.* London: Quartet Books.

Maxwell, David. 1891. *Stepping-Stones to Socialism.* Hull: William Andrews & Co., Hull Press.

Mayr, Ernst. 1972. "Sexual Selection and Natural Selection." In Campbell 1972.

———. 1976. *Evolution and the Diversity of Life: Selected Essays.* Cambridge, Mass.: Belknap Press.

———. 1982. *The Growth of Biological Thought: Diversity, Evolution, and Inheritance.* Cambridge, Mass.: Harvard University Press, Belknap Press.

———. 1985. "Darwin's Five Theories of Evolution." In Kohn 1985, 755–772.

———. 1992. "The Idea of Teleology." *Journal of the History of Ideas* 53:117–135.

McCosh, James. 1890. *The Religious Aspect of Evolution.* Rev. ed. New York: Charles Scribner's Sons.

McElrone, Hugh P. 1886. Review of *Progress and Poverty,* by Henry George. *The Independent* (New York) 38 (1984):1583–1584.

McKinney, H. Lewis. 1969. "Wallace's Earliest Observations on Evolution: 28 December 1845." *Isis* 60:370–373.

———. 1970. "Henry Walter Bates." In Gillispie 1970–1980, 1:500.

———. 1972. *Wallace and Natural Selection.* New Haven, Conn., and London: Yale University Press.

Mello, J. M. 1911. *The Mystery of Life and Mind, with Special Reference to "The World of Life," by A. R. Wallace.* Warwick: Henry H. Lagy.

Menand, Louis. 2001. *The Metaphysical Club.* New York: Farrar, Straus & Giroux.

Millhauser, Milton. 1959. *Just before Darwin: Robert Chambers and Vestiges.* Middletown, Conn.: Wesleyan University Press.

Mills, Eric L. 1984. "A View of Edward Forbes, Naturalist." *Annals of Natural History* 11:365–393.

Milner, Richard. 1990. "Darwin for the Prosecution, Wallace for the Defense." Pt. 1, "How Two Great Naturalists Put the Supernatural on Trial." Pt. 2, "Spirit of a Dead Controversy." *North Country Naturalist* 2:19–37, 38–50.

———. 1996. "Charles Darwin and Associates, Ghostbusters." *Scientific American* 275 (4):96–101.

Minkin, L. 1991. *The Contentious Alliance: Trade Unions and the Labour Party.* Edinburgh: Edinburgh University Press.

Mivart, St. George. 1871. *On the Genesis of Species.* London: Macmillan & Co.

Moore, James R. 1979. *The Post-Darwinian Controversies: A Study of the Protestant Struggle to Come to Terms with Darwin in Great Britain and America, 1870–1900.* Cambridge: Cambridge University Press.

———, ed. 1989. *History, Humanity and Evolution: Essays for John C. Greene.* Cambridge: Cambridge University Press.

———. 1991. "Deconstructing Darwinism: The Politics of Evolution in the 1860s." *Journal of the History of Biology* 24:353–408.

———. 1997. "Wallace's Malthusian Moment: The Common Context Revisited." In Lightman 1997, 290–311.

Morris, William. 1903. *Communism: A Lecture.* [Edited by George Bernard Shaw.] Fabian Tract no. 113. London: Fabian Society.

Moses, William Stainton. 1894. *Spirit Teachings.* Memorial ed. London: London Spiritualist Alliance.

Mossner, Ernest Campbell. 1967. "Deism." In *The Encyclopedia of Philosophy,* ed. Paul Edwards, 326–336. New York: Macmillan Company & Free Press.

Murray, John, and Reverend A. F. Renard. 1891. *Report on Deep-Sea Deposits Based on the Specimens Collected during the Voyage of H.M.S. "Challenger."* London: Eyre & Spottiswoode.

Myers, Frederick W. H. 1895. "Resolute Credulity." *Proceedings of the Society for Psychical Research,* vol. 11 (July).

Myers, Gerald E. 1986. *William James: His Life and Thought.* New Haven, Conn., and London: Yale University Press.

Myers, Greg. 1990. *Writing Biology: Texts in the Social Construction of Scientific Knowledge.* Madison: University of Wisconsin Press.

Nelson, Geoffrey K. 1969. *Spiritualism and Society.* London: Routledge & Kegan Paul.

Nicholson, P. J. 1990. *The Political Philosophy of the British Idealists.* Cambridge: Cambridge University Press.

Noakes, R. J. 1998. "Cranks and Visionaries: Science, Spiritualism, and Transgression in Victorian England." Ph.D. diss., Cambridge University.

Nordhoff, Charles. 1875. *The Communistic Societies of the United States; From Personal Visit and Observation: Including Detailed Accounts of the Economists, Zoarites, Shakers, the Amana, Oneida, Bethel, Aurora, Icarian, and Other Existing Societies, Their Religious Creeds, Social Practices, Numbers, Industries, and Present Condition.* New York: Harper & Brothers.

Northrop, W. B. 1913. "Alfred Russel Wallace." (Interview/obituary.) *Outlook* (New York) 105:618–622.

Numbers, Ronald L. 1998. *Darwinism Comes to America.* Cambridge, Mass.: Harvard University Press.

Oates, David. 1988. "Social Darwinism and Natural Theodicy." *Zygon* 23 (4):439–454.

Offer, Avner. 1981. *Property and Politics, 1870–1914: Land Ownership, Law, Ideology, and Urban Development in England.* Cambridge: Cambridge University Press.

O'Hara, Robert J. 1991. "Representations of the Natural System in the Nineteenth Century." *Biology and Philosophy* 6 (2):255–274.

Oldroyd, David, and Ian Langham, eds. 1983. *The Wider Domain of Evolutionary Thought.* Dordrecht: D. Reidel Publishing.

Oppenheim, Janet. 1985. *The Other World: Spiritualism and Psychical Research in England, 1850–1914.* Cambridge: Cambridge University Press.

Oreskes, Naomi. 1999. *The Rejection of Continental Drift: Theory and Method in American Earth Science.* Oxford: Oxford University Press.

Osborn, Henry Fairfield. 1913. "Alfred Russel Wallace." *Popular Science Monthly* 83:523–537.

Ospovat, Dov. 1978. "Perfect Adaptation and Teleological Explanation: Approaches to the Problem of the History of Life in the Mid-Nineteenth Century." *Studies in the History of Biology* 2:33–56.

Otter, Sandra M. Den. 1996. *British Idealism and Social Explanation: A Study in Late Victorian Thought.* Oxford: Clarendon Press.

Owen, Alex. 1990. *The Darkened Room: Women, Power, and Spiritualism in Late Victorian England.* Philadelphia: University of Pennsylvania Press.

Owen, Robert Dale. 1861. *Footfalls on the Boundary of Another World.*

———. 1871. *The Debatable Land between This World and the Next: With Illustrative Narrations.* London: Trubner & Co.

Palfreman, Jon. 1976. "William Crookes: Spiritualism and Science." *Ethics in Science* 3 (4):211–227.

———. 1979. "Between Scepticism and Credulity: A Study of Victorian Scientific Attitudes to Modern Spiritualism." In *On the Margins of Science: The Social Construction of Rejected Knowledge,* ed. Roy Wallis, 201–236. Keele: University of Keele.

Paradis, James G. 1989. "*Evolution and Ethics* in its Victorian Context." In *Evolution and Ethics,* ed. James G. Paradis and George C. Williams, 3–55. Princeton, N.J.: Princeton University Press.

Parker, Kelly A. 1998. *The Continuity of Peirce's Thought.* Nashville, Tenn., and London: Vanderbilt University Press.

Parker, Percy L., ed. 1912. *Character and Life: A Symposium.* London: Williams & Norgate.

Paul, Diane B. 1988. "The Selection of the 'Survival of the Fittest.'" *Journal of the History of Biology* 21:411–425.

Pearson, M. B. 1995. "A. R. Wallace's Malay Archipelago Journals and Notebooks." *Linnean* 11 (2):12–13.

Pease, Edward R. 1916. *The History of the Fabian Society.* London.

Peck, Steven L. 2003. "Randomness, Contingency, and Faith: Is There a Science of Subjectivity?" *Zygon* 38 (1):5–23.

Peckham, Morse, ed. 1959. *The "Origin of Species" by Charles Darwin: A Variorum Text.* Philadelphia: University of Pennsylvania Press.

Peirce, Charles Sanders. [1903] 1997. *Pragmatism as a Principle and Method of Right Thinking: The 1903 Harvard Lectures on Pragmatism.* Edited by Patricia Ann Turrisi. Albany: State University of New York Press, 1997.

———. 1906. "Review of Wallace's *My Life.*" *Nation* 82:160–161.

Perkin, Harold J. 1989. *The Rise of Professional Society: England since 1880.* London: Routledge.

Pickering, Andrew. 1995. *The Mangle of Practice: Time, Agency, and Science.* Chicago: University of Chicago Press.

Pittenger, Mark. 1993. *American Socialists and Evolutionary Thought, 1870–1920.* Madison: University of Wisconsin Press.

Plowright, John. 1987. "Political Economy and Christian Polity: The Influence of Henry George in England Reassessed." *Victorian Studies* 30 (2):235–252.

Podmore, Frank. 1897. *Studies in Psychical Research.* London: G. P. Putnam's Sons.

———. [1902] 1963. *Modern Spiritualism: A History and a Criticism.* Reprinted as *Mediums of the 19th Century.* 2 Vols. New Hyde Park, N.Y.: University Books.

Porter, Charlotte M. 1995. "The History of Scientific Illustration." *Journal of the History of Biology* 28:545–550.

Porter, Duncan M. 1993. "On the Road to the *Origin* with Darwin, Hooker and Gray." *Journal of the History of Biology* 26:1–38.

Porter, Noah. 1886. "Some Thoughts upon Evolution." *Independent* (New York) 38, no. 1957 (June 3): 683–686.

Porter, Roy. 1997. *The Greatest Benefit to Mankind: A Medical History of Humanity from Antiquity to the Present.* London: HarperCollins Publishers.

Potter, Vincent G. 1996. *Peirce's Philosophical Perspectives.* Edited by Vincent M. Colapietro. New York: Fordham University Press.

Poulton, Edward B. 1890. *The Colours of Animals: Their Meaning and Use, Especially Considered in the Case of Insects.* London: Kegan Paul, Trench, Trubner, & Co.

———. 1923–1924. "Alfred Russel Wallace, 1823–1913." *Proceedings of the Royal Society of London,* ser. B, 95:i–xxxv.

Powell, Baden. 1855. *Essays on the Spirit of the Inductive Philosophy, the Unity of Worlds, and the Philosophy of Creation.* London: Longman.

Prance, Ghillean. 1999. "Alfred Russel Wallace." *Linnean* 15 (1):18–36.

Pratt, Mary Louise. 1992. *Imperial Eyes: Travel Writing and Transculturation.* New York: Routledge.

Prete, F. R. 1990. "The Conundrum of the Honey Bees: One Impediment to the Publication of Darwin's Theory." *Journal of the History of Biology* 23:271–290.

Provine, William B. 1985. "Adaptation and Mechanisms of Evolution after Darwin: A Study in Persistent Controversies." In Kohn 1985, 825–866.

Prynn, David. 1976. "The Clarion Clubs, Rambling and the Holiday Associations in Britain since the 1890s." *Journal of Contemporary History* 11: 65–77.

Pyenson, Lewis, and Susan Sheets-Pyenson. 1999. *Servants of Nature: A History of Scientific Institutions, Enterprises and Sensibilities.* New York and London: W. W. Norton & Co.

Quammen, David. 1997. *The Song of the Dodo: Island Biogeography in an Age of Extinctions.* New York: Touchstone Books.

Quinn, Philip L., and Charles Taliaferro, eds. 1997. *A Companion to Philosophy of Religion.* Oxford: Blackwell Publishers.

Raby, Peter. 1997. *Bright Paradise: Victorian Scientific Travellers.* London: Pimlico.

———. 2001. *Alfred Russel Wallace: A Life.* Princeton, N.J.: Princeton University Press.

Rainger, Ronald. 1991. *An Agenda for Antiquity: Henry Fairfield Osborn and Vertebrate Paleontology at the American Museum of Natural History, 1890–1935.* Tuscaloosa: University of Alabama Press.

Rann, Ernest H. 1909. "Dr. Alfred Russel Wallace at Home." *Pall Mall Magazine* 43:274–284.

Raposa, Michael L. 1989. *Peirce's Philosophy of Religion.* Bloomington: Indiana University Press.

Rayher, Edward S. 1996. "Confusion and Cohesion in Emerging Sciences: Darwin, Wallace, and Social Darwinism." Ph.D. diss., University of Massachusetts Amherst.

Rehbock, Philip F. 1983. *The Philosophical Naturalists: Themes in Early Nineteenth-Century British Biology.* Madison: University of Wisconsin Press.

Richards, Evelleen, 1983. "Darwin and the Descent of Woman." In Oldroyd and Langham 1983, 57–111.

———. 1989a. "Huxley and Woman's Place in Science: 'The Woman Question' and the Control of Victorian Anthropology." In Moore 1989, 253–284.

———. 1989b. "The 'Moral Anatomy' of Robert Knox: The Interplay between Biological and Social Thought in Victorian Scientific Naturalism." *Journal of the History of Biology* 22:373–436.

———. 1997. "Redrawing the Boundaries: Darwinian Science and Victorian Women Intellectuals." In Lightman 1997, 119–142.

Richards, Robert J. 1987. *Darwin and the Emergence of Evolutionary Theories of Mind and Behavior.* Chicago: University of Chicago Press.

———. 1992. *The Meaning of Evolution: The Morphological Construction and Ideological Reconstruction of Darwin's Theory.* Chicago: University of Chicago Press.

Ridley, Mark. 1982. "Coadaptation and the Inadequacy of Natural Selection." *British Journal for the History of Science* 15:45–68.

Ritchie, David George. 1889. *Darwinism and Politics.* London: Swan Sonnenschein & Co.

Ritvo, Harriet. 1997. "Zoological Nomenclature and the Empire of Victorian Science." In Lightman 1997, 334–353.

Robbins, Peter. 1982. *The British Hegelians, 1875–1925.* New York and London: Garland Publishing.

Roberts, Jon H. 1988. *Darwinism and the Divine in America: Protestant Intellectuals and Organic Evolution, 1859–1900*. Madison: University of Wisconsin Press.

Robinson, Gloria. 1976. "August Weismann." In Gillispie 1970–1980, 14:232–239.

Rodgers, Daniel T. 1998. *Atlantic Crossings: Social Politics in a Progressive Age*. Cambridge, Mass.: Harvard University Press, Belknap Press.

Romanes, George John. 1888. *Mental Evolution in Man: Origin of Human Faculty*. London: Kegan Paul, Trench & Co.

———. 1897. *Post-Darwinian Questions: Heredity and Utility*. Vol. 2 of *Darwin, and after Darwin*. Chicago: Open Court Publishing.

Ross, John M. 1989. "Alfred Russel Wallace: Theistic Darwinian." *King's Theological Review* 12 (Autumn): 46–48.

Royle, Edward. 1998. *Robert Owen and the Commencement of the Millennium: A Study of the Harmony Community*. Manchester and New York: Manchester University Press.

Rubino, Carl A. 1997. "Journeys, Maps, and Territories: Charting Uncertain Terrain in Science and Literature." *Intertexts* 1:118–130.

Rudwick, Martin J. S. 1982. "Charles Darwin in London: The Integration of Public and Private Science." *Isis* 73:186–206.

———. 1985. *The Great Devonian Controversy: The Shaping of Scientific Knowledge by Gentlemanly Scientists*. Chicago: University of Chicago Press.

Rupke, Nicolaas A. 1994. *Richard Owen: Victorian Naturalist*. New Haven, Conn.: Yale University Press.

———. 1996. "Eurocentric Ideology of Continental Drift." *History of Science* 34:251–272.

Ruse, Michael. 1974. "The Darwin Industry." *History of Science* 12:43–58.

———. 1979. *The Darwinian Revolution: Science Red in Tooth and Claw*. Chicago: University of Chicago Press.

———. 1998. "Darwinism and Atheism: Different Sides of the Same Coin." *Endeavour* 22:17–20.

———. 2000. "Teleology: Yesterday, Today, and Tomorrow?" *Studies in History and Philosophy of Biological and Biomedical Sciences* 31C (1):213–232.

Russett, Cynthia Eagle. 1989. *Sexual Science: The Victorian Construction of Womanhood*. Cambridge, Mass.: Harvard University Press.

Sarkar, Sahotra. 1998. "Wallace's Belated Revival." *Journal of Biosciences* 23 (1):3–7.

Scarpelli, Giacomo. 1992. " 'Nothing in Nature That Is Not Useful': The Anti-Vaccination Crusade and the Idea of *Harmonia Naturae* in Alfred Russel Wallace." *Nuncius: Annali di Storia Dell Scienza* 7:109–130.

Schabas, Margaret. 1997. "Victorian Economics and the Science of the Mind." In Lightman 1997, 72–93.

Schiebinger, Londa. 1993. *Nature's Body: Gender in the Making of Modern Science*. Boston: Beacon Press.

———. 1999. *Has Feminism Changed Science?* Cambridge, Mass.: Harvard University Press.

Schwartz, Joel S. 1984. "Darwin, Wallace, and the *Descent of Man*." *Journal of the History of Biology* 17:271–289.

———. 1990. "Darwin, Wallace, and Huxley, and Vestiges of the Natural History of Creation." *Journal of the History of Biology* 23:127–153.

———. 1995. "George John Romanes's Defense of Darwinism." *Journal of the History of Biology* 28:281–316.

Schwartzman, Jack. 1997. "The Death of Henry George: Scholar or Statesman?" *American Journal of Economics and Sociology* 56 (4):391–405.

Sclater, Philip Lutley. 1858. "On the General Distribution of the Members of the Class Aves." *Journal of the Proceedings of the Linnean Society of London (Zoology)* 2:130.

Sclater, William Lutley, and Philip Lutley Sclater. 1899. *The Geography of Mammals.* London: Kegan Paul, Trench, Trubner.

Secord, Anne. 1994a. "Corresponding Interests: Artisans and Gentlemen in Nineteenth-Century Natural History." *British Journal for the History of Science* 27:383–408.

————. 1994b. "Science in the Pub: Artisan Botanists in Early Nineteenth-Century Lancashire." *History of Science* 22:269–315.

Secord, James A. 1986. *Controversy in Victorian Geology: The Cambrian-Silurian Dispute.* Princeton, N.J.: Princeton University Press.

————. 1989. "Behind the Veil: Robert Chambers and *Vestiges.*" In Moore 1989, 165–194.

————. 1994. Introduction to *Vestiges of the Natural History of Creation and Other Evolutionary Writings,* by Robert Chambers. Edited by James A. Secord, ix–xlv. Chicago: University of Chicago Press.

————. 2000. *Victorian Sensation: The Extraordinary Publication, Reception, and Secret Authorship of "Vestiges of the Natural History of Creation."* Chicago and London: University of Chicago Press.

Shapin, Steven. 1992. "Discipline and Bounding: The History and Sociology of Science as Seen through the Externalism-Internalism Debate." *History of Science* 30:333–369.

————. 1994. *A Social History of Truth: Civility and Science in Seventeenth Century England.* Chicago: University of Chicago Press.

Shaw, George Bernard, ed. 1920. *Fabian Essays in Socialism.* London: Fabian Society and G. Allen & Unwin.

Sheets-Pyenson, Susan. 1989. *Cathedrals of Science: The Development of Colonial Natural History Museums in the Late Nineteenth Century.* Montreal: McGill-Queens University Press.

Shermer, Michael. 1994. "A Heretic-Scientist among the Spiritualists: Alfred Russel Wallace and 19th-Century Spiritualism—Part I." *Skeptic* 3 (1):70–83.

————. 1996. "Heretic-Personality: Alfred Russel Wallace and the Nature of Heretical Science." *Skeptic* 4 (3):84–93.

————. 2001. *The Borderlands of Science: Where Sense Meets Nonsense.* New York: Oxford University Press.

————. 2002. *In Darwin's Shadow: The Life and Science of Alfred Russel Wallace: A Biographical Study on the Psychology of History.* New York: Oxford University Press.

Short, T. L. 2002. "Darwin's Concept of Final Cause: Neither New nor Trivial." *Biology and Philosophy* 17:323–340.

Shortland, Michael, ed. 1996. *Hugh Miller and the Controversies of Victorian Science.* Oxford: Clarendon Press.

Shortland, Michael, and Richard Yeo, eds. 1996. *Telling Lives in Science: Essays on Scientific Biography.* New York: Cambridge University Press.

Siegfried, Charlene Haddock. 1990. *William James's Radical Reconstruction of Philosophy.* Albany: State University of New York Press.

Silagi, Michael. 1989. "Henry George and Europe: As Dissident Economist and Path-Breaking Philosopher, He Was a Catalyst for British Social Reform." *American Journal of Economics and Sociology* 48 (1):113–122.

Simon, Linda. 1998. *Genuine Reality: A Life of William James*. New York: Harcourt Brace & Co.

Skrupskelis, Ignas K., and Elizabeth M. Berkeley, eds. 1992–. *The Correspondence of William James*. Charlottesville: University Press of Virginia.

Smith, C. U. M. 1997. "Worlds in Collision: Owen and Huxley on the Brain." *Science in Context* 10:343–365.

Smith, Charles H., ed. 1991. *Alfred Russel Wallace: An Anthology of His Shorter Writings*. Oxford: Oxford University Press.

———. 1992. *Alfred Russel Wallace on Spiritualism, Man, and Evolution: An Analytical Essay*. Torrington, Conn.: C. H. Smith.

———. 1999. "Alfred Russel Wallace on Evolution: A Change of Mind?" Paper presented at the Symposium on the History of Medicine and Science, Hattiesburg Miss., 26–27 February.

———. 2002. "Wallace, Alfred Russel." In *Dictionary of Nineteenth-Century British Philosophers*. Edited by W. J. Mander and A. P. F. Sell, 2:1156–1160. Bristol: Thoemmes Press.

Smith, Crosbie. 1998. *The Science of Energy: A Cultural History of Energy Physics in Victorian Britain*. Chicago: University of Chicago Press.

Smith, Crosbie, and M. Norton Wise. 1989. *Energy and Empire: A Biographical Study of Lord Kelvin*. Cambridge: Cambridge University Press.

Smith, Robert W. 1999. "Essay Review: Martians and Other Aliens." *Studies in History and Philosophy of Biological and Biomedical Sciences* 30 (2):237–254.

Smith, Roger. 1972. "Alfred Russel Wallace: Philosophy of Nature and Man." *British Journal for the History of Science* 6:177–199.

Smocovitis, Vassiliki Betty. 1991. "Essay Review: The Politics of *Writing Biology*." *Journal of the History of Biology* 24:521–527.

Sober, Elliott. 1980. "Evolution, Population Thinking, and Essentialism." *Philosophy of Science* 47:350–383.

Soderqvist, Thomas. 1996. "Existential Projects and Existential Choice in Science: Science Biography as an Edifying Genre." In Shortland and Yeo 1996, 45–84.

Sowan, Paul W. 2001. "Alfred Russel Wallace and the Grass Roots of Natural History and Social Justice." *Linnean* 17 (1):17–18.

Spencer, Herbert. 1851. *Social Statics*. London: Chapman.

———. 1879. *The Data of Ethics*. London: Williams & Norgate.

———. 1884. *The Man versus the State*. London: Williams & Norgate.

———. 1904. *An Autobiography*. 2 vols. London: Williams & Norgate.

Stack, David A. 2000. "The First Darwinian Left: Radical and Socialist Responses to Darwin, 1859–1914." *History of Political Thought* 21 (4):682–710.

Stafford, Barbara Maria. 1984. *Voyage into Substance: Art, Science, Nature and the Illustrated Travel Account, 1760–1840*. Cambridge, Mass.: MIT Press.

Stafford, Robert. 1989. *Scientist of Empire: Sir Roderick Murchison, Scientific Exploration, and Victorian Imperialism*. Cambridge: Cambridge University Press.

Stamos, David N. 1999. "Darwin's Species Category Realism." *History and Philosophy of the Life Sciences* 21 (2):137–186.

Stape, J. H. 2000. "Wallace, Alfred Russel." In *Oxford Reader's Companion to Conrad*, ed. Owen Knowles and Gene M. Moore, 396–397. Oxford: Oxford University Press.

Stepan, Nancy Leys. 2001. *Picturing Tropical Nature*. Ithaca, N.Y.: Cornell University Press.

Stern, Bernard J. 1935. "Letters of Alfred Russel Wallace to Lester F. Ward." *Scientific Monthly* 40 (April): 375–379.

Stevens, P. F. 1995. "George Bentham and the Darwin/Wallace Papers of 1858: More Myths Surrounding the Origin and Acceptance of Evolutionary Ideas." *Linnean* 11 (2):14–16.

Stigler, George J. 1969. "Alfred Marshall's Lectures on Progress and Poverty." *Journal of Law and Economics* 12:184–226.

Still, Arthur. 1995. Introduction to special issue on William James. *History of the Human Sciences* 8:1–7.

Stocking, George W., Jr. 1962. "Lamarckianism in American Social Science: 1890–1915." *Journal of the History of Ideas* 23:239–56.

———. 1987. *Victorian Anthropology*. New York: Free Press.

Strauss, D. 1998. "Reflections on *Lowell and Mars*." *Annals of Science* 55:95–103.

Strauss, Sylvia. 1988. "Gender, Class, and Race in Utopia." In *Looking Backward, 1988–1888: Essays on Edward Bellamy*, ed. Daphne Patai, 68–90. Amherst: University of Massachusetts Press.

Strick, James. 1999. "Darwinism and the Origin of Life: The Role of H. C. Bastian in the British Spontaneous Generation Debates, 1868–1873." *Journal of the History of Biology* 32:51–92.

———. 2000. *Sparks of Life: Darwinism and the Victorian Debates over Spontaneous Generation*. Cambridge, Mass.: Harvard University Press.

Swedenborg, Emanuel. 1843–1844. *The Animal Kingdom, Considered Anatomically, Physically and Philosophically*. Edited by James Garth Wilkinson. 2 vols. London: W. Newberry, H. Bailliere.

———. 1894. *The Earth in Our Solar System Which Are Called Planets and the Earths in the Starry Heaven, Their Inhabitants, and the Spirits and Angels There, from Things Heard and Seen*. London: Swedenborg Society. (Wallace's annotated copy is housed in ARWL.)

———. 1905. *God, Creation, Man*. London. (Wallace's annotated copy is housed in ARWL.)

Swinburne, Richard. 1993. *The Coherence of Theism*. Rev. ed. Oxford: Clarendon Press.

Tait, Peter Guthrie, and Balfour Sewart. [1875] 1889. *The Unseen Universe; or, Physical Speculation on a Future State*. London: Macmillan & Co.

Taylor, Eugene. 1990. "William James on Darwin: An Evolutionary Theory of Consciousness." *Annals of the New York Academy of Science* 602:7–33.

———. 1996. *William James on Consciousness beyond the Margin*. Princeton, N.J.: Princeton University Press.

Taylor, Jean G. 1983. *The Social World of Batavia: European and Eurasian in Dutch Asia*. Madison: University of Wisconsin Press.

Teall, J. J. H. 1892. "Deep-Sea Deposits: A Review of the Work of the 'Challenger' Expedition." *Natural Science* 1:17–27.

Thomas, H. W. 1905. Introduction to *In the World Celestial*, by T. A. Bland. 4th ed. Chicago: T. A. Bland & Co.

Thomas, John L. 1983. *Alternative America: Henry George, Edward Bellamy, and Henry Demarest Lloyd and the Adversary Tradition*. Cambridge, Mass.: Harvard University Press, Belknap Press.

Todes, D. P. 1989. *Darwin without Malthus*. Oxford: Oxford University Press.

Toledo-Piza Ragazzo, Mônica de. 2002. *Peixes do Rio Negro/Fishes of the Rio Negro.* São Paulo: Editora da Universidade de São Paulo, Imprensa Oficial do Estado.

Trusted, Jennifer. 1991. *Physics and Metaphysics: Theories of Space and Time.* London and New York: Routledge.

Turner, Frank M. 1974. *Between Science and Religion: The Reaction to Scientific Naturalism in Late Victorian England.* New Haven, Conn., and London: Yale University Press.

————. 1993. *Contesting Cultural Authority: Essays in Victorian Intellectual Life.* Cambridge: Cambridge University Press.

Tyndall, John. 1874. *Address Delivered before the British Association, Assembled at Belfast.* London: Longmans, Green.

Tyndall, John, A. Henfrey, Thomas H. Huxley, et al. 1867. *The Culture Demanded by Modern Life.* New York: Appleton.

Van Oosterzee, Penny. 1997. *Where Worlds Collide: The Wallace Line.* Ithaca, N.Y., and London: Cornell University Press.

Van Riper, A Bowdoin. 1993. *Men Among the Mammoths: Victorian Science and the Discovery of Human Prehistory.* Chicago: University of Chicago Press.

Varila, Armi. 1977. *The Swedenborgian Background of William James' Philosophy.* Helsinki: Suomalainen Tiedeakatemia.

Vetter, Jeremy. 1999. *Contemplating Man under All His Varied Aspects: The Anthropological Work of Alfred Russel Wallace, 1843–70.* M.Phil. thesis, University of Oxford.

Vorzimmer, Peter J. 1970. *Charles Darwin: The Years of Controversy; The Origin of Species and Its Critics, 1859–1882.* Philadelphia: Temple University Press.

Wallace, Alfred Russel. 1853a. *A Narrative of Travels on the Amazon and Rio Negro, with an Account of the Native Tribes, and Observations on the Climate, Geology, and Natural History of the Amazon Valley.* London: Reeve & Co.

————. 1853b. "On the Rio Negro." *Journal of the Royal Geographical Society* 23:212–217.

————. 1855. "On the Law Which Has Regulated the Introduction of New Species." *Annals and Magazine of Natural History,* 2d ser., 16:184–196.

————. 1855–1859. "Notebook, 1855–1859." MS., Library of the Linnean Society of London.

————. 1856a. "Attempts at a Natural Arrangement of Birds." *Annals and Magazine of Natural History,* 2d ser., 18:193–216.

————. 1856b. "On the Habits of the Orang-utan of Borneo." *Annals and Magazine of Natural History,* 2d ser., 18:26–32.

————. 1857. "On the Natural History of the Aru Islands." *Annals and Magazine of Natural History,* 2d ser., suppl. to vol. 20, 473–485.

————. 1858. "Note on the Theory of Permanent and Geographical Varieties." *Zoologist* 16:5887–5888.

————. [1858] 1969. "On the Tendency of Varieties to Depart Indefinitely from the Original Type." *Journal of the Proceedings of the Linnean Society, Zoology* 3 (20 August): 53–62. Reprinted in Wallace's *Natural Selection and Tropical Nature: Essays on Descriptive and Theoretical Biology.* Westmead: Gregg International Publishers, 1969, 20–30.

————. 1860. "On the Zoological Geography of the Malay Archipelago." *Journal of the Linnean Society of London, Zoology* 4:172–184.

————. 1863. "On the Physical Geography of the Malay Archipelago." *Journal of the Royal Geographical Society* 33:217–234.

————. 1864a. "On Some Anomalies in Zoological and Botanical Geography." *Natural History Review* 4:111–123.

————. 1864b. "The Origin of Human Races and the Antiquity of Man Deduced from the Theory of 'Natural Selection.'" *Journal of the Anthropological Society of London* 2:clviii–clxx (followed by an account of related discussion, clxx–clxxxvii).

————. 1865a. "On the Phenomena of Variation and Geographical Distribution as Illustrated by the Papilionidae of the Malayan Region." *Transactions of the Linnean Society of London* 25:1–71.

————. 1865b. "On the Varieties of Man in the Malay Archipelago." *Transactions of the Ethnological Society of London* 3:196–215.

————. [1866] 1875. *The Scientific Aspect of the Supernatural: Indicating the Desirableness of an Experimental Enquiry by Men of Science into the Alleged Powers of Clairvoyants and Mediums.* London: F. Farrah. Expanded and reprinted in *On Miracles and Modern Spiritualism: Three Essays.* London: James Burns.

————. 1867a. "Mimicry, and Other Protective Resemblances among Animals." *Westminster Review* 32:1–43. Reprinted in Wallace [1891] 1969, 34–90.

————. 1867b. "On Birds' Nests and Their Plumage; or, The Relation between Sexual Differences of Colour and the Mode of Nidification in Birds." Paper presented at the British Association for the Advancement of Science meeting, sec. D, Biology, 9 September, Dundee. An expanded version was published as "A Theory of Birds' Nests: Shewing the Relation of Certain Sexual Differences of Colour in Birds to Their Mode of Nidification," *Journal of Travel and Natural History* 1, no. 2 (March 1868): 73–89, and reprinted as "A Theory of Birds' Nests, Showing the Relation of Certain Differences of Colour in Female Birds to Their Mode of Nidification" in Wallace [1891] 1969, 118–140.

————. [1869] 1962. *The Malay Archipelago; The Land of the Orang-Utan and the Bird of Paradise: A Narrative of Travel with Studies of Man and Nature.* New York: Dover Publications. (Reprint of the 10th ed., London: Macmillan & Co., 1891.)

————. 1869a. "Museums for the People." *Macmillan's Magazine* 19 (111):244–250.

————. 1869b. "The Origin of Moral Intuitions." Letter to the editor. *Scientific Opinion* 2, no. 46 (15 September): 336b–337a.

————. 1869c. "Sir Charles Lyell on Geological Climates and the Origin of Species." *Quarterly Review* 126:359–394.

————. [1870] 1891. *Contributions to the Theory of Natural Selection: A Series of Essays.* London: Macmillan. Reprinted (with alterations) in *Natural Selection and Tropical Nature: Essays on Descriptive and Theoretical Biology.* London: Macmillan & Co.

————. 1870a. "An Answer to the Arguments of Hume, Lecky, and Others, against Miracles." *Spiritualist* (London) 1 (15 November): 113c–116b.

————. 1870b. "Government Aid to Science." *Nature* 1:288–289.

————. 1870c. "The Measurement of Geological Time." *Nature* 1:399–401, 452–455.

————. 1871. Review of *The Descent of Man, and Selection in Relation to Sex,* by Charles Darwin. *Academy* 2:177–183.

————. 1872a. "Ethnology and Spiritualism." *Nature* 5:363–364.

————. 1872b. Review of *The Beginnings of Life: Being Some Account of the Nature, Modes of Origin, and Transformations of Lower Organisms,* by H. Charlton Bastian. Pts. 1, 2. *Nature* 6 (8 and 15 August): 284–287, 299–303.

————. 1872c. Review of *The Debatable Land between This World and the Next,* by Robert Dale Owen. *Quarterly Journal of Science* 2:237–247.

————. 1872d. Review of *Primitive Culture,* by Edward B. Tylor. *Academy* 3, no. 42 (15 February): 69–71.

————. 1874. "A Defence of Modern Spiritualism," pt. 1. Pt. 2, "Spirit-Photographs." *Fortnightly Review* 15 (1874): 630–657, 785–807.

————. 1875a. "Notes of Personal Evidence." Postscript to expanded version of "The Scientific Aspect of the Supernatural" (1866) reprinted in Wallace 1875b.

————. 1875b. *On Miracles and Modern Spiritualism: Three Essays.* London: James Burns.

————. [1876] 1962. *The Geographical Distribution of Animals: With a Study of the Relations of Living and Extinct Faunas as Elucidating the Past Changes of the Earth's Surface.* 2 vols. New York and London: Hafner Publishing.

————. 1876a. "Opening Address by the President, Alfred Russel Wallace." Given in Glasgow on 6 September 1876 to sec. D, Biology, of the BAAS. *Nature* 14:403–412.

————. 1876b. "A Sitting with Dr. Slade." *Spiritualist* (London) 9, no. 4 (25 August): 42.

————. 1878a. "The Curiosities of Credulity." *Athenaeum,* no. 2620 (12 January), 54–55.

————. 1878b. "Epping Forrest." *Fortnightly Review* 24:628–645.

————. 1878c. *Tropical Nature, and Other Essays.* London: Macmillan & Co.

————. [1879] 1883. *Australasia,* ed. and extended by Alfred R. Wallace, with ethnological app. by A. H. Keane. In *Stanford's Compendium of Geography and Travel.* 3d ed. London: Edward Stanford.

————. 1880. "How to Nationalize the Land: A Radical Solution of the Irish Land Problem." *Contemporary Review* 38:716–736.

————. [1880] 1892. *Island Life; or, the Phenomena and Causes of Insular Faunas and Floras, Including a Revision and Attempted Solution of the Problem of Geological Climates.* 2d rev. ed. London: Macmillan & Co.

————. [1882] 1906. *Land Nationalisation: Its Necessity and Its Aims; Being a Comparison of the System of Landlord and Tenant with That of Occupying Ownership in Their Influence on the Well-being of the People.* 4th ed. London: Swan Sonnenschein and Co.

————. [1884] 1969. "The Morality of Interest—the Tyranny of Capital." *Christian Socialist,* no. 10 (March), 150–151. Reprinted in Wallace [1905] 1969, 2:244–249.

————. 1884–1885. *To Members of Parliament and Others: Forty-five Years of Registration Statistics, Proving Vaccination to Be Both Useless and Dangerous: In Two Parts.* London: E. W. Allen.

————. 1885a. "The *Journal of Science* on Spiritualism." *Light* (London) 5, no. 236 (11 July): 327–328.

————. 1885b. "Modern Spiritualism: Are Its Phenomena in Harmony with Science?" *Sunday Herald* (Boston), 26 April, 9c–d.

————. 1885c. "President's Address". In *Report of the Land Nationalisation Society, 1884–1885: Fourth Annual Meeting of the Land Nationalisation Society,* 5–15. London: Land Nationalisation Society.

————. 1886–1887. "American Journal: 1886–1887." Ms. in Library of Linnean Society of London.

————. [1889]. 1975. *Darwinism: An Exposition of the Theory of Natural Selection with Some of Its Applications.* 2d ed. New York: AMS Press.

————. 1889. "Letter" (in which Wallace clarifies his views on socialism and land nationalization). *Land and Labor,* no. 1 (November), 7–8.

————. [1890] 1900c. "Human Selection." *Fortnightly Review* 48 (1 September): 325–337. Reprinted in Wallace 1900c, 1:509–526.

————. [1891] 1969. *Natural Selection and Tropical Nature: Essays on Descriptive and Theoretical Biology.* London: Macmillan & Co.; reprint, Westmead: Gregg International.

————. 1891a. "English and American Flowers," pt. 1. Pt. 2, "Flowers and Forests of the Far West." *Fortnightly Review* 50 (1 October and 1 December): 525–534, 796–810.

————. 1891b. "An English Nationalist: Alfred Russell [*sic*] Wallace Converted by Bellamy's Book." Note containing extracts from two Wallace letters concerning Richard T. Ely's book *An Introduction to Political Economy* (1889). *New York Times,* 1 February, 9d. Extracts reprinted as "A British Convert," *New Nation* 1, no. 3 (14 February 1891): 50c, and as "A Distinguished Convert," *New Nation* 1, no. 9 (28 March 1891): 135a–b.

————. 1891c. "Presidential Address of Alfred Russel Wallace." Tenth annual meeting of the Land Nationalisation Society. In *Report of the Land Nationalisation Society, 1890–1891,* 16–23. London: Land Nationalisation Society.

————. [1892] 1900c. "Human Progress: Past and Future." *Arena* 5 (January): 145–159. Reprinted in Wallace 1900c, 2:493–509.

————. 1892a. "H. W. Bates, the Naturalist of the Amazons." Obituary. *Nature* 45:398–399.

————. 1892b. "Note on Sexual Selection." *Natural Science* 1:749–750.

————. August 1892c. "The Permanence of the Great Oceanic Basins." *Natural Science* 1, no.6 (August): 418–426.

————. 1893a. "The Ice Age and Its Work." Pt. 1, "Erratic Blocks and Ice-Sheets." Pt. 2, "Erosion of Lake Basins." *Fortnightly Review* 54 (1 November and 1 December): 616–633, 750–774.

————. 1893b. "President's Address, 1893: The Conditions Essential to the Success of Small Holdings." Read by William Volckman, chairman, twelfth annual meeting of the Land Nationalisation Society. In *Report of the Land Nationalisation Society, 1892–1893,* 15–23. LNS Tract No. 50. London: Land Nationalisation Society.

————. 1894a. "The Future of Civilization." *Nature* 49:549–551.

————. 1894b. "The Social Economy of the Future." In *The New Party Described by Some of Its Members,* ed. Andrew Reid, 177–211. London: Hodder Brothers.

————. 1895. "The Method of Organic Evolution." Pts. 1 and 2. *Fortnightly Review* 57 (1 February and 1 March), 211–224, 435–445.

————. 1896a. *Miracles and Modern Spiritualism.* 3d rev. ed. London: George Redway.

————. 1896b. "The Problem of Utility: Are Specific Characters Always or Generally Useful?" *Journal of the Linnean Society of London, Zoology* 25:481–496.

————. 1896c. "The Proposed Gigantic Model of the Earth." *Contemporary Review* 69:730–740.

————. 1898a. "Re-occupation of the Land." In William Morris et al., *Hand and Brain: A Symposium of Essays on Socialism,* 25–43. East Aurora, N.Y.: Roycroft Printing Shop. Revised and reprinted in Wallace 1900c, 2:478–492.

————. 1898b. "Spiritualism and Social Duty." *Light* (London) 18 (9 July): 334–337. Reprinted with "verbal modifications" in Wallace 1900c, 2:521–528.

————. [1898] 1970. *The Wonderful Century: Its Successes and Failures.* Westmead: Gregg International Publishers.

———. 1899. "The Inefficiency of Strikes: Is There Not a Better Way?" In *The Labour Annual: 1899,* 105. Edited by Joseph Edwards. Manchester: Edwards.

———. 1900a. "Evolution." *Sun* (New York) 68, no. 114 (23 December): 4a–g, 5a.

———. 1900b. "Imperial Might and Human Right." *Clarion* (London), no. 450 (21 July): 230b.

———. 1900c. *Studies Scientific and Social.* 2 vols. London: Macmillan & Co.

———. [1903] 1907. *Man's Place in the Universe; A Study of the Results of Scientific Research in Relation to the Unity or Plurality of Worlds.* 6th ed. London: Chapman & Hall, Ltd.

———. 1903a. "Man's Place in the Universe: As Indicated by the New Astronomy." *Fortnightly Review* 73:395–411.

———. 1903b. "Man's Place in the Universe: A Reply to Criticisms." *Fortnightly Review* 74:380–390.

———. 1905. "If There Were a Socialist Government—How Should It Begin?" *Clarion* (London), no. 715 (18 August): 5a–f.

———. [1905] 1969. *My Life: A Record of Events and Opinions.* 2 vols. Facsimile reprint. Westmead: Gregg International Publishers.

———. 1907. *Is Mars Habitable? A Critical Examination of Professor Percival Lowell's Book "Mars and Its Canals," with an Alternative Explanation.* London: Macmillan & Co.

———. 1908a. "The Present Position of Darwinism." *Contemporary Review* 94:129–141.

———. 1908b. "Darwinism *versus* Wallaceism." *Contemporary Review* 94:716–717.

———. 1910a. *The World of Life: A Manifestation of Creative Power, Directive Mind and Ultimate Purpose.* London: Chapman & Hall, Ltd.

———. 1910b. "New Thoughts on Evolution." Interview by Harold Begbie. 2 pts. *Daily Chronicle* (London), no. 15197, 3 November, 4d; no. 15198, 4 November, 4d–e.

———. 1911. "Scientist's 88th Birthday: Interview with Dr. Russel Wallace." *Daily News* (London), no. 20227, 9 January, 1e–f.

———. 1912a. "Evolution and Character." In Parker 1912, 3–50.

———. 1912 b. "The Great Strike—and After: Hopes of a National Peace." Interview by Harold Begbie. *Daily Chronicle* (London), no. 15622, 13 March, 4d–e.

———. 1912c. "Mr. Blatchford's Dogmatism." *Christian Commonwealth* 32:815b.

———. 1912d. "The Origin of Life: A Reply to Dr. Schaefer." *Everyman* 1:5–6.

———. 1912e. "The Problem of Life." Interview. *Daily News and Leader* (London and Manchester), no. 20748, 7 September, 1a–b.

———. 1913a. *The Revolt of Democracy: With the Life Story of the Author by James Marchant, F.R.S. Edin.* London: Cassell & Co.

———. 1913b. *Social Environment and Moral Progress.* New York: Cassell & Co.

Waller, John C. 2001. "Gentlemanly Men of Science: Sir Francis Galton and the Professionalization of the British Life-Sciences." *Journal of the History of Biology* 34:83–114.

Ward, Lester F. 1903. *Pure Sociology: A Treatise on the Origin and Spontaneous Development of Society.* New York: Macmillan Co.

Ward, Paul. 1998. *Red Flag and Union Jack: Englishness, Patriotism and the British Left, 1881–1924.* Suffolk: Boydell Press for the Royal Historical Society.

West, Shearer, ed. 1996. *The Victorians and Race.* Aldershot: Scolar Press.

Williams, Wesley C. 1971. "Robert Chambers." In Gillispie 1970–1980, 3:192.

Wilson, David B. 1996. "On the Importance of Eliminating *Science* and *Religion* from the History of Science and Religion: The Cases of Oliver Lodge, J. H. Jeans and A. S. Eddington." In *Facets of Faith and Science.* Vol. 1, *Historiography and Modes of Interaction,* ed. Jitse M. Van der Meer. Lanham, Md.: University Press of America.

Wilson, John G. 2000. *The Forgotten Naturalist: In Search of Alfred Russel Wallace.* Melbourne: Australia Scholarly Publishing.

Wilson, Leonard G. 1970. *Sir Charles Lyell's Scientific Journals on the Species Question.* New Haven, Conn.: Yale University Press.

Winsor, Mary P. 1991. *Reading the Shape of Nature: Comparative Zoology at the Agassiz Museum.* Chicago: University of Chicago Press.

———. 2003. "Non-essentialist Methods in Pre-Darwinian Taxonomy." *Biology and Philosophy,* vol. 18 (in press).

Winter, Alison. 1994. "Mesmerism and Popular Culture in Early Victorian England." *History of Science* 32:317–343.

———. 1997. "The Construction of Orthodoxies and Heterodoxies in the Early Victorian Life Sciences." In Lightman 1997, 24–50.

———. 1998. *Mesmerized: Powers of Mind in Victorian Britain.* Chicago: University of Chicago Press.

Wright, Margaret R. 1997. "Marcella O'Grady Boveri (1863–1950): Her Three Careers in Biology." *Isis* 88:627–652.

Wykstra, Stephen J. 1996. "Have Worldviews Shaped Science? A Reply to Brooke." In *Facets of Faith and Science.* Vol. 1, *Historiography and Modes of Interaction,* ed. Jitse M. van der Meer, 91–111. Lanham, Md.: University Press of America.

———. 1997. "Should Worldviews Shape Science? Toward an Integrationist Account of Scientific Theorizing." In *Facets of Faith and Science.* Vol. 2, *The Role of Beliefs in Mathematics and the Natural Sciences,* ed. Jitse M. van der Meer, 123–171. Lanham, Md.: University Press of America.

Wynne, Brian. 1979. "Physics and Psychics: Science, Symbolic Action, and Social Control in Late Victorian England." In *Natural Order: Historical Studies of Scientific Culture,* ed. Barry Barnes and Steven Shapin, 167–186. Beverly Hills, Calif.: Sage.

Yanni, Carla. 1999. *Nature's Museums: Victorian Science and the Architecture of Display.* Baltimore: Johns Hopkins University Press.

Yeo, Richard. 1984. "Science and Intellectual Authority in Mid-Nineteenth Century Britain: Robert Chambers and *Vestiges of the Natural History of Creation.*" *Victorian Studies* 28:5–31.

———. 1993. *Defining Science: William Whewell, Natural Knowledge, and Public Debate in Early Victorian Britain.* Cambridge: Cambridge University Press.

Yeo, Stephen. 1977. "A New Life: The Religion of Socialism in Britain, 1883–1896." *History Workshop* 4:5–56.

Young, Robert M., 1971. " 'Non-Scientific' Factors in the Darwinian Debates." In *Actes du XIIe Congres Internationale d'Histoire des Sciences (1968),* 8:221–226. Paris: Blanchard.

———. 1985a. "Darwinism *Is* Social." In Kohn 1985, 609–638.

———. 1985b. *Darwin's Metaphor: Nature's Place in Victorian Culture.* Cambridge: Cambridge University Press.

INDEX

Note: An italicized page number immediately following a person's name indicates an illustration of that person. Wallace's relationships with others are dealt with extensively here; to locate that information in the index, look under the other person's name; for example, under Darwin, Charles, is the subhead "Wallace's relationship with." Wallace's various essays and books are alphabetized under the name of the work, rather than under Wallace.